从入门到实战·微课视频

Java Web 开发从入门到实战

◎ 陈恒 姜学 主编

清华大学出版社
北京

内 容 简 介

本书详细讲解了 Java Web 开发中 JSP、Servlet 以及 Ajax 技术的基础知识和实际应用,同时介绍了 Struts 2 框架以及 Spring MVC 框架的基础知识。为了让读者做到"学中做,做中学",本书配备了相应的实践环节。

全书共分 15 章,内容包括 Web 前端基础、Java Web 开发环境的构建、JSP 语法、JSP 内置对象、JSP 与 JavaBean、Java Servlet 基础、会话跟踪技术、过滤器与监听器、EL 与 JSTL、文件的上传与下载、JDBC 数据库访问、Ajax、JSON 以及 jQuery 技术、电子商务平台的设计与实现、Struts 2 框架基础以及 Spring MVC 框架基础等重要内容。书中实例侧重实用性和启发性,趣味性强,分布合理,通俗易懂,使读者能够快速掌握 Java Web 开发的基础知识、编程技巧以及完整的开发体系,为适应实战应用打下坚实的基础。

本书附有教学视频、源代码、课件、教学大纲、教学日历、教案、习题与实践环节答案等配套资源,可以作为大学计算机及相关专业的教材或教学参考书,也可以作为 Java 技术的培训教材,同时也适合广大 Java Web 开发人员阅读与使用。

本书封面贴有清华大学出版社防伪标签,无标签者不得销售。

版权所有,侵权必究。举报: 010-62782989,beiqinquan@tup.tsinghua.edu.cn。

图书在版编目(CIP)数据

Java Web 开发从入门到实战 / 陈恒,姜学主编. —北京:清华大学出版社,2019(2024.1重印)
(从入门到实战·微课视频)
ISBN 978-7-302-52911-8

Ⅰ. ①J… Ⅱ. ①陈… ②姜… Ⅲ. ①JAVA 语言-程序设计 Ⅳ. ①TP312.8

中国版本图书馆 CIP 数据核字(2019)第 083533 号

策划编辑:魏江江
责任编辑:王冰飞
封面设计:刘 键
责任校对:时翠兰
责任印制:刘海龙

出版发行:清华大学出版社
网　　址:https://www.tup.com.cn, https://www.wqxuetang.com
地　　址:北京清华大学学研大厦 A 座　　邮　编:100084
社 总 机:010-83470000　　邮　购:010-62786544
投稿与读者服务:010-62776969, c-service@tup.tsinghua.edu.cn
质 量 反 馈:010-62772015, zhiliang@tup.tsinghua.edu.cn

印 装 者:三河市君旺印务有限公司
经　　销:全国新华书店
开　　本:185mm×260mm　　印　张:28.75　　字　数:695 千字
版　　次:2019 年 7 月第 1 版　　印　次:2024 年 1 月第 13 次印刷
印　　数:22001~24000
定　　价:69.80 元

产品编号:082526-01

前言

本书使用 Tomcat 9.0+Eclipse IDE for Java EE 平台详细讲解了 JSP、Servlet、Ajax 技术、Struts 2 框架以及 Spring MVC 框架的基础知识和实际应用。本书的重点不是简单地介绍基础知识，而是精心设计了大量实例。读者通过本书可以快速地掌握 JSP、Servlet、Ajax 技术、Struts 2 框架以及 Spring MVC 框架等 Java Web 开发技术的实战应用，提高 Java Web 开发能力。

全书共 15 章，各章的具体内容如下：

第 1 章概括地介绍 Web 前端基础知识，包括 HTML、CSS 与 JavaScript。

第 2 章主要介绍 Java Web 开发环境的构建，并通过一个简单的 Web 应用讲解了 Java Web 开发的基本步骤。

第 3 章主要讲解 JSP 语法，包括 JSP 脚本元素以及常用的 JSP 标记。

第 4 章介绍常见的 JSP 内置对象，包括 request、response、out、session、application、pageContext、page、config 和 exception。

第 5 章介绍 JSP 与 JavaBean，JSP 和 JavaBean 技术相结合不仅可以实现数据的表示和处理分离，而且可以提高代码重用的程度。

第 6 章讲述 Java Servlet 的基础知识，包括部署 Servlet、Servlet 的生命周期、通过 JSP 页面访问 Servlet、重定向与转发、基于 Servlet 的 MVC 模式等内容，是本书的重点内容之一。

第 7 章重点介绍 Cookie、session、URL 重写以及隐藏表单域等会话跟踪技术。

第 8 章详细地讲述了过滤器与监听器的概念、运行原理以及实际应用。过滤器可以过滤浏览器对服务器的请求，也可以过滤服务器对浏览器的响应。监听器可以监听 Java Web 应用中某些对象信息的创建、销毁、增加、修改、删除等动作的发生，然后做出相应的响应处理。

第 9 章主要介绍 EL 与 JSTL 核心标签库的基本用法。

第 10 章介绍如何使用 Servlet 的 HttpServletRequest 对文件上传的支持。

第 11 章重点介绍在 JSP 中如何访问关系数据库，例如 MySQL、Oracle 和 SQL Server 等数据库，同时还介绍了连接池、分页查询、Excel 与数据库互访以及使用 CallableStatement 语句访问数据库存储子程序等内容。

第 12 章重点介绍 Ajax、JSON 以及 jQuery 技术，重点掌握 JSON 在 Ajax 中的应用以及 jQuery 对 Ajax 的实现。

第 13 章是本书的重点内容之一，将前面章节的知识进行大综合，详细讲解一个基于

MVC（JSP+JavaBean+Servlet）模式的电子商务平台的设计与实现过程。

第 14 章简述 Struts 2 框架基础知识，包括 Struts 2 的体系结构、环境构建以及基本流程。

第 15 章简述 Spring MVC 框架基础知识，包括 Spring MVC 的工作原理、环境构建、基于注解的控制器、表单标签库与数据绑定以及 Spring 的数据库编程。

为便于教学，本书配有教学课件、教学大纲、电子教案、程序源码、习题与实践环节答案等配套资源，读者可以扫描封底课件二维码免费下载。另外，本书还配有教学视频，读者可以扫描书中的二维码在线观看。

本书获得辽宁省向应用型转变示范专业"计算机科学与技术"项目支持。

由于编者水平有限，书中难免会有不足之处，敬请广大读者批评指正。

编　者

2019 年 3 月

目 录

源码下载

第 1 章　Web 前端基础 ············· 1

1.1　HTML ······················ 1
1.1.1　HTML 文件的基本结构 ······· 1
1.1.2　编写 HTML 页面 ············ 2
1.1.3　常用 HTML 标签 ············ 2
1.1.4　实践环节——制作调查问卷 ··· 13

1.2　CSS ······················ 13
1.2.1　CSS 基本语法 ·············· 13
1.2.2　在网页中添加 CSS 的方法 ···· 14
1.2.3　选择符的分类 ············· 16
1.2.4　伪类及伪对象 ············· 17
1.2.5　常见的 DIV+CSS 布局类型 ··· 18
1.2.6　实践环节——页面布局 ······ 25

1.3　JavaScript ················ 28
1.3.1　在网页中添加 JavaScript 的方法 ··· 28
1.3.2　JavaScript 基本语法 ········ 29
1.3.3　流程控制与函数 ············ 32
1.3.4　JavaScript 对象 ············ 36
1.3.5　JavaScript 对象模型 ········ 39
1.3.6　操作 HTML ················ 43
1.3.7　表单验证 ················· 48
1.3.8　实践环节——表单验证 ······ 51

1.4　本章小结 ···················· 51
习题 1 ························· 51

第 2 章　Java Web 开发环境的构建 ···· 54

2.1　构建开发环境 ················ 54
2.1.1　开发工具 ················· 54

III

2.1.2　工具的集成 ··· 55
　2.2　使用 Eclipse 开发 Web 应用 ··· 58
　　　2.2.1　JSP 文件和 JSP 运行原理 ·· 58
　　　2.2.2　一个简单的 Web 应用 ··· 59
　　　2.2.3　实践环节——使用 Eclipse 开发 Web 应用 ······················· 61
　2.3　本章小结 ·· 62
　习题 2 ··· 62

第 3 章　JSP 语法 ·· 63

　3.1　JSP 页面的基本构成 ·· 63
　　　3.1.1　一个 JSP 页面 ··· 63
　　　3.1.2　JSP 页面注释 ·· 64
　　　3.1.3　实践环节——识别 JSP 页面元素 ······································ 65
　3.2　JSP 脚本元素 ·· 65
　　　3.2.1　Java 程序片 ··· 65
　　　3.2.2　成员变量与方法的声明 ··· 67
　　　3.2.3　Java 表达式 ··· 68
　　　3.2.4　实践环节——在 JSP 页面中输出英文字母表 ····················· 68
　　　3.2.5　实践环节——网站访问量的统计 ······································ 69
　　　3.2.6　实践环节——打印表格 ··· 69
　3.3　JSP 指令标记 ·· 69
　　　3.3.1　page 指令 ··· 69
　　　3.3.2　include 指令 ·· 71
　　　3.3.3　实践环节——制作导航栏 ·· 72
　3.4　JSP 动作标记 ·· 73
　　　3.4.1　include 动作标记 ··· 73
　　　3.4.2　forward 动作标记 ·· 74
　　　3.4.3　param 动作标记 ·· 76
　　　3.4.4　实践环节——include 和 param 动作标记的应用 ··············· 77
　　　3.4.5　实践环节——登录验证 ··· 77
　3.5　本章小结 ··· 77
　习题 3 ··· 78

第 4 章　JSP 内置对象 ·· 79

　4.1　request 对象 ·· 80
　　　4.1.1　常用方法 ·· 80
　　　4.1.2　使用 request 对象存取数据 ··· 83

| | 4.1.3 | 中文乱码问题 ··· 84 |
| | 4.1.4 | 实践环节——获取客户端的基本信息 ······································· 85 |

4.2 response 对象 ··· 85
    4.2.1 动态改变 contentType 属性值 ·· 85
    4.2.2 设置响应表头 ··· 86
    4.2.3 使用 response 对象重定向 ·· 87
    4.2.4 实践环节——登录验证 ·· 89

4.3 out 对象 ·· 90

4.4 session 对象 ·· 91
    4.4.1 session 对象的 ID ·· 91
    4.4.2 使用 session 对象存取数据 ··· 94
    4.4.3 session 对象的生命周期 ··· 97
    4.4.4 实践环节——购物车 ···100

4.5 application 对象 ··101
    4.5.1 什么是 application ··101
    4.5.2 使用 application 对象存取数据 ···101
    4.5.3 实践环节——网站访客计数器 ··103

4.6 pageContext 对象 ···103

4.7 page 对象 ···104

4.8 config 对象 ···105

4.9 exception 对象 ··105

4.10 JSP 的 4 种作用域 ··107

4.11 本章小结 ···109

习题 4 ··109

第 5 章 JSP 与 JavaBean ·· 111

5.1 编写 JavaBean ··111

5.2 在 JSP 中使用 JavaBean ··113
    5.2.1 动作标记 useBean ··113
    5.2.2 动作标记 getProperty ···115
    5.2.3 动作标记 setProperty ···117
    5.2.4 实践环节——使用 JavaBean 计算三角形的
           面积和周长 ···120

5.3 本章小结 ··121

习题 5 ··121

第 6 章 Java Servlet 基础 ·· 122

6.1 Servlet 类与 Servlet 对象 ··122

6.2 Servlet 对象的创建与运行 ······124
    6.2.1 在 web.xml 中部署 Servlet ······124
    6.2.2 基于注解的方式部署 Servlet ······126
    6.2.3 实践环节——@WebServlet 的应用 ······128
6.3 Servlet 的生命周期 ······128
6.4 通过 JSP 页面访问 Servlet ······129
6.5 doGet()和 doPost()方法 ······132
6.6 重定向与转发 ······134
    6.6.1 重定向 ······134
    6.6.2 转发 ······135
    6.6.3 实践环节——登录验证 ······137
6.7 在 Java Servlet 中使用 session ······138
6.8 基于 Servlet 的 MVC 模式 ······141
    6.8.1 MVC 模式 ······141
    6.8.2 使用 JSP、Servlet、JavaBean 实现 MVC ······142
    6.8.3 模型周期 ······144
    6.8.4 实践环节——四则运算 ······147
6.9 本章小结 ······147
习题 6 ······147

第 7 章 会话跟踪技术 ······149

7.1 Cookie 技术 ······149
7.2 session 技术 ······152
7.3 URL 重写技术 ······153
7.4 隐藏表单域技术 ······157
7.5 实践环节——登录注销系统 ······159
7.6 本章小结 ······160
习题 7 ······160

第 8 章 过滤器与监听器 ······161

8.1 Filter 类与 Filter 对象 ······162
8.2 Filter 对象的部署与运行 ······163
    8.2.1 在 web.xml 中部署过滤器 ······163
    8.2.2 基于注解的方式部署过滤器 ······165
    8.2.3 过滤器的执行顺序 ······166
    8.2.4 实践环节——过滤器的初始化参数 ······168
8.3 过滤器的实际应用 ······168

	8.3.1 字符编码过滤器	168
	8.3.2 登录验证过滤器	169
8.4	监听器的分类与创建	173
	8.4.1 监听对象自身创建和销毁的监听器	173
	8.4.2 监听对象属性变化的监听器	176
	8.4.3 监听 session 中某个对象状态变化的监听器	177
8.5	监听器的实际应用——统计在线用户	180
8.6	本章小结	184
习题 8		184

第 9 章 EL 与 JSTL 185

9.1	EL	185
	9.1.1 EL 简介	185
	9.1.2 EL 语法	186
	9.1.3 EL 运算符	187
	9.1.4 EL 隐含对象	188
9.2	JSTL	191
	9.2.1 JSP 标准标签库 JSTL	191
	9.2.2 实践环节——使用 JSTL 标签打印九九乘法表	201
9.3	本章小结	202
习题 9		202

第 10 章 文件的上传与下载 203

10.1	基于 Servlet 的文件上传	203
	10.1.1 文件上传的表单设置	203
	10.1.2 @MultipartConfig 注解	204
	10.1.3 Part 接口	204
	10.1.4 单文件上传	205
	10.1.5 多文件上传	208
	10.1.6 实践环节——上传文件的限定	210
10.2	文件的下载	211
	10.2.1 文件下载的实现方法	211
	10.2.2 文件下载示例	211
	10.2.3 文件名乱码问题	214
10.3	本章小结	215
习题 10		215

第 11 章　JDBC 数据库访问 ································· 216

- 11.1　使用纯 Java 数据库驱动程序连接 MySQL ··············· 216
- 11.2　数据库操作 ··············· 220
  - 11.2.1　查询操作 ··············· 220
  - 11.2.2　新增、更新与删除操作 ··············· 221
  - 11.2.3　实践环节——使用 Statement 语句实现多条件查询 ··· 225
- 11.3　游动查询 ··············· 225
- 11.4　使用连接池 ··············· 229
- 11.5　其他典型数据库的连接 ··············· 232
  - 11.5.1　连接 SQL Server 数据库 ··············· 232
  - 11.5.2　连接 Oracle 数据库 ··············· 233
- 11.6　使用 PreparedStatement 语句访问数据库 ··············· 233
  - 11.6.1　预处理语句的优点 ··············· 233
  - 11.6.2　使用通配符 ··············· 233
  - 11.6.3　实践环节——使用 PreparedStatement 语句实现多条件查询 ··············· 236
- 11.7　分页查询 ··············· 236
- 11.8　使用 CallableStatement 语句访问存储子程序 ··············· 240
- 11.9　Excel 与数据库互访 ··············· 244
  - 11.9.1　向数据库导入 Excel ··············· 245
  - 11.9.2　从数据库导出 Excel ··············· 250
- 11.10　本章小结 ··············· 256
- 习题 11 ··············· 257

第 12 章　Ajax、JSON 以及 jQuery 技术 ································· 258

- 12.1　Ajax 技术 ··············· 258
  - 12.1.1　Ajax 简介 ··············· 259
  - 12.1.2　XMLHttpRequest 对象 ··············· 259
  - 12.1.3　Ajax 向服务器发送请求 ··············· 260
  - 12.1.4　Ajax 获得服务器的响应 ··············· 261
  - 12.1.5　Ajax 示例 ··············· 262
- 12.2　JSON 技术 ··············· 265
  - 12.2.1　JSON 的数据结构 ··············· 265
  - 12.2.2　JSON 在 JavaScript 中的使用 ··············· 266
  - 12.2.3　JSON 在 Ajax 中的使用 ··············· 267
- 12.3　jQuery 技术 ··············· 270
  - 12.3.1　jQuery 简介 ··············· 270

12.3.2 jQuery 的安装 271
12.3.3 jQuery 语法 271
12.3.4 jQuery 对 Ajax 的实现 271
12.4 实践环节——注册邮箱的唯一性验证 276
12.5 本章小结 276
习题 12 277

第 13 章 电子商务平台的设计与实现 278

13.1 系统设计 278
    13.1.1 系统功能需求 278
    13.1.2 系统模块划分 279
13.2 数据库设计 280
    13.2.1 数据库概念结构设计 280
    13.2.2 数据库逻辑结构设计 281
    13.2.3 创建数据表 283
13.3 系统管理 283
    13.3.1 导入相关的 jar 包 283
    13.3.2 JSP 页面管理 283
    13.3.3 组件与 Servlet 管理 288
13.4 组件设计 288
    13.4.1 过滤器 289
    13.4.2 验证码 291
    13.4.3 实体模型 294
    13.4.4 数据库操作 294
    13.4.5 业务模型 311
    13.4.6 工具类 318
13.5 后台管理子系统的实现 319
    13.5.1 管理员登录 319
    13.5.2 添加商品 320
    13.5.3 查询商品 324
    13.5.4 修改商品 329
    13.5.5 删除商品 330
    13.5.6 订单管理 332
    13.5.7 用户管理 334
    13.5.8 公告管理 335
    13.5.9 退出系统 338
13.6 前台电子商务子系统的实现 338
    13.6.1 导航栏 338

13.6.2	销售排行	341
13.6.3	人气排行	341
13.6.4	最新商品	342
13.6.5	公告栏	342
13.6.6	用户注册	342
13.6.7	用户登录	343
13.6.8	商品详情	345
13.6.9	关注商品	346
13.6.10	购物车	347
13.6.11	下单	350
13.6.12	用户中心	352

13.7 本章小结 ······ 353

第 14 章 Struts 2 框架基础 ······ 354

14.1 Struts 2 的体系结构 ······ 354
14.2 Struts 2 的工作环境 ······ 355
    14.2.1 Struts 2 的下载 ······ 355
    14.2.2 使用 Eclipse 开发 Struts 2 的 Web 应用 ······ 356
    14.2.3 第一个 Struts 2 应用 ······ 357
14.3 Struts 2 的 Action ······ 359
    14.3.1 Action 的创建与配置 ······ 359
    14.3.2 Action 接收请求参数 ······ 362
    14.3.3 Action 访问 Servlet API ······ 365
    14.3.4 Action 中常见的结果类型 ······ 371
14.4 基于 Struts 2 框架的注册与登录 ······ 376
14.5 本章小结 ······ 387
习题 14 ······ 388

第 15 章 Spring MVC 框架基础 ······ 390

15.1 Spring MVC 的工作原理 ······ 390
15.2 Spring MVC 的工作环境 ······ 392
    15.2.1 Spring 的下载与安装 ······ 392
    15.2.2 使用 Eclipse 开发 Spring MVC 的 Web 应用 ······ 393
15.3 基于注解的控制器 ······ 397
    15.3.1 Controller 注解类型 ······ 397
    15.3.2 RequestMapping 注解类型 ······ 397
    15.3.3 编写请求处理方法 ······ 398

　　　　15.3.4　Controller 接收请求参数的常见方式 ·················· 399
　　　　15.3.5　Spring MVC 框架中的重定向与转发 ················· 404
　　　　15.3.6　应用@Autowired 进行依赖注入 ····················· 405
　　　　15.3.7　@ModelAttribute ································· 408
　15.4　表单标签库与数据绑定 ······································· 409
　　　　15.4.1　表单标签库 ······································· 409
　　　　15.4.2　数据绑定 ··· 413
　15.5　Spring 的数据库编程 ·· 421
　　　　15.5.1　Spring JDBC 的配置 ······························· 421
　　　　15.5.2　Spring JdbcTemplate 的常用方法 ···················· 422
　　　　15.5.3　基于@Transactional 注解的声明式事务管理 ············ 426
　15.6　基于 Spring MVC 框架的注册与登录 ··························· 427
　　　　15.6.1　JSON 数据转换 ····································· 427
　　　　15.6.2　注册与登录 ······································· 427
　15.7　本章小结 ·· 437
　习题 15 ··· 437

附录 A　案例 1：基于 Struts 2 框架的名片管理系统的设计与实现 ············· 438

附录 B　案例 2：基于 Spring MVC 框架的名片管理系统的设计与实现 ············· 439

附录 C　案例 3：Spring MVC 整合 MyBatis ··············· 440

附录 D　案例 4：在 Eclipse 中使用 Maven 整合 Spring MVC+MyBatis 框架 ··············· 441

参考文献 ·· 442

第 1 章 Web 前端基础

学习目的与要求

本章对 HTML、CSS 与 JavaScript 进行简要讲述，包括 HTML 的常用标签、CSS 的基本语法、JavaScript 的语法基础和 JavaScript 对象等内容。通过本章的学习，读者能够掌握 HTML 的常用标签、CSS 的使用方法以及 JavaScript 的语法基础，能够设计与开发静态 Web 页面，并编写页面中的 JavaScript 代码。

本章主要内容

- HTML
- CSS
- JavaScript

HTML 的英文全称是 HyperText Markup Language，即超文本标记语言，它是 Internet 上用于编写网页的主要标记语言。

CSS 是英文 Cascading Style Sheet 的缩写，又称为"层叠样式表"，简称为样式表。它是 W3C 定义的标准，是一种用来为结构化文档（例如 HTML 文档）添加样式（字体、间距和背景等）的计算机语言。CSS 是对 HTML 处理样式的补充，能将内容和样式处理相分离，大大降低了编程者的工作量。

JavaScript 是一种描述性的脚本语言（Script Language），它由客户端浏览器解释执行，在执行期间无需 Web 服务器，减轻了 Web 服务器的负担。JavaScript 可以向 HTML 页面添加交互行为、读写元素、验证表单以及进行事件处理。

1.1 HTML

1.1.1 HTML 文件的基本结构

一个完整的 HTML 文件由各种元素与标记组成，包括标题、段落、表格、文本和超链接等。下面是一个 HTML 文件的基本结构。

```
<html>
    <head>
        …
    </head>
    <body>
        …
    </body>
</html>
```

从上面的代码段可以看出，HTML 文件的基本结构分为 3 个部分，其中各部分的含义如下。

- <html>…</html>：表示 HTML 文件开始和结束的位置，里面包括<head>和<body>等标记。HTML 文件中所有的内容都应该在这两个标记之间。
- <head>…</head>：HTML 文件的头部标记，通常将这两个标记之间的内容统称为 HTML 的头部。
- <body>…</body>：用来指明文档的主体区域，网页所要显示的内容都要放在这个标记内，通常将这两个标记之间的内容统称为 HTML 的主体。

1.1.2　编写 HTML 页面

编写 HTML 页面有两种常用方法，一种是利用操作系统自带的记事本编写，一种是利用可视化网页制作软件（例如 Dreamweaver）编写。本书从第 2 章开始使用集成开发环境（IDE）Eclipse 编写 Web 程序。

HTML、CSS 与 JavaScript 并不需要特殊的开发环境，它们都由客户端的浏览器执行。HTML 文件的扩展名为.html 或.htm，CSS 文件的扩展名为.css，JavaScript 文件的扩展名为.js。

1.1.3　常用 HTML 标签

常用 HTML 标签简单地划分为以下 4 种格式。

- <标记名称>：单一型，无设置值。例如
。
- <标记名称 属性="属性值">：单一型，有设置值。例如<hr color="red">。
- <标记名称>…</标记名称>：对称型，无设置值。例如<title>…</title>。
- <标记名称 属性="属性值">…</标记名称>：对称型，有设置值。例如<body bgcolor="red">…</body>。

下面介绍常用的 HTML 标签。

❶ 标题

HTML 将和文本相关的标题分成 6 个级别，1～6 级标题的语法格式如下：

```
<h1>…</h1>
```

```
<h2>…</h2>
<h3>…</h3>
<h4>…</h4>
<h5>…</h5>
<h6>…</h6>
```

<h1>到<h6>作为标题标记依据重要性递减，字号从<h1>到<h6>由大变小。为了更好地理解，请看下面的代码段：

```
<h1>学习标题标记</h1>
    <h2>第 1 章 Web 前端基础</h2>
        <h3>1.1HTML</h3>
            <h4>1.1.3 常用 HTML 标签</h4>
```

<h1>（一级）标题是重中之重，一般运用于网站标题或者头条新闻上；<h2>（二级）标题主要出现在页面的主体内容的文章标题和栏目标题上；<h3>（三级）标题一般出现在页面的边侧栏上。由于页面层级关系不能太深，所以<h4>、<h5>和<h6>一般较少出现。

❷ 段落

在 HTML 网页中，使用<p>标记实现一个新段落，语法格式如下：

```
<p>段落的内容</p>
```

<p>标记中的属性 align 能够设置段落中文字的对齐方式，对齐方式分为左对齐、居中和右对齐，语法格式如下：

```
<p align="对齐方式"></p>
```

其中，align 取值为 left 时，文字显示左对齐；align 取值为 right 时，文字显示右对齐；align 取值为 center 时，文字显示居中对齐。

【例】1-1 有 3 段文字，对齐方式依次为左、中、右。

代码如下：

```
<!DOCTYPE html>
<html>
<head>
<meta charset="UTF-8">
<title>Insert title here</title>
</head>
<body>
    <p align="left">居左文字</p>
    <p align="center">居中文字</p>
    <p align="right">居右文字</p>
</body>
</html>
```

❸ 滚动（marquee）

在 HTML 页面中可以使用<marquee>标记让文字滚动，该标记有滚动方向（direction）、滚动方式（behavior）、滚动次数（loop）、滚动速度（scrollamount）、滚动延迟（scrolldelay）、背景颜色（bgcolor）、宽度和高度等常用属性。其语法格式如下：

```
<marquee direction="滚动方向" behavior ="滚动方式">滚动的文字</marquee>
```

其中，direction 的值有 up、down、left 和 right，分别表示向上、向下、向左和向右滚动，向左滚动是默认情况；behavior 的值有 scroll、slide 和 alternate，分别表示循环滚动、只滚动一次和来回交替滚动；loop 的值为整数；scrollamount 的值为文字每次移动的长度，以像素为单位；scrolldelay 的单位是毫秒。

【例】1-2 编写一个网页，网页中有一段滚动的文字，文字滚动方向为默认方向，文字滚动的背景色为蓝色，文字滚动方式为来回交替滚动。网页运行效果如图 1.1 所示。

《赠汪伦》
李白
李白乘舟将欲行，
忽闻岸上踏歌声。
桃花潭水深千尺，
不及汪伦送我情。

大家好，我正在学习《古诗三百首》，哈哈，羡慕嫉妒恨吧！

图 1.1　example1_2.html 运行效果

页面文件 example1_2.html 的代码如下：

```html
<!DOCTYPE html>
<html>
<head>
<meta charset="UTF-8">
<title>Insert title here</title>
</head>
<body>
    <p align="center">《赠汪伦》<br><font size="2">李白</font><br>
        李白乘舟将欲行，<br>
        忽闻岸上踏歌声。<br>
        桃花潭水深千尺，<br>
        不及汪伦送我情。<br>
    </p>
    <marquee bgcolor="blue" behavior="alternate">
        <font color="white">大家好，我正在学习《古诗三百首》,哈哈,羡慕嫉妒恨吧!</font>
    </marquee>
</body>
</html>
```

❹ 列表

1）无序列表标记<ul>

<ul>标记用于设置无序列表，在每个列表项目文字之前，以项目符号作为每条列表项的前缀，各个列表没有级别之分。无序列表的语法格式如下：

```
<ul>
    <li>列表项</li>
    <li>列表项</li>
    …
</ul>
```

无序列表的项目符号在默认情况下是●，而通过<ul>标记的 type 属性可以改变无序列表的项目符号，避免项目符号单调。type 的可取值为 disc、circle 和 square，分别代表●、○和■。

2）有序列表标记<ol>

有序列表中的项目以数字或英文字母开头，通常各项目之间是有先后顺序的。有序列表的语法格式如下：

```
<ol>
    <li>列表项</li>
    <li>列表项</li>
    …
</ol>
```

有序列表和无序列表一样，也有项目类型，也可以通过 type 属性设置自己的项目类型。在默认情况下，有序列表的项目序号是数字。有序列表的 type 属性的取值如表 1.1 所示。用户也可以通过 start 属性改变项目序号的起始值，虽然起始值只能是数字，但同样对字母或罗马数字起作用。例如项目类型为 a、起始值为 5，那么项目序号就从英文字母 e 开始编号。

表 1.1 有序列表的 type 属性的取值

属 性 值	项 目 序 号
1	1、2、3、4…
a	a、b、c、d…
A	A、B、C、D…
i	i、ii、iii、iv…
I	I、II、III、IV…

例 1-3 编写一个网页（运行效果如图 1.2 所示），在网页中分别定义一个无序列表和一个有序列表，无序列表的项目符号为○，有序列表的项目序号为 a、b、c、d…。

无序列表--车类

○ 小轿车
○ 小货车
○ 重型卡车

计算机网络专业的学生应该具备的能力

a. 办公自动化能力
b. 计算机硬件选购与测试能力
c. 计算机组装与维护能力
d. 网站建设与维护能力
e. 动态网页设计能力

图 1.2 example1_3.html 运行效果

页面文件 example1_3.html 的代码如下：

```
<!DOCTYPE html>
<html>
<head>
<meta charset="UTF-8">
<title>Insert title here</title>
</head>
<body>
    <h2>无序列表--车类</h2>
    <ul type="circle">
        <li>小轿车</li>
        <li>小货车</li>
        <li>重型卡车</li>
    </ul>
    <h2>计算机网络专业的学生应该具备的能力</h2>
    <ol type="a">
        <li>办公自动化能力</li>
        <li>计算机硬件选购与测试能力</li>
        <li>计算机组装与维护能力</li>
        <li>网站建设与维护能力</li>
        <li>动态网页设计能力</li>
    </ol>
</body>
</html>
```

❺ 图像与多媒体

HTML 图像通过<img>标记进行插入。<img>标记有很多属性，其中 src 属性是必需的，它指定要插入图像文件的位置与名称。img 标记的语法格式如下：

```
<img src="图像文件的路径及名称">
```

在网页中可以使用<bgsound>标记给网页添加背景音乐，语法格式如下：

```
<bgsound src="音乐文件的路径及名称" loop="播放次数">
```

在网页中可以使用<embed>标记将多媒体文件（例如 Flash 动画、MP3 音乐、ASF 视频等）添加到网页中，语法格式如下：

```
<embed src="多媒体文件的路径及名称" width="播放器的宽度" height="播放器的高度">
</embed>
```

图像与多媒体文件的路径可以是相对路径，也可以是绝对路径。绝对路径是完全路径，是文件在硬盘上的真正路径；相对路径是以当前文件所在的路径为基准进行相对文件的查找。

❻ 超链接

超链接的作用是建立从一个位置到另一个位置的链接。利用超链接不仅可以进行网页间的相互访问，还可以使网页链接到其他相关的多媒体文件上。

超链接标记<a>是一个非常重要的标记，它可以成对出现在文档的任何位置，其语法格式如下：

```
<a href="链接路径" target="目标窗口的打开方式">链接内容</a>
```

其中，"链接内容"可以是文字内容，也可以是一张图片。target 属性的值可以为_self、_blank、_top 以及_parent，_self 是 target 属性的默认值。_blank 表示目标页面会在一个新的空白窗口中打开。_top 表示目标页面会在顶层框架中打开。_parent 表示目标页面会在当前框架的上一层打开。

【例】1-4　假设有 3 个文件，分别为 index.html、addGoods.html 和 updateGoods.html。其中 index.html 是起始页面，addGoods.html 和 updateGoods.html 在 goods 文件夹下，goods 文件夹和 index.html 在同一目录下。在 index.html 中可以链接到后面两个页面上。

index.html 的代码如下：

```
<!DOCTYPE html>
<html>
<head>
<meta charset="UTF-8">
<title>电子商务后台首页</title>
</head>
<body>
    <ul>
        <li><a href="index.html">首页</a></li>
        <li><a href="goods/addGoods.html">添加商品</a></li>
        <li><a href="goods/updateGoods.html">修改商品</a></li>
    </ul>
    <p>首页</p>
</body>
</html>
```

在例 1-4 中，由于 addGoods.html、updateGoods.html 和 index.html 的相对路径为 goods/，在 addGoods.html 中需要跳转到 index.html 和 updateGoods.html 两个页面上。

addGoods.html 的代码如下：

```
<!DOCTYPE html>
<html>
<head>
<meta charset="UTF-8">
<title>添加商品</title>
</head>
<body>
    <ul>
        <li><a href="../index.html">首页</a></li>
        <li><a href="addGoods.html">添加商品</a></li>
        <li><a href="updateGoods.html">修改商品</a></li>
```

```
        </ul>
        <p>添加商品页面</p>
</body>
</html>
```

由于起始页面 index.html 位于当前页面（addGoods.html）的上一级，所以从当前页面到起始页面的链接写为../index.html；而 updateGoods.html 和 addGoods.html 在同一级别目录下，可以省略路径名。

❼ **表格**

一个表格由行、列和单元格构成，可以有多行，每行可以有多个单元格。创建表格要以<table>标记开始、以</table>标记结束，语法格式如下：

```
<table>
    <tr>
        <td>单元格中的内容</td>
        <td>单元格中的内容</td>
        ...
    </tr>
    <tr>
        <td>单元格中的内容</td>
        <td>单元格中的内容</td>
        ...
    </tr>
    ...
</table>
```

在一个表格中包含几组<tr>和</tr>标记就表示该表格有几行，在一行中包含几组<td>和</td>标记就表示该行中有几个单元格。

在制作表格时可能需要某个单元格占据多列的位置，这时就要使用单元格的 colspan 属性设置单元格所跨的列数，语法格式如下：

```
<td colspan="跨的列数值">
```

例如下表第 1 行的单元格的水平跨度为 5。

同样，当需要某个单元格占据多行的位置时就要使用 rowspan 属性设置该单元格所跨的行数。

```
<td rowspan="跨的行数值">
```

例如下表第 1 个单元格的垂直跨度为 2。

例1-5 编写网页 example1_5.html,在网页中有一个表格,表格标题为"个人简历"、边框宽度为 1、边框颜色为 green。网页运行效果如图 1.3 所示。

个人简历				
基本资料	姓名	李小白	性别	女
	政治面貌	群众	出生日期	1988-12-09
业余爱好				
替身	各种球类（包括中国足球）	看言情	爬山	压马路

图 1.3　example1_5.html 运行效果

页面文件 example1_5.html 的代码如下：

```html
<!DOCTYPE html>
<html>
<head>
<meta charset="UTF-8">
<title>example1_5.html</title>
</head>
<body>
    <table border="1" style="border-color: green" >
        <caption>个人简历</caption>
        <tr>
            <th rowspan="2" align="left">基本资料</th>
            <th align="left">姓名</th>
            <td>李小白</td>
            <th align="left">性别</th>
            <td>女</td>
        </tr>
        <tr>
            <th align="left">政治面貌</th>
            <td>群众</td>
            <th align="left">出生日期</th>
            <td>1988-12-09</td>
        </tr>
        <tr>
            <th colspan="5">业余爱好</th>
        </tr>
        <tr>
            <td>替身</td>
            <td>各种球类（包括中国足球）</td>
            <td>看言情</td>
            <td>爬山</td>
            <td>压马路</td>
        </tr>
    </table>
</body>
</html>
```

❽ 表单

表单是将用户输入的信息封装提交给服务器，实现与用户的交互。表单是用<form>标记定义的，它类似于一个容器，表单对象必须在表单中才有效，例如 input。定义表单的语法格式如下：

```
<form action="表单的处理程序">
...
    input 元素
...
</form>
```

<form>标记有很多属性，如表 1.2 所示。

表 1.2 <form>标记的属性及含义

属 性 名	含 义 描 述
action	action 属性值是表单提交的地址，即表单收集到信息后传递到的程序地址，例如某页面
name	name 属性用于给表单命名。表单名称可以控制表单与后台程序之间的关系
method	method 属性用于指定使用哪种提交方法将表单数据提交到服务器。在默认情况下，提交方法为 get。get 方法将表单内容附在 URL 地址后面，因此有长度限制（最大 8192 个字符），而且不安全。post 方法将用户在表单中输入的数据包含在表单主体中，一起提交给服务器，该方法没有信息长度的限制，也比较安全
enctype	enctype 属性用于指定表单信息提交的编码方式，这个编码方式通常情况下采用默认的（application/x-www-form-urlencoded）即可，但上传文件时必须选择 mime 编码（multipart/form-data）

<input>标记是最常用的表单标记，该标记允许用户在表单中（文本框、单选框、复选框等）输入信息，输入类型是由类型属性（type）定义的。常用的输入类型如下：

1）文本框与密码框

<input>标记的 type 属性值为 text，代表的是单行文本框，在其中可以输入任何类型的文本、数字或字母，输入的内容以单行显示。<input>标记的 type 属性值为 password，代表的是密码框，在其中输入的字符都以星号（*）或圆点（●）显示。例如下面的原始代码：

```
<form>
    姓名：<input type="text" value="" name="userName"/>
    <br>
    密码：<input type="password" value="123456" name="pwd"/>
</form>
```

上述原始代码呈现的结果如图 1.4 所示。

2）单选按钮与复选框

单选按钮用来让用户进行单一选择，例如让用户选择性别。单选按钮在页面中以圆圈（〇）显示，语法格式如下：

图 1.4 文本框与密码框

```
<input type="radio" value="单选按钮的值" name="单选按钮的名称" checked/>
```

其中，name 代表单选按钮的名称，一组单选按钮的名称都相同，action 处理程序通过

name 获取被选中的单选按钮的 value 值。checked 表示该单选按钮被选中，在一组单选按钮中只有一个单选按钮被设置为 checked。例如下面的原始代码：

```
<form>
        <input type="radio" value="male" name="sex" checked/>男
        <input type="radio" value="female" name="sex" />女
</form>
```

上述原始代码呈现的结果如图 1.5 所示。

大家经常会看到"请选择您喜欢的歌手：□张三 □李四 □王五"，这样的网页就是使用复选框实现的。复选框与单选按钮不同的是能够实现选项的多选，它以一个方框（□）表示，语法格式如下：

图 1.5　单选按钮

```
<input type="checkbox" value="复选框的值" name="复选框的名称" checked/>
```

其中，当用户选中复选框后，value 属性的值传递给处理程序。name 代表的是复选框的名称，一组复选框的名称都相同，处理程序通过 name 获取被选中的复选框的 value 值（以数组的形式返回，数组元素为被选中的复选框的 value 值）。checked 表示该复选框被选中，在一组复选框中可以同时有多个被选中。例如下面的原始代码：

```
<form>
    <input type="checkbox" value="zhangsan" name="lover" checked/>张三
    <input type="checkbox" value="lisi" name="lover" checked/>李四
    <input type="checkbox" value="wangwu" name="lover"/>王五
</form>
```

上述原始代码呈现的结果如图 1.6 所示。

3）按钮

在网页的表单中，按钮起到至关重要的作用。没有按钮，网页很难和用户进行交互。单击按钮可以激发提交表单的动作（提交按钮）；也可以将表单恢复到初始的状态（重置按钮）；还可以根据程序的要求发挥其他的作用（普通按钮）。

图 1.6　复选框

普通按钮主要是配合脚本语言（JavaScript）来进行表单的处理，语法格式如下：

```
<input type="button" value="按钮的值" name="按钮的名称"/>
```

其中，value 的取值就是显示在按钮上的文字，在普通按钮中可以添加 onclick、onfocus 等 JavaScript 事件实现特定的功能。

当用户在表单中输入信息后，想清除输入的信息，将表单恢复成初始状态时，需要使用重置按钮，语法格式如下：

```
<input type="reset" value="按钮的值" name="按钮的名称"/>
```

当用户单击提交按钮时可以实现表单内容的提交，语法格式如下：

```
<input type="submit" value="按钮的值" name="按钮的名称"/>
```

例如下面的原始代码：

```
<form>
```

```
姓名：<input type="text" name="userName"/><br>
<input type="submit" value="提交"/>
<input type="reset" value="重置"/>
<input type="button" value="关闭" onclick="window.close()"/>
</form>
```

上述原始代码呈现的结果如图 1.7 所示。

4）文件域

大家经常会碰到上传照片或文件的功能，这就要用到文件域。文件域是由一个文本框和一个"浏览"按钮组成的。当用户上传文件时，可以直接在文本框中输入要上传文件的路径，也可以单击"浏览"按钮选择文件，语法格式如下：

图 1.7　按钮

```
<input type="file" name="文件域的名称"/>
```

在使用文件域上传文件时，一定不要忘记设置 form 表单信息提交的编码方式为 enctype="multipart/form-data"。例如下面的原始代码：

```
<form action="" enctype="multipart/form-data">
    你的靓照：<input type="file" name="fileName"/>
</form>
```

上述原始代码呈现的结果如图 1.8 所示。

5）下拉列表

下拉列表的语法格式如下：

图 1.8　文件域

```
<select name="下拉列表的名称" size="显示的项数" multiple >
    <option value="选项值 1" selected>选项 1 显示内容
    <option value="选项值 2" >选项 2 显示内容
    …
</select>
```

其中，选项值是提交给服务器的值，而选项显示内容才是真正在页面中要显示的。selected 表示此选项在默认状态下是选中的，size 用来设定列表在页面中最多显示的项数，当超出这个值时就会出现滚动条。multiple 表示列表可以进行多项选择。例如下面的原始代码：

```
<select name="cities" size="2" multiple >
    <option value="beijing" selected>北京
    <option value="shanghai" selected>上海
    <option value="guangzhou" >广州
    <option value="shenzhen" >深圳
</select>
```

上述原始代码呈现的结果如图 1.9 所示。

6）文本区

文本区用来输入多行文本。文本区和其他表单控件不一样，它使用的是

图 1.9　列表

<textarea>标记而不是<input>标记,语法格式如下:

```
<textarea name="文本区的名称" cols="列数" rows="行数"></textarea>
```

其中,cols 用于设定文本区的列数,也就是其宽度;rows 用于设定文本区的行数,也就是高度,当文本区的内容超出这一范围时就会出现滚动条。

1.1.4 实践环节——制作调查问卷

编写网页 practice1_1.html,具体要求如下:

① 网页中有个 form 表单,表单处理程序为本页面程序,表单提交方式为"post",表单提交编码方式为"multipart/form-data"。

② 页面运行效果如图 1.10 所示。

图 1.10　practice1_1.html 页面运行效果

1.2　CSS

用户不需要使用复杂的工具来创建 CSS 文件,可以使用文本编辑器或者 Web 开发工具来创建。无论采用哪种方式,都需要创建一个以.css 为扩展名的文件。

1.2.1　CSS 基本语法

CSS 的语法由 3 个部分构成,即选择符(selector)、属性(property)和属性值(value)。其语法格式如下:

```
选择符 {
    属性:值
}
```

其中，选择符用来指定针对哪个 HTML 标签应用样式表，任何一个 HTML 标签都可以是一个 CSS 的选择符。例如：

```
body {
    color: blue
}
```

其中，body 就是选择符，color 就是属性，blue 就是属性值。该规则表示在网页 body 标签里的内容为蓝色。为选择符指定多个样式，需要在属性之间用分号加以分隔。下面的选择符 p 就包含两个样式，一个是对齐方式居中，一个是字体颜色为红。

```
p {
    text-align: center;
    color: red
}
```

用户可以将相同的属性和属性值赋予多个选择符（组合），选择符之间用逗号分隔。例如：

```
h1,h2,h3,h4,h5,h6
{
    /*字体颜色为蓝色*/
    color: blue
}
```

该规则是将所有正文标题（<h1>到<h6>）的字体颜色都变成蓝色。"/*"和"*/"之间的内容为 CSS 的注释，但要注意不要将注释嵌入到选择符语句里面。

1.2.2　在网页中添加 CSS 的方法

CSS 在网页中按位置可以分为 3 种，即内嵌样式、内部样式和外部样式。

❶ 内嵌样式

内嵌样式是将样式代码写在标记里面，使用 style 作为属性，样式语句作为属性值。内嵌样式只对所在标记有效。例如：

```
<p style="font-size:20pt; color:red">
这个 style 定义<p></p>里面的文字是 20pt 字号，字体颜色是红色。
</p>
```

❷ 内部样式

内部样式是使用<style>标记将样式代码写在 HTML 的<head></head>里面，内部样式只对所在网页有效。例如：

```
<!DOCTYPE html>
<html>
<head>
```

```
<meta charset="UTF-8">
<style type="text/css">
    h1 {
        border-width:1;
        text-align:center;
        color:red
    }
</style>
</head>
<body>
    <h1>这个标题使用了 style。</h1>
</body>
</html>
```

❸ **外部样式**

1）链接样式表

将样式代码写在一个以.css 为扩展名的 CSS 文件中，然后在每个需要用到这些样式的网页中引用这个 CSS 文件。通过 HTML 的 link 元素将外部的样式文件链接到网页中。例如：

```
<!DOCTYPE html>
<html>
<head>
<meta charset="UTF-8">
<title>Insert title here</title>
<link rel="stylesheet" href="mystyle.css" type="text/css" />
</head>
<body>
    <h1>标题</h1>
    <p>段落内容</p>
</body>
</html>
```

其中，rel 和 type 属性表明这是一个样式文件，href 属性指定了外部样式文件的相对地址。外部的样式文件不能含有任何像<head>或<style>这样的 HTML 标记，样式表文件仅由样式规则或声明组成。mystyle.css 文件的内容如下：

```
p{
    background: yellow;
}
```

2）导入样式表

在<style>标签内使用@import 导入外部样式文件。例如：

```
<!DOCTYPE html>
<html>
<head>
```

```
<meta charset="UTF-8">
<title>Insert title here</title>
<style type="text/css">
    <!--
    @import url("mystyle.css");
    h1{color:red}
    -->
</style>
</head>
<body>
    <h1>标题</h1>
    <p>段落内容</p>
</body>
</html>
```

使用外部样式相对于使用内嵌样式和内部样式有以下优点。

① 样式代码可以复用：一个外部 CSS 文件可以被很多网页共用。

② 便于修改：如果要修改样式，只需要修改 CSS 文件，而不需要修改每个网页。

③ 提高网页显示的速度：如果样式写在网页里，会降低网页显示的速度，如果网页引用一个 CSS 文件，这个 CSS 文件已经在缓存区（其他网页早已经引用过它），网页显示的速度就比较快。

因此，在实际开发中一般使用外部样式，不推荐使用内嵌样式和内部样式。

1.2.3 选择符的分类

❶ 普通选择符

任意的 HTML 标记都可以作为选择符，这样的选择符称为普通选择符。其样式仅仅作用在选择符指定的 HTML 标记上。例如：

```
p {
    color:red
}
```

❷ 类选择符

HTML 标记名加上类名，中间用"."号分开，类名供该 HTML 标记的 class 属性使用。如果希望<p>有两种样式，一种是居中对齐，一种是居右对齐，那么可以写成如下的样式：

```
p.right {
    text-align:right
}
p.center {
    text-align:center
}
```

其中，right 和 center 就是两个 class。在网页中可以引用这两个 class 设置段落的对齐方式。

示例代码如下：

```
<p class="center">这一段内容是居中显示。</p>
<p class="right">这一段内容是居右显示。</p>
```

如果省略 HTML 标记名只写".类名"，表示这个类名适用于所有的 HTML 标记的 class 属性，这种选择符称为通用类选择符。

❸ id 选择符

HTML 标记名加上 id 名，中间用"#"号分开，id 名供该 HTML 标记的 id 属性使用。例如：

```
p#svp {
    font-size:12pt
}
```

其中，svp 是个 id 选择符的名字，在网页中可以引用这个 id 选择符设置<p>的样式。示例代码如下：

```
<p id="svp">这一段话的字体大小为 12pt。</p>
```

如果省略 HTML 标记名只写"#id 名"，表示这个 id 名适用于所有的 HTML 标记的 id 属性，这种选择符称为通用 id 选择符。

1.2.4 伪类及伪对象

伪类及伪对象由 CSS 自动支持，属于 CSS 的一种扩展类型。其名称不能被用户自定义，在使用时只能按照标准格式进行应用。

❶ 超链接伪类

超链接伪类共有 4 个，它们是 a:link、a:visited、a:hover 和 a:active。a:link 表示未被访问的链接，a:visited 表示已被访问过的链接，a:hover 表示鼠标指针悬浮在其上的链接，a:active 表示鼠标点中激活链接。由于优先级的关系，在写超链接<a>的 CSS 时一定要按照 a:link、a:visited、a:hover、a:active 的顺序书写。例如：

```
a:link {color: red}           /*未被访问的链接，红色*/
a:visited {color: green}      /*已被访问过的链接，绿色*/
a:hover {color: yellow}       /*鼠标指针悬浮在其上的链接，黄色*/
a:active {color: blue}        /*鼠标点中激活的链接，蓝色*/
```

❷ 常用伪对象

:first-letter 设置对象内的第一个字符的样式表属性，例如设置段落标记的第一个字符的样式代码如下：

```
p:first-letter {
    color: red;
    font-size: 16px
}
```

:first-line 设置对象内的第一行的样式表属性,例如设置 body 对象里第一行的样式代码如下:

```css
body:first-line {
    color: red;
    font-size: 16px
}
```

1.2.5 常见的 DIV+CSS 布局类型

DIV+CSS 布局是当前网页布局中最流行的类型之一。

❶ DIV

DIV 是一个放置内容的容器,用于大面积、大区域的块状排版,样式需要编写 CSS 来实现。

例 1-6 DIV 示例。

```html
<!DOCTYPE html>
<html>
<head>
    <meta charset="UTF-8">
    <title>div_css1</title>
    <style type="text/css">
    .mainBox {
        border: 1px dashed #0099CC;
        margin: 3px;
        padding: 0px;
        float: left;
        height: 300px;
        width: 192px;
    }
    .mainBox h3 {
        float: left;
        height: 20px;
        width: 179px;
        color: #FFFFFF;
        padding: 6px 3px 3px 10px;
        background-color: #0099CC;
        font-size: 16px;
    }
    .mainBox p {
        line-height: 1.5em;
        text-indent: 2em;
        margin: 35px 5px 5px 5px;
    }
```

```
        </style>
    </head>
    <body>
        <div class="mainBox">
            <h3>前言</h3>
            <p>正文内容</p>
        </div>
        <div class="mainBox">
            <h3>CSS 盒子模式</h3>
            <p>正文内容</p>
        </div>
        <div class="mainBox">
            <h3>转变思想</h3>
            <p>正文内容</p>
        </div>
    </body>
</html>
```

上述代码的运行效果如图 1.11 所示。

图 1.11 DIV 示例

❷ 一列固定

宽度的属性值是固定像素，无论怎样改变浏览器窗口的大小，DIV 的宽度都不改变。

【例】1-7 一列固定的示例。

```
<!DOCTYPE html>
<html>
<head>
  <meta charset="UTF-8">
  <title>div_css2</title>
  <style type="text/css">
  .oneFixed {
    border: 1px dashed #0099CC;
    background-color: cyan;
```

```
        height: 300px;
        width: 300px
    }
    </style>
</head>
<body>
    <div class="oneFixed">
        一列固定宽度
    </div>
</body>
</html>
```

上述代码的运行效果如图 1.12 所示。

图 1.12 一列固定宽度

❸ 两列固定宽度

两列的布局需要用到两个 DIV，宽度的属性值是固定像素。

例1-8 两列固定宽度示例。

```
<!DOCTYPE html>
<html>
<head>
    <meta charset="UTF-8">
    <title>div_css3</title>
    <style type="text/css">
    #left {
        border: 1px dashed #0099CC;
        background-color: cyan;
        height: 200px;
        width: 300px;
        float: left
    }
    #right{
        border: 1px dashed #0099CC;
```

```
            background-color: LightSkyBlue;
            height: 200px;
            width: 300px;
            float: left
        }
    </style>
</head>
<body>
    <div id="left">
        左边
    </div>
    <div id="right">
        右边
    </div>
</body>
</html>
```

上述代码的运行效果如图 1.13 所示。

图 1.13 两列固定宽度

❹ 三列浮动中间宽度自适应

三列浮动中间宽度自适应就是要求左边 DIV 固定宽度且左显示,右边 DIV 固定宽度且右显示,中间 DIV 根据左右 DIV 的间距变化宽度自适应。

例1-9 三列浮动中间宽度自适应示例。

```
<!DOCTYPE html>
<html>
<head>
    <meta charset="UTF-8">
    <title>div_css4</title>
    <style type="text/css">
    body{
        margin: 0px
    }
    #left {
        border: 1px solid #0099CC;
        background-color: cyan;
        height: 300px;
        width: 100px;
```

```
        position: absolute;
        top: 0px;
        left: 0px
    }
    #center{
        border: 1px solid #0099CC;
        background-color: #7FFFAA;
        margin-left: 100px;
        margin-right: 100px;
        height: 300px
    }
    #right{
        border: 1px solid #0099CC;
        background-color: cyan;
        height: 300px;
        width: 100px;
        position: absolute;
        right: 0px;
        top: 0px
    }
    </style>
</head>
<body>
    <div id="left">
        左边
    </div>
    <div id="center">
        中间自适应
    </div>
    <div id="right">
        右边
    </div>
</body>
</html>
```

上述代码的运行效果如图 1.14 所示。

图 1.14　三列浮动中间宽度自适应

❺ 三行两列居中高度自适应

三行两列居中高度自适应就是要求整个网页内容居中，第一行 DIV 固定高度且居顶端显示，第三行 DIV 固定高度且居底端显示，中间 DIV 根据内容的变化高度自适应。

例 1-10　三行两列居中高度自适应示例。

```html
<!DOCTYPE html>
<html>
<head>
    <meta charset="UTF-8">
    <title>div_css5</title>
    <style type="text/css">
    body {
        background: #999;
        text-align: center;
        color: #333;
        font-family: arial, verdana, sans-serif;
    }

    #header {
        width: 776px;
        margin-right: auto;
        margin-left: auto;
        padding: 0px;
        background: #EEE;
        height: 60px;
        text-align: left;
    }

    #contain {
        margin-right: auto;
        margin-left: auto;
        width: 776px;
    }

    #mainbg {
        width: 776px;
        padding: 0px;
        background: #60A179;
        float: left;
    }

    #right {
        float: right;
        margin: 2px 0px 2px 0px;
        padding: 0px;
```

```css
        width: 574px;
        background: #ccd2de;
        text-align: left;
}

#left {
    float: left;
    margin: 2px 2px 0px 0px;
    padding: 0px;
    background: #F2F3F7;
    width: 200px;
    text-align: left;
}

#footer {
    clear: both;
    width: 776px;
    margin-right: auto;
    margin-left: auto;
    padding: 0px;
    background: #EEE;
    height: 60px;
}

.text {
    margin: 0px;
    padding: 20px;
}
</style>
</head>
<body>
    <div id="header">header</div>
    <div id="contain">
        <div id="mainbg">
            <div id="right">
                <div class="text">
                    right
                    <p>1</p>
                    <p>1</p>
                    <p>1</p>
                    <p>1</p>
                    <p>1</p>
                </div>
            </div>
            <div id="left">
```

```
                <div class="text">left</div>
            </div>
        </div>
    </div>
    <div id="footer">footer</div>
</body>
</html>
```

上述代码的运行效果如图 1.15 所示。

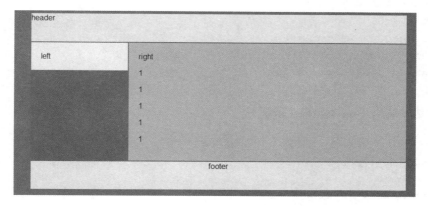

图 1.15　三行两列居中高度自适应

1.2.6　实践环节——页面布局

按照下面的步骤编写网页 practice1_2.html。

❶ 使用 DIV 定义结构

一个典型的版面分栏结构为页头、导航栏、内容、版权。
结构代码如下：

```
<div id="header"></div>
<div id="navigator"></div>
<div id="content"></div>
<div id="footer"></div>
```

将这 4 个盒子装进一个更大的盒子——body 中，如下：

```
<body>
    上面 4 行代码
</body>
```

❷ 定义 body 的属性

```
body {
    font-family: Arial, Helvetica, sans-serif;
    font-size: 12px;
```

```css
    margin: 0px auto;
    height: auto;
    width: 800px;
    border: 1px solid #006633;
}
```

❸ 定义页头（header）的属性

```css
#header {
    height: 100px;
    width: 800px;
    background-image: url(plane.jpg);
    background-repeat: no-repeat;
    margin: 0px 0px 3px 0px;
}
```

❹ 定义导航栏（navigator）的属性

```css
#navigator {  /*定义一个导航栏的长盒子*/
    height: 25px;
    width: 800px;
    font-size: 14px;
    list-style-type: none;   /*让 navigator 这个大盒子下面的小盒子 li 列表样式不显
                               示，这对于标准浏览器很重要*/
}

#navigator li {
    float: left;             /*让 li 这些小盒子左对齐*/
}

#navigator li a {
    color: #000000;
    text-decoration: none;   /*让 li 盒子里面的链接样式没有下画线*/
    padding-top: 4px;
    display: block;          /*让 li 里面的链接以块状呈现，就像一个按钮，而不必一定
                               要点中链接文字才起作用*/
    width: 131px;
    height: 22px;
    text-align: center;
    background-color: #009966;
    margin-left: 2px;
}

#navigator li a:hover {
    background-color: #006633;  /*鼠标指针移到链接盒子上面改变盒子的背景色*/
    color: #FFFFFF;
}
```

❺ 定义内容部分（content）的属性

```css
#content {
    height: auto;
    width: 780px;
    line-height: 1.5em;
    padding: 10px;
}

#content p {
    text-indent: 2em;
}

#content h3 {
    font-size: 16px;
    margin: 10px;
}
```

❻ 定义页脚（footer）的属性

```css
#footer {
    height: 50px;
    width: 780px;
    line-height: 2em;
    text-align: center;
    background-color: #009966;
    padding: 10px;
}
```

❼ 定义各标记的边界和填充（开头处）

```css
* {
    margin: 0px;
    padding: 0px;
}
```

❽ 结构代码

```html
<body>
    <div id="header"></div>
    <div id="navigator">
        <ul id="navigator">
            <li><a href="#">首页</a></li>
            <li><a href="#">文章</a></li>
            <li><a href="#">相册</a></li>
            <li><a href="#">Blog</a></li>
            <li><a href="#">论坛</a></li>
            <li><a href="#">帮助</a></li>
        </ul>
```

```
        </div>
        <div id="content">
            <h3>前言</h3>
            <p>CSS 是英文 Cascading Style Sheet 的缩写,又称为"层叠样式表",简称为样
            式表。它是 W3C 定义的标准,一种用来为结构化文档(如 HTML 文档)添加样式(字体、
            间距和背景等)的计算机语言。</p>
            <h3>理解 DIV+CSS 布局</h3>
            <p>简单地说 DIV+CSS(DIV CSS)被称为"Web 标准"中常用术语之一。首先认识 DIV
            是用于搭建 html 网页结构(框架)标签,再认识 CSS 是用于创建网页表现(样式/美化)
            样式表的统称,通过 CSS 来设置 DIV 标签样式,这一切我们常常称之为 DIV+CSS。
            </p>
        </div>
        <div id="footer">
            <p>关于| 广告服务| 招聘| 客服中心| Q Q 留言| 网站管理| 会员登录| 购物车</p>
            <p>Copyright ©清华大学出版社</p>
        </div>
</body>
```

❾ 运行页面

页面运行效果如图 1.16 所示。

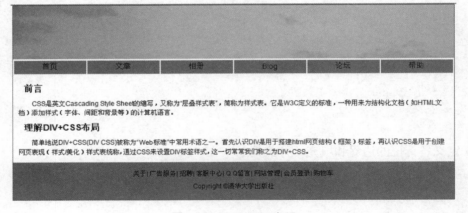

图 1.16 DIV+CSS 布局

1.3 JavaScript

JavaScript 和 CSS 一样,不需要使用复杂的工具来编写,可以使用文本编辑器或者 Web 开发工具来编写。

1.3.1 在网页中添加 JavaScript 的方法

❶ 嵌入使用

在网页代码中的任何位置都可以嵌入 JavaScript 代码,但建议嵌入到<head>标记中。

示例代码如下:

```html
<!DOCTYPE html>
<html>
<head>
<meta charset="UTF-8">
<title>JavaScript 嵌入学习</title>
<script type="text/javascript">
    alert("第一次看到警告框很兴奋!");
</script>
</head>
<body>
    好好学习JavaScript 知识。
</body>
</html>
```

❷ 引入使用

当多个页面使用相同的 JavaScript 代码时,可以将共用的代码保存在以.js 为扩展名的文件中,然后在页面中使用 src 属性引入外部 JS 文件。例如:

1) myFirst.js

```
alert("被引入到页面中!");
```

2) 引入外部 JS 文件学习.html

```html
<!DOCTYPE html>
<html>
<head>
<meta charset="UTF-8">
<title>引入外部 JS 文件</title>
<script type="text/javascript" src="myFirst.js" charset="GBK">
</script>
</head>
<body>
    好好学习JavaScript 知识。
</body>
</html>
```

1.3.2　JavaScript 基本语法

❶ 变量

使用 var 可以声明任意类型的变量,例如:

```
var firstNumber = 10;
```

❷ 类型转换

JavaScript 是弱类型语言,变量的类型对应于其值的类型。用户可以对不同类型的变量执行运算，解释器强制转换数据类型，然后进行处理。例如数值与字符串相加，将数值强制转换为字符串；布尔值与字符串相加，将布尔值强制转换为字符串；数值与布尔值相加，将布尔值强制转换为数值。

字符串到数值的转换。例如 parseInt(s)将字符串转换为整数，parseFloat(s)将字符串转换为浮点数，Number(s)将字符串转换为数字。注意，parseInt()和 parseFloat()方法只对 string 类型有效，且需要是以数字开头的字符串。

❸ 运算符

1）赋值运算符

赋值运算符的运算规则及说明如表 1.3 所示。

表 1.3　赋值运算符

运　算　符	示　　例	说　　明
=	x=y;	将变量 y 的值赋给 x
+=	x+=y;	x=x+y;
-=	x-=y;	x=x-y;
*=	x*=y;	x=x*y;
/=	x/=y;	x=x/y;

2）数学运算符

数学运算符的运算规则及说明如表 1.4 所示。

表 1.4　数学运算符

运　算　符	示　　例	说　　明
+	A = 5 + 8　　//结果是 13 A ="5"+ 8　　//结果是"58"	如果操作数都是数字，执行加法运算，如果其中的操作数有字符串，会执行连接字符串的运算
-	A = 8 - 5	减法
*	A = 8 * 5	乘法
/	A = 8 / 5　　//结果是 1.6	除法
%	10 % 3 = 1	取余
++	++x 返回递增后的 x 值 x++返回递增前的 x 值	递增
--	--x 返回递减后的 x 值 x--返回递减前的 x 值	递减
-	如果 a 等于 5，则-a=-5	此运算符返回操作数的相反数

3）逻辑运算符

逻辑运算符的运算规则及说明如表 1.5 所示。

表 1.5　逻辑运算符

运　算　符	示　　例	说　　明
&&	expr1 && expr2	逻辑与（表达式 1 错误，表达式 2 不再运算）
||	expr1 || expr2	逻辑或（表达式 1 正确，表达式 2 不再运算）
!	!expr	逻辑非

4）typeof 运算符

对变量或值调用 typeof 运算符将返回对应的值，typeof 运算符的运算规则及说明如

表 1.6 所示。

表 1.6 typeof 运算符

示　　例	返 回 结 果	说　　明
typeof(true)	boolean	变量或值是 boolean 类型
typeof(300)	number	变量或值是 number 类型
typeof('abc')	string	变量或值是 string 类型
typeof(null)	object	变量或值是一种引用类型或 null 类型
typeof(f)	function	变量是一个函数

❹ 注释

为了保证程序的可读性，以及便于日后修改和维护代码，可以在 JavaScript 程序中为代码写注释。在 JavaScript 程序里用两个斜杠"//"来表示单行注释；多行注释用"/*"表示开始，用"*/"表示结束。注释的示例代码如下：

```
aGoodIdea = "Comment your code thoroughly."; //这是单行注释
/*
这是多行注释 行一。
这是多行注释 行二。
*/
```

❺ 变量的命名规则

以字母、下画线（_）或美元符号（$）开头；余下的字符可以是下画线、美元符号，或任何的字母、数字；不能有空格，大小写敏感；不能使用 JavaScript 中的关键字或保留字命名。

❻ 部分保留字

break、delete、function、return、typeof、case、do、if、switch、var、catch、else、in、this、void、continue、false、instanceof、throw、while、finally、new、true、with、default、for、null、try。

例 1-11 编写网页 example1_11.html，在网页中嵌入使用 JavaScript。

JavaScript 程序的具体要求如下：

（1）声明变量 firstNumber，并将"I am a String"赋值给该变量。

（2）声明变量 secondNumber，并将 300 赋值给该变量。

（3）声明变量 thirdNumber，并将 firstNumber+secondNumber 赋值给该变量。

（4）使用 alert()弹出变量 thirdNumber 的值。

（5）使用 typeof 运算符判断变量 firstNumber、secondNumber 以及 thirdNumber 的值类型，并使用 alert()分别弹出 typeof 的返回结果。

页面文件 example1_11.html 的代码如下：

```
<!DOCTYPE html>
<html>
<head>
<meta charset="UTF-8">
<title>example1_11.html</title>
```

```
<script type="text/javascript">
    var firstNumber = "I am a String";
    var secondNumber = 300;
    var thirdNumber = firstNumber + secondNumber;
    alert(thirdNumber);
    alert("firstNumber 的值类型: " + typeof(firstNumber));
    alert("secondNumber 的值类型: " + typeof(secondNumber));
    alert("thirdNumber 的值类型: " + typeof(thirdNumber));
</script>
</head>
<body>
</body>
</html>
```

1.3.3 流程控制与函数

❶ 流程控制语句

1）if 条件语句

```
if(表达式){
    语句
}
```

或

```
if(表达式)){
    语句
}else{
    语句
}
```

2）switch 条件语句

```
switch(表达式){
    case case1:
        语句
        break;
    case case2:
        语句
        break;
    ...
    default:
        default 语句
}
```

3）for 循环语句

```
for(表达式1;表达式2;表达式3){
```

```
    语句
}
```

4）while 循环语句

```
while(表达式){
    语句
}
```

5）do-while 循环语句

```
do{
    语句
}while(表达式)
```

6）break/continue 语句

break 语句让执行语句从循环语句或其他程序块中跳出。continue 语句让执行语句跳过本次循环的剩余语句进入下一次循环。

❷ 函数

通常将完成某个功能的一组语句写成一个函数，函数的定义格式如下：

```
function 函数名([参数,参数]){
    函数体
}
```

其中，function 是关键字，函数没有类型，参数也没有类型。例如：

```
function gogo(obj){
    document.write("函数没有类型，参数也没有类型");
}
```

❸ arguments 对象

函数可以接收任意个数的参数，一般通过 arguments 对象来访问。示例代码如下：

```
function say(){
    if(arguments[1] != "你好"){
        alert(arguments[0]);
    }else{
        alert(arguments[1]);
    }
    alert(arguments.length);          //返回参数的个数
}
```

调用函数如下：

```
say("How are you?","你好");
```

❹ 系统函数

JavaScript 提供了与任何对象无关的系统函数，使用这些函数不需要创建任何对象就可以直接使用。

1）eval(字符串表达式)

该函数的功能是返回字符串表达式的值。例如：

```
var test = eval("2+3");              //test 的值为 5
```

2）parseInt(字符串)

该函数的功能是将以数字开头的字符串转换为整数。例如：

```
var test = parseInt("200.5abc");     //test 的值为 200
```

3）parseFloat(字符串)

该函数的功能是将以数字开头的字符串转换为浮点数。例如：

```
var test = parseFloat("200.5abc");   //test 的值为 200.5
```

4）Number(字符串)

该函数的功能是将数字字符串转换为数字，若字符串中有非数字字符，则返回 NaN。例如：

```
var test = Number("200.5abc");       //test 的值为 NaN
```

例1-12 编写网页 example1_12.html，在网页中嵌入使用 JavaScript 程序打印出九九乘法表，网页运行效果如图 1.17 所示。

```
1*1=1
2*1=2 2*2=4
3*1=3 3*2=6 3*3=9
4*1=4 4*2=8 4*3=12 4*4=16
5*1=5 5*2=10 5*3=15 5*4=20 5*5=25
6*1=6 6*2=12 6*3=18 6*4=24 6*5=30 6*6=36
7*1=7 7*2=14 7*3=21 7*4=28 7*5=35 7*6=42 7*7=49
8*1=8 8*2=16 8*3=24 8*4=32 8*5=40 8*6=48 8*7=56 8*8=64
9*1=9 9*2=18 9*3=27 9*4=36 9*5=45 9*6=54 9*7=63 9*8=72 9*9=81
```

图 1.17 九九乘法表

页面文件 example1_12.html 的代码如下：

```
<!DOCTYPE html>
<html>
<head>
<meta charset="UTF-8">
<title>example1_12.html</title>
<script type="text/javascript">
for(var i = 1;i <= 9; i++){
    for(var j = 1; j <= i; j++){
        document.write( i + "*" + j + "=" + i*j + "  ");//在页面输出
    }
    document.write("<br>");
}
</script>
</head>
```

```
<body>
</body>
</html>
```

❺ 函数调用

1）在链接中调用函数

当用户单击链接时调用函数，格式如下：

```
<a href="javascript:函数">…</a>
```

在链接中调用函数的示例代码如下：

```
<!DOCTYPE html>
<html>
<head>
<meta charset="UTF-8">
<title>链接调用函数</title>
<script type="text/javascript">
function gogo(){
    alert("被链接调用的函数");
}
</script>
</head>
<body>
    <a href="javascript:gogo()">链接调用函数</a>
</body>
</html>
```

2）由事件触发调用函数

触发事件调用函数的格式如下：

```
事件="函数"
```

触发事件调用函数的示例代码如下：

```
<!DOCTYPE html>
<html>
<head>
<meta charset="UTF-8">
<title>事件触发调用函数</title>
<script type="text/javascript">
function gogo(param){
    alert(param);
}
</script>
</head>
<body>
    <form action="">
```

```
            <input type="button" value="点击我" onclick="gogo('O(∩_∩)O哈哈~')"/>
        </form>
    </body>
</html>
```

1.3.4　JavaScript 对象

在一个 JavaScript 对象中可以包含若干属性和方法。属性是描述对象的状态，对象用"."运算符访问属性；方法是描述对象的行为动作，对象用"."运算符调用方法。

❶ **对象的创建**

使用 new 关键字来创建对象，例如：

```
var oStringObject = new String();
```

如果构造函数无参数，则不必加括号。

❷ **JavaScript 内部对象**

1）数组（Array）对象

（1）创建数组。

创建数组的示例代码如下：

```
var myArray = new Array();              //新建一个长度为 0 的数组
var myArray = new Array(100);           //新建一个长度为 100 的数组
var myArray = new Array(1,2,3);         //新建一个指定长度的数组并赋初值
```

数组长度不固定，赋值即可改变长度。数组的主要属性 length 返回数组长度。

（2）数组的常用方法。

- reverse()方法：将 JavaScript 数组对象内容反转。
- concat()方法：将两个或更多数组组合在一起。例如：

```
var newArray = tmpArray.concat(tmpArray);
```

- join()方法：返回一个将数组中的所有元素用指定符号连在一起的字符串。例如：

```
var newString = tmpArray.join(".");
```

- pop()方法：移除数组中的最后一个元素并返回该元素。
- shift()方法：移除数组中的第一个元素并返回该元素。
- slice()方法：返回数组的一部分。例如：

```
var newArray = tmpArray.slice(1,3);
```

（3）数组的使用。

使用数组的示例代码如下：

```
<script type="text/javascript">
    var myArray = new Array();
    for(var i = 0; i < 5; i++){
        myArray[i] = i;
```

```
        }
        for(var j = 0; j < myArray.length; j++){
            alert(myArray[j]);
        }
</script>
```

2）日期（Date）对象

Date 对象可以用来表示任意的日期和时间。

（1）创建 Date 对象。

用户必须使用 new 运算符创建一个 Date 对象。示例代码如下：

```
var date = new Date("July 8,2012");      //2012年7月8日
var date = new Date(2012,7,8);           //2012年7月8日
var date = new Date("2012/7/8") ;        //2012年7月8日
var date = new Date(Milliseconds);       //用自1970年1月1日以来的毫秒数创建的日
                                         //期对象
var date = new Date();                   //当前系统的时间对象
```

（2）获取日期的时间方法。

- getYear()：返回年数。
- getMonth()：返回当月号数（比实际值小 1）。
- getDate()：返回当日号数。
- getDay()：返回星期几（0 表示周日）。
- getHours()：返回小时数。
- getMinutes()：返回分钟数。
- getSeconds()：返回秒数。
- getTime()：返回毫秒数。

（3）设置日期和时间的方法。

- setYear()：设置年数。
- setMonth()：设置当月号数（set6 表示 7 月）。
- setDate()：设置当日号数。
- setHours()：设置小时数。
- setMinutes()：设置分钟数。
- setSeconds()：设置秒数。
- setTime()：设置毫秒数。

（4）Date 对象的使用。

使用 Date 对象的示例代码如下：

```
<script type="text/javascript">
    var date = new Date("2050/12/25");
    document.write("2050的圣诞节是星期"+date.getDay()+"<br>");
    var datenow = new Date();      //得到当前日期对象
    var mills = date - datenow;//两个 Date 对象相减得到两个日期的时间间隔（单位
                               //是毫秒）
```

```
            document.write("2050的圣诞节距离现在还有"+mills+"毫秒<br>");
</script>
```

3）String 对象

（1）创建 String 对象。

创建 String 对象的示例代码如下：

```
var firstString = "This is a string";
var secondString = new String("This is a string");
```

String 对象的主要属性 length 返回字符串的长度。

（2）String 对象的常用方法。

- charAt(i)：返回指定索引位置处的字符，索引从 0 开始。
- concat(str)：连接字符串。
- indexOf(str)：返回 String 对象内第一次出现子字符串的字符位置（从左到右查找）。
- lastIndexOf(str)：返回 String 对象中子字符串最后出现的位置。
- replace(str1,str2)：返回将 str1 替换为 str2 后的字符串。
- split(separator, limit)：将字符串以 separator 作为分割符切割成多个子字符串，并将它们作为一个数组返回；如果有 limit 参数，则返回数组的 limit 个元素。
- substring(start,end)：返回一个指定位置之间的子字符串，不包括 end。
- toLowerCase()：返回一个字符串，字符串中的字母被转换为小写字母。
- toUpperCase()：返回一个字符串，字符串中的字母被转换为大写字母。

（3）String 对象的使用。

使用 String 对象的示例代码如下：

```
<script type="text/javascript">
    var firstString = "This is a string";
    for(var i = 0; i < firstString.length; i++){
        alert(firstString.charAt(i));
    }
</script>
```

4）Math 对象

Math 对象是全局对象，在使用时不需要创建。

（1）Math 对象的属性。

- LN10：10 的自然对数。
- LN2：2 的对数。
- PI：圆周率。
- SQRT1_2：1/2 的平方根。
- SQRT2：2 的平方根。

（2）Math 对象的常用方法。

- abs(x)：返回 x 的绝对值。
- ceil(x)：返回大于或等于 x 的最小整数。

- floor(x)：返回小于或等于 x 的最大整数。
- round(x)：舍入到最近整数。
- sqrt(x)：返回 x 的平方根。
- random()：返回 0～1 的随机数。

（3）Math 对象的使用。

使用 Math 对象的示例代码如下：

```
<script type="text/javascript">
    alert(Math.SQRT2);
    alert(Math.random());
</script>
```

例1-13 编写网页 example1_13.html，在网页中定义一个 JavaScript 函数，功能是去除字符串开头及末尾的空格，并使用超链接调用该函数。

页面文件 example1_13.html 的代码如下：

```
<!DOCTYPE html>
<html>
<head>
    <meta charset="UTF-8">
    <title>example1_13.html</title>
    <script type="text/javascript">
        function print99(x){
            while((x.length>0) && (x.charAt(0) == ' '))
                x = x.substring(1,x.length);
            while(x.length>0&&(x.charAt(x.length-1) == ' '))
                x = x.substring(0,x.length-1);
            alert("Kill=" + x + "==")
            return x;
        }
    </script>
</head>
<body>
    <a href="javascript:print99('   abc def   ')">点击我啊！</a>
</body>
</html>
```

1.3.5 JavaScript 对象模型

❶ 浏览器对象模型

浏览器对象是提供独立于内容而与浏览器窗口进行交互的对象。浏览器对象模型如图 1.18 所示。

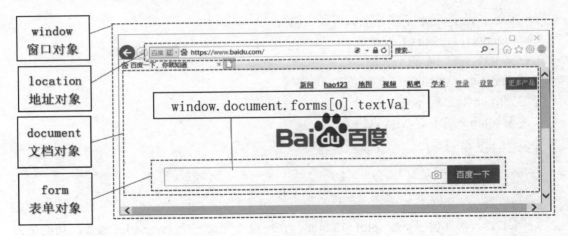

图 1.18 浏览器对象模型

❷ 窗口（window）对象

1）打开新窗口

使用 open() 方法可以打开一个新窗口，示例代码如下：

```
var winObj=open("target.html", "target_1","width=500,height=300,scrollbars=no");
```

open() 方法有 3 个参数，第一个代表要载入新窗口页面的 URL，第二个代表新窗口的名称，第三个代表新窗口的属性，多个属性间用逗号隔开。

2）对话框（与用户交互）方法

（1）alert()。

该方法的功能是弹出一个提示框。示例代码如下：

```
<script type="text/javascript">
        alert("请点击确定按钮！");
</script>
```

其运行效果如图 1.19 所示。

（2）prompt(message,defaultText)。

该方法的功能是弹出可以输入信息的文本框，第一个参数代表用户输入信息的提示，第二个参数代表文本框的默认值。示例代码如下：

```
<script type="text/javascript">
        prompt("What's your name?","chenheng");
</script>
```

其运行效果如图 1.20 所示。

图 1.19 具有确定按钮的对话框　　　　图 1.20 用户输入对话框

（3）confirm(message)。

该方法的功能是弹出对话框，提示确认信息。示例代码如下：

```
<script type="text/javascript">
        if(confirm("真的删除吗？")){
            //单击"确定"按钮后的操作
        }else{
            //单击"取消"按钮后的操作
        }
</script>
```

其运行效果如图 1.21 所示。

❸ history 对象

history 对象记录着浏览器浏览过的每一个页面，这些页面组成了一个历史记录列表。它有下面 3 个主要方法。

- forward()：将历史记录向前移动一个页面。
- back()：将历史记录向后移动一个页面。在网页中经常使用该方法提供"返回"的功能。
- go()：转向历史记录中的指定地址。使用此方法需要一个参数，参数可以是正/负整数或字符串。如果参数是字符串，那么浏览器就会搜索列表，找到最接近当前页面位置且 URL 地址中含有此字符串的页面，然后转到该页面。

图 1.21　确认对话框

使用 history 对象的示例代码如下：

```
history.go(-3);        //向后返回 3 个访问过的页面
histroy.go(3);         //向前返回 3 个访问过的页面
history.back();        //与 history.go(-1);的功能相同
history.forward();     //与 history.go(1);的功能相同
```

❹ location 对象

使用 window 对象的 location 属性可以直接改变 URL 地址，例如：

```
window.location="http://www.baidu.com";
```

或

```
location="http://www.baidu.com";
```

用户还可以使用 location 对象的 href 属性或 replace(URL)方法改变 URL 地址，例如：

```
location.href="http://www.baidu.com";
```

或

```
location.replace("http://www.baidu.com");
```

用户可以使用 location 对象的 href 属性清空页面，例如：

```
location.href = "about:blank";          //清空页面
```

❺ document 对象

1）forms 集合

在同一个页面上有多个表单，通过 document.forms[]数组获得这些表单对象要比使用表单名称方便很多。

2）getElementById(id)方法

该方法的功能是获得指定 id 值的表单控件对象。

3）getElementsByName(name)方法

该方法的功能是获得指定 name 值的表单控件对象，返回的是对象数组。

4）获取表单对象的方法

获取表单对象的方法如下：

```
document.forms[0];              //通过 forms 对象的索引
document.forms["myForm"];       //通过 forms 对象和表单名称
document.myForm;                //通过表单名称
```

5）document 对象的使用

使用 document 对象的示例代码如下：

```html
<!DOCTYPE html>
<html>
<head>
<meta charset="UTF-8">
<title>The Document Object</title>
<script type="text/javascript">
    function testMethod() {
        //1. 通过表单获取控件
        var name_1 = document.forms[0].nameTest;
        alert("name_1=" + name_1.value);
        var name_2 = document.forms["myForm"].nameTest;
        alert("name_2=" + name_2.value);
        var name_3 = document.myForm.nameTest;
        alert("name_3=" + name_3.value);
        var name_4 = document.myForm.elements["nameTest"];
        alert("name_4=" + name_4.value);
        //2. 通过 name 获取控件
        var name_5 = document.getElementsByName("nameTest");
        alert("name_5[0]=" + name_5[0].value);
        //3. 通过 id 获取控件
        var id_1 = document.getElementById("idTest");
        alert("id_1=" + id_1.value);
    }
</script>
</head>
<body>
    <form name="myForm">
```

```
            通过name 获取:<input type="text" name="nameTest" value="firstTextValue"> <br>
            <br> 通过 id 获取:<input type="text" id="idTest" value="secondTextValue">
            <br> <input type="button" value="Click Me" onclick="testMethod()">
        </form>
    </body>
</html>
```

1.3.6 操作 HTML

❶ 事件处理

通常将鼠标或热键的动作称为事件。由鼠标或热键引发的一连串程序的动作称为事件驱动。对事件进行处理的程序或函数称为事件处理程序。

1）窗口或页面的事件处理

窗口或页面的事件处理如表 1.7 所示。

表 1.7 窗口或页面的事件处理

事 件	说　　明
onBlur	当前元素失去焦点时触发
onFocus	某个元素获得焦点时触发
onLoad	页面内容完成装载时触发
onUnload	当前页面被退出或重置时触发

2）键盘或鼠标的事件处理

键盘或鼠标的事件处理如表 1.8 所示。

表 1.8 键盘或鼠标的事件处理

事 件	说　　明
onClick	当鼠标单击时触发
onDblClick	当鼠标双击时触发
onMouseDown	当按下鼠标时触发
onMouseMove	当鼠标移动时触发
onMouseOut	当鼠标离开某对象范围时触发
onMouseOver	当鼠标移动到某对象范围的上方时触发
onMouseUp	当鼠标按下后松开鼠标时触发
onKeyPress	当键盘上的某个键被按下并且释放时触发
onKeyDown	当键盘上的某个键被按下时触发
onKeyUp	当键盘上的某个键被放开时触发

3）表单元素的事件处理

表单元素的事件处理如表 1.9 所示。

表 1.9 表单元素的事件处理

表 单 元 素	主 要 事 件
button（按钮）	onClick、onBlur、onFocus
checkbox（复选框）	onClick、onBlur、onFocus

续表

表单元素	主要事件
file（上传文件）	onClick、onBlur、onFocus
password（密码框）	onBlur、onFocus、onSelect
radio（单选按钮）	onClick、onBlur、onFocus
select（列表）	onFocus、onBlur、onChange
text（文本框）	onClick、onBlur、onFocus、onChange
textarea（文本区）	onClick、onBlur、onFocus、onChange

4）事件处理模型

在 JavaScript 中对事件的处理程序通常由函数完成，即事件="函数名"，例如：

```html
<!DOCTYPE html>
<html>
<head>
<meta charset="UTF-8">
<title>The Document Object</title>
<script type="text/javascript">
    function testMethod() {
        alert("事件处理模型");
    }
 </script>
 </head>
 <body>
    <form name="myForm">
        <input type="button" value="Click Me" onclick="testMethod()" >
    </form>
 </body>
</html>
```

❷ 表单元素

1）表单元素的通用属性与方法

- form 属性：获取该表单控件所属的表单对象。
- name 属性：获取或设置表单控件的名称。
- type 属性：获取表单控件的类型。
- value 属性：获取和设置表单控件的值。
- focus()方法：让表单控件对象获得焦点。
- blur()方法：让表单控件对象失去焦点。

表单元素的通用属性及方法的示例代码如下：

```html
<!DOCTYPE html>
<html>
<head>
<meta charset="UTF-8">
<title>form</title>
```

```
<script type="text/javascript">
    //显示属性
    function test() {
        var text_name = document.myForm.text_name;
        alert("text_name.form=" + text_name.form.name
                + "\ntext_name.name=" + text_name.name
                + "\ntext_name.type=" + text_name.type
                + "\ntext_name.value=" + text_name.value
                + "\ntext_name.defaultValue=" + text_name.defaultValue);
    }
    //获得焦点
    function do_focus() {
        document.myForm.text_name.focus();
    }
    //失去焦点
    function do_blur() {
        document.myForm.text_name.blur();
    }
</script>
</head>
<body>
    <form name="myForm">
        <input type="text" name="text_name" value="textValue"><br>
        <input type="button" name="button_1" value="显示属性" onclick="test()"> <br>
        <input type="button" name="button_2" value="获得焦点" onclick="do_focus()"> <br>
        <input type="button" name="button_3" value="失去焦点" onclick="do_blur()">
    </form>
</body>
</html>
```

2）文本框
- value 属性：获得文本框的值，值是字符串类型。
- defaultValue 属性：获得文本框的默认值，值是字符串类型。
- readonly 属性：只读，文本框中的内容不能修改。
- focus()方法：获得焦点，即获得鼠标光标。
- blur()方法：失去焦点。
- select()方法：选中文本框内容，突出显示输入区域。

文本框的示例代码如下：

```
<!DOCTYPE html>
<html>
<head>
```

```
<meta charset="UTF-8">
<title>文本框求和</title>
<script type="text/javascript">
function add(){
    var sum=0;
    var text_1Value=document.forms[0].text_1.value;
    var text_2Value=document.forms[0].text_2.value;
    sum=Number(text_1Value)+Number(text_2Value);
    document.forms[0].text_3.value=sum;
}
</script>
</head>
<body>
    <form name="form1" method="post" action="">
        <input type="text" name="text_1" value=""><br>
        <input type="text" name="text_2" value=""><br>
        <input type="text" name="text_3" value=""><br>
        <input type="button" value="求和" onclick="add()">
    </form>
</body>
</html>
```

3）复选框

- checked 属性：复选框是否被选中，选中为 true，未选中为 false。
- value 属性：设置或获取复选框的值。

复选框的示例代码如下：

```
<!DOCTYPE html>
<html>
<head>
<meta charset="UTF-8">
<title>复选框反选</title>
<script type="text/javascript">
    function unSelect() {
        var n = document.forms[0].check.length;//得到复选框的个数
        for ( var i = 0; i < n; i++) {
            if (document.forms[0].check[i].checked) {
                document.forms[0].check[i].checked = false;
            } else {
                document.forms[0].check[i].checked = true;
            }
        }
    }
</script>
</head>
<body>
```

```html
        <form action="">
            <input type="checkbox" name="check" value="0" />aaa<br>
            <input type="checkbox" name="check" value="1" />bbb<br>
            <input type="checkbox" name="check" value="2" />ccc<br>
            <input type="checkbox" name="check" value="3" />ddd<br>
            <input type="checkbox" name="check" value="4" />eee<br>
            <input type="checkbox" name="check" value="5" />fff<br>
            <input type="button" value="反选" onclick="unSelect()">
        </form>
</body>
</html>
```

4）单选按钮

- checked 属性：单选按钮是否被选中，选中为 true，未选中为 false。
- value 属性：设置或获取单选按钮的值。

单选按钮的示例代码如下：

```html
<!DOCTYPE html>
<html>
<head>
<meta charset="UTF-8">
<title>弹出单选按钮的值</title>
<script type="text/javascript">
    function gg() {
        var n = document.forms[0].sex.length;
        for ( var i = 0; i < n; i++) {
            if (document.forms[0].sex[i].checked) {
                alert(document.forms[0].sex[i].value);
            }
        }
    }
</script>
</head>
<body>
    <form action="">
        <input type="radio" name="sex" value="male" onclick="gg()" /> 男
        <input type="radio" name="sex" value="female" onclick="gg()" /> 女
    </form>
</body>
</html>
```

5）下拉列表

- length 属性：选项个数。
- selectedIndex 属性：当前被选中选项的索引。
- options 属性：所有的选项组成一个数组，options 表示整个选项数组，第一个选项为

options[0],第二个为 options[1],其他以此类推。
- option 的 value 属性:<option>标记中 value 所指定的值。
- option 的 text 属性:显示于界面中的文本,即<option>和</option>之间的部分。
- 增加选项:每个选项都是一个 option 对象,用户可以创建 option 对象,然后添加到 select 的末尾。例如 "select.options[select.length] = new Option(text,value);"。

下拉列表的示例代码如下:

```html
<!DOCTYPE html>
<html>
<head>
<meta charset="UTF-8">
<title>下拉列表的应用</title>
<script type="text/javascript">
    function gg() {
        var opObject = document.forms[0].elements["cities"];//获得列表对象
        for ( var i = 0; i < opObject.options.length; i++) {
                                                            //使用 options 属性
            if (opObject.options[i].selected) {             //找到被选中的选项
                alert(opObject.options[i].value);           //弹出选中选项的值
                alert(opObject.options[i].text);            //弹出选中选项的文本
            }
        }
        //为列表新增选项
        opObject.options[opObject.length] = new Option("新增" + 1, "new" + 1);
    }
</script>
</head>
<body>
    <form action="">
        <select name="cities" onchange="gg()">
            <option value="daLian">大连</option>
            <option value="beiJing">北京</option>
            <option value="shangHai">上海</option>
            <option value="guangZhou">广州</option>
        </select>
    </form>
</body>
</html>
```

1.3.7 表单验证

当用户在表单中输入的内容提交到服务器之前,在客户端通过表单控件产生的事件,运用 JavaScript,验证用户输入数据的有效性。

❶ 正则表达式

表单验证常用的正则表达式如下。

验证邮政编码：/^\d{6}$/
验证身份证号码：/^\d{15}$|^\d{18}$|^\d{17}[xX]$/
验证电子邮箱地址：/^\w+((-\w+)|(\.\w+))*\@[A-Za-z0-9]+((\.|-)[A-Za-z0-9]+)*\.[A-Za-z0-9]+$/
验证数字或英文字母：/^[a-z0-9]+$/
验证日期格式：
/((^((1[8-9]\d{2})|([2-9]\d{3}))([-\/\._])(10|12|0?[13578])([-\/\._])(3[01]|[12][0-9]|0?[1-9])$)|(^((1[8-9]\d{2})|([2-9]\d{3}))([-\/\._])(11|0?[469])([-\/\._])(30|[12][0-9]|0?[1-9])$)|(^((1[8-9]\d{2})|([2-9]\d{3}))([-\/\._])(0?2)([-\/\._])(2[0-8]|1[0-9]|0?[1-9])$)|(^([2468][048]00)([-\/\._])(0?2)([-\/\._])(29)$)|(^([3579][26]00)([-\/\._])(0?2)([-\/\._])(29)$)|(^([1][89][0][48])([-\/\._])(0?2)([-\/\._])(29)$)|(^([2-9][0-9][0][48])([-\/\._])(0?2)([-\/\._])(29)$)|(^([1][89][2468][048])([-\/\._])(0?2)([-\/\._])(29)$)|(^([2-9][0-9][2468][048])([-\/\._])(0?2)([-\/\._])(29)$)|(^([1][89][13579][26])([-\/\._])(0?2)([-\/\._])(29)$)|(^([2-9][0-9][13579][26])([-\/\._])(0?2)([-\/\._])(29)$))/

应用正则表达式的示例代码如下：

```html
<!DOCTYPE html>
<html>
<head>
<meta charset="UTF-8">
<title>正则表达式的应用</title>
<script type="text/javascript">
//验证E-mail
function checkEmail(){
    var exp=/^\w+((-\w+)|(\.\w+))*\@[A-Za-z0-9]+((\.|-)[A-Za-z0-9]+)*\.[A-Za-z0-9]+$/;
    if(!exp.test(document.forms[0].email.value)){
        alert("E-mail 格式错误!");
        document.forms[0].email.focus();
        return false;
    }else{
        alert("E-mail 格式正确!");
        return true;
    }
}
</script>
</head>
<body>
    <form action="">
        Mail: <input type="text" name="email" /><br>
        <input type="button" value="提交" onclick="return checkEmail()" />
    </form>
```

```
</body>
</html>
```

❷ 表单验证实例

验证密码域不能为空，长度要大于或等于6，只能是字母或数字。代码如下：

```html
<!DOCTYPE html>
<html>
<head>
<meta charset="UTF-8">
<title>表单验证实例</title>
<script type="text/javascript">
function valid(form){
    //验证非空
    if(form.pass.value.length == 0){
        alert("Please enter a password");
        form.pass.focus();
        return false;
    }
    //验证长度
    if(form.pass.value.length < 6){
        alert("Password must be at least 6 characters");
        form.pass.focus();
        return false;
    }
    var exp=/^[a-z0-9]+$/;
    //验证格式
    if(!exp.test(form.pass.value)){
        alert("Password contains illegal characters");
        form.pass.focus();
        return false;
    }
    alert("OK password");
    return true;
}
</script>
</head>
<body>
    <form action="">
        Enter your password:
        <input name="pass" type="password"/>
        <input type="button" value="submit password" onclick="return valid
        (this.form)"/>
    </form>
</body>
</html>
```

1.3.8 实践环节——表单验证

制作用户注册页面 practice1_3.html，具体要求如下：
① 有常用的登录账号、密码、确认密码、姓名、身份证号码（只考虑 18 位的身份证）、出生年月日、住址、邮编、E-mail 等输入区域（自己设定）。
② 自己设定验证规则，在提交时检验是否符合要求，弹出非法的输入，并将焦点返回要输入的控件对象。
③ 根据出生年月日判断身份证号码是否合法（只考虑 18 位的身份证）。
④ 其他验证：登录账号只能是字母或数字且以字母开头；密码要在 8 位以上且需要有字母和数字之外的字符；出生年月日的格式为 yyyy-mm-dd；邮编为 6 位数字；E-mail 的基本格式验证。

1.4 本章小结

本章简要介绍了 HTML、CSS 和 JavaScript，对于更多内容，请读者参考其他教材或访问 "http://www.w3school.com.cn" 或 "http://www.runoob.com"。

习题 1

1. 在以下标记中，用于设置页面标题的是（　　）。
   A．<title>　　　　B．<caption>　　　　C．<head>　　　　D．<html>
2. 若设计网页的背景图形为 bg.png，在以下标记中正确的是（　　）。
   A．<body background="bg.png">　　　　B．<body bground="bg.png">
   C．<body image="bg.png">　　　　　　D．<body bgcolor="bg.png">
3. 下列关于 HTML 文件的说法正确的是（　　）。
   A．HTML 标记都必须配对使用
   B．在<title>和</title>标记之间的是头信息
   C．HTML 标记是区分大小写的，<B>跟<b>表示的意思是不一样的
   D．<!-- -->标记是注释标记
4. 下列（　　）表示的不是按钮。
   A．type="submit"　　B．type="reset"　　C．type="myButton"　　D．type="button"
5. 若产生一个 4 行 30 列的多行文本域，在以下方法中，正确的是（　　）。
   A．<input type="text" rows="4" cols="30" name="txtintrol">
   B．<textarea rows="4" cols="30" name="txtintro">
   C．<textarea rows="4" cols="30" name="txtintro"></textarea>
   D．<textarea rows="30" cols="4" name="txtintro"></textarea>

6. 下列属性中用于设置文本框显示宽度的属性是（   ）。
   A．size   B．maxlength   C．value   D．length
7. 下面对表单的说法错误的是（   ）。
   A．表单在 Web 页面中用来给访问者填写信息，从而能采集客户端信息，使页面具有交互信息的功能
   B．当用户填写完信息后单击普通按钮做提交（submit）操作
   C．一个表单用<form></form>标签来创建
   D．action 属性的值是指处理程序的程序名（包括网络路径、网址或相对路径）
8. 下面说法中错误的是（   ）。
   A．在 HTML 语言中，<input>标记具有重要的地位，它能够将浏览器中的控件加载到 HTML 文档中，该标记既有开始标记，又有结束标记
   B．<input type="text">是设定一个单行的文本输入区域
   C．size 属性指定控件宽度，表示该文本输入框所能显示的最大字符数
   D．maxlength 属性表示该文本输入框允许用户输入的最大字符数
9. 下面对于按钮的说法中正确的是（   ）。
   A．按钮可分为普通按钮、提交按钮和重置按钮
   B．<input type="button">表示这是一个提交到服务器的按钮
   C．<input type="reset">表示这是一个普通按钮
   D．name 属性用来指定按钮页面显示的名称
10. 下面说法中错误的是（   ）。
    A．<input type="checkbox " checked>中的 checked 属性用来设置该复选框默认是否被选中
    B．<input type="hidden">表示一个隐藏区域，用户可以在其中输入某些要传送的信息
    C．<input type="password">表示这是一个密码区域，当用户输入密码时区域内将会显示 "*" 号
    D．<input type="radio">表示这是一个单选按钮
11. 下列说法中错误的是（   ）。
    A．<select></select>标记对用来创建一个菜单下拉列表框
    B．下拉列表框中的 multiple 属性不用赋值，直接加入到标记中即可使用，加入此属性后列表框就可以多选
    C．<option>标记用来指定列表框中的一个选项
    D．不可以设定输入多行的文本区的大小
12. 下面对于样式表的说法中错误的是（   ）。
    A．CSS 就是 Cascading Style Sheets，中文翻译为 "层叠样式表"，简称样式表
    B．将 CSS 指定的格式加入到 HTML 中的方法有两种
    C．外部链接 CSS 时需要在 HTML 文件里加一个超链接，连接到外部的 CSS 文档
    D．内定义 CSS 时需要在 HTML 文件内的<head></head>标记对之间加一段 CSS 的描述内容

13. 下面对于样式表的说法中错误的是（　　）。
    A．CSS 的定义由 3 个部分构成，即选择符（selector）、属性（properties）和属性的取值（value）
    B．选择符可以是多种形式，一般是要定义样式的 HTML 标记，用户可以通过此方法定义它的属性和值，属性和值要用逗号隔开
    C．CSS 可以定义字体属性
    D．CSS 可以定义颜色和背景属性
14. 下列说法中错误的是（　　）。
    A．p {font-family:"sans serif"}定义段落字体为 sans serif
    B．p {text-align: center; color: red}定义段落居中排列，并且段落中的文字为红色
    C．background-image 属性用来设置背景图片
    D．background-color 属性用来设置背景颜色
15. （　　）对象表示浏览器的窗口，可用来检索关于该窗口状态的信息。
    A．document　　　B．window　　　C．location　　　D．history
16. （　　）对象表示给定浏览器窗口中的 HTML 文档，用来检索关于文档的信息。
    A．document　　　B．window　　　C．screen　　　D．history
17. （　　）方法要求窗口显示刚刚访问的前一个页面。
    A．back()　　　B．go()　　　C．display()　　　D．view()
18. 有关变量的命名规则，下列说法中错误的是（　　）。
    A．以字母、下画线（_）或美元符号（$）开头
    B．首字符之外的字符可以是下画线、美元符号，或任何的字母、数字
    C．不能有空格，不区分大小写
    D．不能使用 JavaScript 中的关键字或保留字命名
19. 单击按钮触发的事件是（　　）。
    A．onClick　　　B．onFocus　　　C．onChange　　　D．onLoad
20. 页面加载时产生的事件是（　　）。
    A．onClick　　　B．onFocus　　　C．onChange　　　D．onLoad
21. 用 HTML 编写网页版个人简历，具体要求如下：
（1）页面内容以 DIV+CSS 形式体现。
（2）信息内容包括个人基础信息、近期照片、学习经历、兴趣爱好、求职意向。

第 2 章 Java Web 开发环境的构建

学习目的与要求

本章对开发工具和环境配置进行简要介绍。通过本章的学习，读者能够掌握使用 Eclipse 平台开发 Java Web 应用的基本步骤。

本章主要内容
- 构建开发环境
- 使用 Eclipse 开发 Web 应用

JSP 是 Java Server Pages（Java 服务器页面）的简称，是基于 Java 语言的一种 Web 应用开发技术，是由 Sun 公司倡导，多家公司共同参与建立的一种动态网页技术标准。

在学习 JSP 之前，读者应具有 Java 语言基础以及 HTML 语言方面的知识。本章第 2 节通过一个简单的 Web 应用介绍了 JSP 项目开发的基本步骤，这些基本步骤对后续章节的学习是极其重要的。

2.1 构建开发环境

所谓"工欲善其事，必先利其器"，在开发 JSP 应用程序之前需要构建其开发环境。

视频讲解

2.1.1 开发工具

❶ Java 开发工具包（JDK）

JSP 引擎需要 Java 语言的核心库和相应编译器，在安装 JSP 引擎之前需要安装 Java 标准版（Java SE）提供的开发工具包 JDK。登录"http://www.oracle.com/technetwork/java"，在 Software Downloads 中选择 Java SE 提供的 JDK，根据操作系统的位数下载相应的 JDK，例如 32 位的系统使用 32 位的 JDK。本书采用的 JDK 是 jdk-8u152-windows-x64.exe。

❷ **JSP 引擎**

运行包含 JSP 页面的 Web 项目还需要一个支持 JSP 的 Web 服务软件,该软件也称为 JSP 引擎或 JSP 容器,通常将安装了 JSP 引擎的计算机称为一个支持 JSP 的 Web 服务器。目前比较常用的 JSP 引擎有 Tomcat、JRun、Resin、WebSphere、WebLogic 等,本书采用的是 Tomcat 9.0.56。登录 Apache 软件基金会的官方网站"http://jakarta.Apache.org/tomcat",下载 Tomcat 9.0 的免安装版(本书采用 apache-tomcat-9.0.2-windows-x64.zip)。登录网站后,首先在 Download 里选择 Tomcat 9.0,然后在 Binary Distributions 的 Core 中选择 zip 即可。

❸ **Eclipse**

为了提高开发效率,通常需要安装 IDE(集成开发环境)工具,在本书中使用的 IDE 工具是 Eclipse。Eclipse 是一个可用于开发 JSP 程序的 IDE 工具。登录"http://www.eclipse.org/ide",选择 Java EE,根据操作系统的位数下载相应的 Eclipse。本书采用的 Eclipse 是 eclipse-jee-oxygen-2-win32-x86_64.zip。

2.1.2 工具的集成

❶ **JDK 的安装与配置**

1)安装 JDK

双击下载后的 jdk-8u152-windows-x64.exe 文件图标将出现安装向导界面,选择接受软件安装协议。建议采用默认的安装路径"C:\Program Files\Java\jdk1.8.0_152"。需要注意的是,在安装 JDK 的过程中,JDK 还额外提供一个 Java 运行环境 JRE(Java Runtime Environment),并提示是否修改 JRE 默认的安装路径"C:\Program Files\Java\jre1.8.0_152",建议采用该默认的安装路径。

2)配置系统环境变量

安装 JDK 后需要配置"环境变量"的"系统变量"Java_Home 和 Path。在 Windows 10 系统下,系统变量示例如图 2.1 和图 2.2 所示。

图 2.1 新建系统变量 Java_Home

❷ **Tomcat 的安装与启动**

在安装 Tomcat 之前需要事先安装 JDK 并配置系统环境变量 Java_Home。将下载的 apache-tomcat-9.0.2-windows-x64.zip 解压到某个目录下,比如解压到 E 盘的 Java soft 下,解压缩后将出现如图 2.3 所示的目录结构。

图 2.2　编辑环境变量 Path

图 2.3　Tomcat 目录结构

通过执行 Tomcat 根目录下 bin 文件夹中的 startup.bat 来启动 Tomcat 服务器。执行 startup.bat 启动 Tomcat 服务器会占用一个 MS-DOS 窗口，出现如图 2.4 所示的界面，如果关闭当前 MS-DOS 窗口将关闭 Tomcat 服务器。

图 2.4　执行 startup.bat 启动 Tomcat 服务器

Tomcat 服务器启动后，在浏览器的地址栏中输入"http://localhost:8080"将出现如图 2.5 所示的 Tomcat 测试页面。

第 2 章　Java Web 开发环境的构建

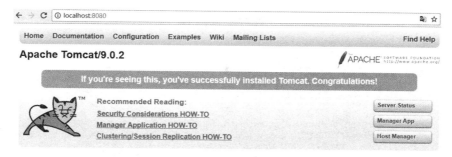

图 2.5　Tomcat 测试页面

8080 是 Tomcat 服务器默认占用的端口。用户可以通过修改 Tomcat 的配置文件修改端口号。用记事本打开 conf 文件夹下的 server.xml 文件，找到以下代码：

```
<Connector port="8080" protocol="HTTP/1.1"
          connectionTimeout="20000"
          redirectPort="8443" />
```

将 port="8080"中的 8080 更改为新的端口号，保存 server.xml 文件后重新启动 Tomcat 服务器即可，比如将 8080 修改为 9090 等。如果修改为 9090，那么在浏览器的地址栏中要输入 "http://localhost:9090" 才能打开 Tomcat 的测试页面。

需要说明的是，一般情况下不要修改 Tomcat 默认的端口号，除非 8080 已经被占用。在修改端口时应避免与公用端口冲突，一旦冲突会造成其他程序不能正常使用。

❸ 安装 Eclipse

在 Eclipse 下载完成后，解压到自己设置的路径下，即可完成安装。在 Eclipse 安装后，双击 Eclipse 安装目录下的 eclipse.exe 文件启动 Eclipse。

❹ 集成 Tomcat

启动 Eclipse，选择 Window/Preferences 命令，在弹出的对话框中选择 Server/Runtime Environments 选项。然后在弹出的窗口中单击 Add 按钮，弹出如图 2.6 所示的 New Server Runtime Environment 界面，在此可以配置各种版本的 Web 服务器。

图 2.6　Tomcat 配置界面

在图 2.6 中选择 Apache Tomcat v9.0 服务器版本，单击 Next 按钮，进入如图 2.7 所示的界面。

图 2.7　选择 Tomcat 目录

在图 2.7 中单击 Browse 按钮，选择 Tomcat 的安装目录，单击 Finish 按钮即可完成 Tomcat 的配置。

至此，可以使用 Eclipse 创建 Dynamic Web Project，并在 Tomcat 下运行。

2.2　使用 Eclipse 开发 Web 应用

视频讲解

2.2.1　JSP 文件和 JSP 运行原理

❶ **JSP 文件**

在一个 JSP 文件中可以有普通的 HTML 标记、JSP 规定的标记以及 Java 程序。JSP 文件的扩展名是.jsp，文件名必须符合标识符规定，即名字可以由字母、下画线、美元符号和数字组成。

❷ **JSP 运行原理**

当 Web 服务器上的一个 JSP 页面第一次被客户端请求执行时，Web 服务器上的 JSP 引擎首先将 JSP 文件转译成一个 Java 文件，并将 Java 文件编译成字节码文件，字节码文件在服务器端创建一个 Servlet 对象，然后执行该 Servlet 对象，同时发送一个 HTML 页面到客户端响应客户端的请求。当这个 JSP 页面再次被请求时，JSP 引擎为每个客户端启动一个线程并直接执行对应的 Servlet 对象响应客户端的请求，这也是 JSP 响应速度比较快的原因之一。

JSP 引擎用如下方式处理 JSP 页面：

① 将 JSP 页面中的静态元素（HTML 标记）直接交给客户端浏览器执行显示。

② 对 JSP 页面中的动态元素（Java 程序和 JSP 标记）进行必要的处理，将需要显示的结果发送给客户端浏览器。

2.2.2 一个简单的 Web 应用

使用 Eclipse 开发一个 Web 应用需要如下 3 个步骤：
- 创建项目。
- 创建 JSP 文件。
- 发布项目到 Tomcat 并运行。

❶ 创建项目

① 启动 Eclipse，进入 Eclipse 开发界面。

② 选择 File/New/Project 命令，打开 New Project 对话框，在该对话框中选择 Web 节点下的 Dynamic Web Project 子节点，如图 2.8 所示。

③ 单击 Next 按钮，打开 New Dynamic Web Project 对话框，在该对话框的 Project name 文本框中输入项目名称，这里为 firstProject。然后选择 Target runtime 区域中的服务器，如图 2.9 所示。

图 2.8　New Project 对话框

图 2.9　New Dynamic Web Project 对话框

④ 单击 Finish 按钮，完成项目 firstProject 的创建。此时在 Eclipse 平台左侧的 Project Explorer 中将显示项目 firstProject，依次展开各节点，可显示如图 2.10 所示的目录结构。

❷ 创建 JSP 文件

在 firstProject 项目创建完成后，可以根据实际需要创建类文件、JSP 文件或者其他文件。对于这些文件的创建会在需要时介绍，下面将创建一个名字为 myFirst.jsp 的 JSP 文件。

图 2.10　项目 firstProject 的目录结构

① 选中 firstProject 项目的 WebContent 节点，右击，在弹出的快捷菜单中选择 New/JSP File 命令，打开 New JSP File 对话框，在该对话框的 File name 文本框中输入文件名 myFirst.jsp，其他采用默认设置，单击 Finish 按钮完成 JSP 文件的创建。

② JSP 文件创建完成后，在 firstProject 项目的 WebContent 节点下将自动添加一个名称为 myFirst.jsp 的 JSP 文件，同时 Eclipse 会自动将 JSP 文件在右侧的编辑框中打开。

③ 将 myFirst.jsp 文件中的默认代码修改如下：

```jsp
<%@ page language="java" contentType="text/html; charset=UTF-8"
pageEncoding="UTF-8"%>
<!DOCTYPE html>
<html>
<head>
<meta charset="UTF-8">
<title>Insert title here</title>
</head>
<body>
    <div align="center">真高兴,忙乎半天了,终于要看到人生中第一个JSP页面了。</div>
</body>
</html>
```

④ 将编辑好的 JSP 页面保存（Ctrl+S），至此完成一个简单 JSP 程序的创建。

在创建 JSP 文件时，Eclipse 默认创建的 JSP 文件的编码格式为 ISO-8859-1，为了让页面支持中文，还需要将编码格式修改为 UTF-8、GBK 或 GB2312。建议读者在 Eclipse 中（Window→Preferences→Web→JSP Files→Editor→Templates→New JSP File(html)→Edit）将 JSP 文件模板修改如下：

```jsp
<%@ page language="java" contentType="text/html; charset=${encoding}"
pageEncoding="${encoding}"%>
<!DOCTYPE html>
<html>
<head>
<meta charset="${encoding}">
<title>Insert title here</title>
</head>
<body>
${cursor}
</body>
</html>
```

在一个项目的 WebContent 节点下可以创建多个 JSP 文件，另外 JSP 文件中使用到的图片文件、CSS 文件（层叠样式表）以及 JavaScript 文件都放在 WebContent 节点下。

❸ 发布项目到 Tomcat 并运行

在完成 JSP 文件的创建后，可以将项目发布到 Tomcat 并运行该项目。下面介绍具体的方法。

① 在 firstProject 项目的 WebContent 节点下找到 myFirst.jsp 并选中该 JSP 文件，右击，在弹出的快捷菜单中选择 Run As/Run On Server 命令，打开 Run On Server 对话框，在该对话框中选中 Always use this server when running this project 复选框，其他采用默认设置，如图 2.11 所示。

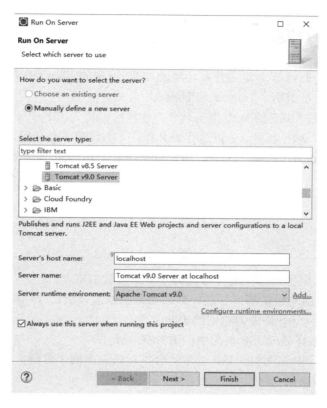

图 2.11 Run On Server 对话框

② 单击 Finish 按钮，即可通过 Tomcat 运行该项目，运行后的效果如图 2.12 所示。如果用户想在浏览器中运行该项目，可以将图 2.12 中的 URL 地址复制到浏览器的地址栏中，并按下 Enter 键运行。

图 2.12 运行 firstProject 项目

注：在 Eclipse 中，默认将 Web 项目发布到 Eclipse 的工作空间的.metadata\.plugins\org.eclipse.wst.server.core\tmp0\wtpwebapps\目录下。

2.2.3 实践环节——使用 Eclipse 开发 Web 应用

① 创建一个名称为 sencondTest 的项目。

② 在 sencondTest 项目中创建一个名称为 yourFirst.jsp 的文件，它将在 JSP 页面中显示"不错！不错！自己能创建 JSP 文件了，并且可以发布运行了"。

③ 发布并运行 sencondTest 项目。

2.3 本章小结

本章主要介绍了 Java Web 集成开发环境的构建，用到的工具有 JDK、Tomcat 和 Eclipse。它们的安装顺序依次是 JDK、Tomcat、Eclipse。

Tomcat 是一个支持 JSP 的 Web 服务软件，该软件也称为 JSP 引擎或 JSP 容器。JSP 引擎是支持 JSP 程序的 Web 容器，负责运行 JSP 程序，并将有关结果发送给客户端。目前流行的 JSP 引擎有 Tomcat、Resin、JRun、WebSphere、WebLogic 等，本书使用的是 Tomcat 服务器。

习题 2

1. 安装 Tomcat 服务器所在的计算机需要事先安装 JDK 吗？
2. Tomcat 服务器的默认端口号是什么？如果想修改该端口号，应该修改哪个文件？
3. First.jsp 和 first.jsp 是否为相同的 JSP 文件名？
4. JSP 引擎是怎样处理 JSP 页面中的 HTML 标记的？
5. 当用户请求 JSP 页面时，JSP 引擎将会执行该页面的字节码文件响应客户的请求，执行字节码文件的结果是（　　）。

　　A．发送一个 JSP 源文件到客户端　　B．发送一个 Java 文件到客户端
　　C．发送一个 HTML 页面到客户端　　D．什么都不做

第 3 章 JSP 语法

学习目的与要求

本章主要介绍 JSP 脚本元素、JSP 指令标记和 JSP 动作标记。通过本章的学习,读者能够理解 JSP 页面的组成部分,掌握 JSP 语法,能够使用 JSP 开发 Web 页面。

本章主要内容
- JSP 页面的基本构成
- JSP 脚本元素
- JSP 指令标记
- JSP 动作标记

一个 JSP 页面通常由 HTML 标记、JSP 注释、Java 脚本元素和 JSP 标记 4 种基本元素组成。这 4 种基本元素在 JSP 页面中如何被使用是本章介绍的重点。

本章涉及的 JSP 页面保存在工程 ch3 的 WebContent 中。

3.1 JSP 页面的基本构成

视频讲解

3.1.1 一个 JSP 页面

在 HTML 静态页面文件中加入和 Java 相关的动态元素就构成了一个 JSP 页面。一个 JSP 页面通常由以下 4 种基本元素组成。

① 普通的 HTML 标记。
② JSP 注释。
③ Java 脚本元素,包括声明、Java 程序片和 Java 表达式。
④ JSP 标记,例如指令标记、动作标记和自定义标记等。

例 3-1 根据 example3_1.jsp 代码中的注释识别 JSP 页面中的基本元素。

页面文件 example3_1.jsp 的代码如下：

```jsp
<%@ page language="java" contentType="text/html; charset=UTF-8" pageEncoding="UTF-8"%>
<!DOCTYPE html>
<jsp:include page="a.jsp"/>    <!-- JSP 动作标记-->
<%!
int i = 0;                     //数据声明
int add(int x, int y) {        //方法声明
    return x + y;
}
%>
<html>        <!--HTML 标记-->
<head>
<meta charset="UTF-8">
<title>Insert title here</title>
</head>
<body>
<%
i ++;                          //Java 程序片
int result = add(1, 2);
%>
i 的值为<%=i%>        <%--Java 表达式--%>
<br>
1+2 的和为<%=result%>
</body>
</html>
```

页面文件 a.jsp 的代码如下：

```jsp
<%@ page language="java" contentType="text/html; charset=UTF-8" pageEncoding="UTF-8"%>
<!DOCTYPE html>
<html>
<head>
<meta charset="UTF-8">
<title>Insert title here</title>
</head>
<body>
被 example3_1.jsp 动态引用。
</body>
</html>
```

3.1.2 JSP 页面注释

在 example3_1.jsp 代码中有许多 JSP 注释。注释能够增强 JSP 文件的可读性，便于 Web 项目的更新和维护。JSP 页面中常见的注释有以下两种。

❶ HTML 注释

格式：<!--HTML 注释-->

在标记符"<!--"和"-->"之间加入注释内容就构成了 HTML 注释。

JSP 引擎对于 HTML 注释也要进行处理，即不将它看作是注释，如果其中有 JSP 代码，也将被 JSP 引擎处理。JSP 引擎将处理之后的 HTML 注释交给客户端，用户通过浏览器查看 JSP 源文件时能够看到 HTML 注释。

❷ JSP 注释

格式：<%--JSP 注释--%>

在标记符"<%--"和"--%>"之间加入注释内容就构成了 JSP 注释。

JSP 引擎将 JSP 注释当作是真正的注释，在编译 JSP 页面时将忽略这部分代码，因此用户通过浏览器查看 JSP 源文件时无法看到 JSP 注释。

3.1.3 实践环节——识别 JSP 页面元素

识别出如下 JSP 页面的基本元素：

```
<%@ page language="java" contentType="text/html; charset=UTF-8"
    pageEncoding="UTF-8"%>
<!DOCTYPE html>
<!--学习JSP页面的基本构成-->
<%!
    String content = "JSP页面基本构成：";
%>
<html>
<head>
<meta charset="UTF-8">
<title>shijian3_1.jsp</title>
</head>
<body>
<%
    content = content + "HTML标记、JSP注释、JSP标记以及Java脚本元素";
%>
<%=content%>
</body>
</html>
```

3.2 JSP 脚本元素

JSP 中的 Java 脚本元素包括声明、Java 程序片以及 Java 表达式。

视频讲解

3.2.1 Java 程序片

在标记符"<%"和"%>"之间插入的 Java 代码被称为 JSP 页面的 Java 程序片。Java

程序片的格式如下：

```
<% Java 代码 %>
```

一个 JSP 页面可以有任意段 Java 程序片，这些程序片将被 JSP 引擎（本书中指 Tomcat 服务器）按顺序执行。在一个程序片中声明的变量称为 JSP 页面的局部变量，它们在 JSP 页面后继的所有程序片以及表达式内都有效。

当多个客户请求一个 JSP 页面时，JSP 引擎为每个客户启动一个线程，不同的线程会分别执行该 JSP 页面中的 Java 程序片，程序片中的局部变量会在不同的线程中被分配不同的内存空间，因此一个客户对 JSP 页面局部变量操作的结果不会影响到其他客户。Java 程序片的执行原理如图 3.1 所示。

图 3.1　Java 程序片的执行原理

例 3-2　编写一个 JSP 页面 example3_2.jsp，页面中存在一段 Java 程序片，该程序片内声明了一个整型的局部变量 n，初始值为 0。

页面文件 example3_2.jsp 的代码如下：

```jsp
<%@ page language="java" contentType="text/html; charset=UTF-8" pageEncoding="UTF-8"%>
<!DOCTYPE html>
<html>
<head>
<meta charset="UTF-8">
<title>example3_2.jsp</title>
</head>
<body>
<%
    int n = 0;
    n++;
    out.print("n = " + n);
%>
</body>
</html>
```

如果有 5 个客户请求 example3_2.jsp 页面，JSP 引擎会启动 5 个线程，页面中的 Java 程序片在每个线程中均会被执行一次，共计执行 5 次；在内存中，局部变量 n 对应 5 处不同的存储空间，初始值都为 0，且都仅执行了一次自加运算，所以 5 个不同的客户看到的页面效果是相同的。

3.2.2 成员变量与方法的声明

成员变量和方法的声明格式如下：

```
<%! 变量或方法定义 %>
```

在标记符"<%!"和"%>"之间声明的变量被称为 JSP 页面的成员变量，它们可以是 Java 语言允许的任何数据类型，例如：

```
<%!
    int n = 0;
    Date date;
%>
```

成员变量在整个 JSP 页面内都有效（与书写位置无关），因为 JSP 引擎将 JSP 页面转译成 Java 文件时把这些变量作为类的成员变量，这些变量的内存空间直到服务器关闭才释放。因此，多个用户共享 JSP 页面的成员变量。任何用户对 JSP 页面成员变量操作的结果都会影响到其他用户。

在标记符"<%!"和"%>"之间声明的方法被称为 JSP 页面的成员方法，该方法在整个 JSP 页面内有效，但是该方法内定义的变量仅在该方法内有效。

例 3-3 编写一个 JSP 页面 example3_3.jsp，页面中声明一个成员变量 n（初始值为 0）和方法 add()（求两个整数的和）。另外，页面中还有一段 Java 程序片，在程序片中声明一个局部变量 m，并且对成员变量 n 和局部变量 m 分别进行自加。

页面文件 example3_3.jsp 的代码如下：

```
<%@ page language="java" contentType="text/html; charset=UTF-8" pageEncoding="UTF-8"%>
<!DOCTYPE html>
<html>
<head>
<meta charset="UTF-8">
<title>example3_3.jsp</title>
</head>
<%!
int n=0;
int add(int x,int y){
    return x+y;
}
%>
<body>
```

```
<%
    int m=0;
    n++;
    m++;
    int result=add(1,2);
    out.print("成员变量 n 的值为: "+n+"<br>");
    out.print("局部变量 m 的值为: "+m+"<br>");
    out.print("1+2="+result+"<br>"+"<br>");
    out.print("第"+n+"个客户");
%>
</body>
</html>
```

在 example3_3.jsp 中，变量 n 是成员变量，被所有客户共享；变量 m 是局部变量，被每个客户独享。如果有 3 个客户请求这个 JSP 页面，则看到的效果如图 3.2 所示。

图 3.2 3 个客户请求 example3_3.jsp 页面的效果

从 example3_3.jsp 中可知 Java 程序片具有如下特点：
① 调用 JSP 页面声明的方法。
② 操作 JSP 页面声明的成员变量。
③ 声明局部变量。
④ 操作局部变量。

3.2.3　Java 表达式

在标记符"<%="和"%>"之间可以插入一个表达式，这个表达式必须能求值。表达式的值由 Web 服务器负责计算，并将计算结果用字符串形式发送到客户端作为 HTML 页面的内容显示。

在 Java 表达式中可以有算术表达式、逻辑表达式或条件表达式等，但用户使用 Java 表达式时应该注意以下两点：
① 不可以在"<%="和"%>"之间插入语句，即所输入内容的末尾不能以分号结束。
② "<%="是一个完整的符号，"<%"和"="之间不能有空格。

3.2.4　实践环节——在 JSP 页面中输出英文字母表

编写一个 JSP 页面，在 JSP 页面中使用 Java 程序片输出 26 个小写的英文字母。

3.2.5 实践环节——网站访问量的统计

利用成员变量被所有客户共享这一性质实现一个简单的计数器，页面效果如图 3.3 所示。

图 3.3 简单的计数器

3.2.6 实践环节——打印表格

在浏览器中输出大小为 15×10 的表格，页面效果如图 3.4 所示。

图 3.4 15×10 的表格

3.3 JSP 指令标记

视频讲解

常用的 JSP 指令有 page 和 include 指令。

3.3.1 page 指令

page 指令标记用来定义整个 JSP 页面的一些属性和这些属性的值。用户可以用一个 page 指令指定多个属性的值，也可以使用多个 page 指令分别为每个属性指定值。page 指令的格式如下：

```
<%@ page 属性1="属性1的值" 属性2="属性2的值" …%>
```

或

```
<%@ page 属性1="属性1的值" %>
<%@ page 属性2="属性2的值" %>
```

```
<%@ page 属性3="属性3的值" %>
...
<%@ page 属性n="属性n的值" %>
```

page 指令的主要属性有 contentType、import、language 和 pageEncoding 等。

❶ 属性 contentType

JSP 页面使用 page 指令标记只能为 contentType 属性指定一个值，用来确定响应的 MIME 类型（MIME 类型就是设定某种文件用对应的一种应用程序来打开的方式类型）。当用户请求一个 JSP 页面时，服务器会告诉浏览器使用 contentType 属性指定的 MIME 类型来解释执行所接收到的信息。例如，浏览器使用 Word 应用程序打开用户请求时可以将 contentType 属性值设置为：

```
<%@page contentType="application/msword;charset=UTF-8"%>
```

常见的 MIME 类型有 text/html（HTML 解析器，所谓的网页形式）、text/plain（普通文本）、application/pdf（PDF 文档）、application/msword（Word 应用程序）、image/jpeg（JPEG 图形）、image/png（PNG 图像）、image/gif（GIF 图形）以及 application/vnd.ms-powerpoint（PowerPoint 应用程序）。

❷ 属性 import

JSP 页面使用 page 指令标记可以为 import 属性指定多个值，import 属性的作用是为 JSP 页面引入包中的类，以便在 JSP 页面的程序片、变量及方法声明或表达式中使用包中的类。

❸ 属性 language

language 属性用来指定 JSP 页面使用的脚本语言，目前该属性的值只能取"java"。

❹ 属性 pageEncoding

contentType 中的 charset 是指服务器发送给客户浏览器时所见到的网页内容的编码；pageEncoding 是指 JSP 文件自身存储时所用的编码。

在 JSP 规范中，如果 pageEncoding 属性存在，那么 JSP 页面的字符编码方式就由 pageEncoding 决定，否则就由 contentType 属性中的 charset 决定，如果 charset 也不存在，JSP 页面的字符编码方式就采用默认的 ISO-8859-1。

【例】3-4 编写一个 JSP 页面 example3_4.jsp，当用户请求该页面时，浏览器启动本地的 PowerPoint 应用程序打开该页面。

页面文件 example3_4.jsp 的代码如下：

```
<%@ page language="java" contentType="application/vnd.ms-powerpoint;
charset=UTF-8" pageEncoding="UTF-8"%>
<!DOCTYPE html>
<html>
<head>
<meta charset="UTF-8">
<title>example3_4.jsp</title>
</head>
<body>
```

在学习 page 指令标记时请牢牢记住只能为 JSP 页面设置一个 contentType 属性值，可为 import 属性设置多个值。

```
</body>
</html>
```

3.3.2 include 指令

一个网站中的多个 JSP 页面有时需要显示同样的信息，比如该网站的 Logo 或导航条等，为了便于网站程序的维护，通常在这些 JSP 页面的适当位置嵌入一个相同的文件。include 指令标记的作用就是将 JSP 文件、HTML 网页文件或其他文本文件静态嵌入到当前的 JSP 网页中。该指令的语法格式如下：

```
<%@include file="文件的URL"%>
```

所谓静态嵌入就是"先包含后处理"，在编译阶段完成对文件的嵌入。也就是先将当前 JSP 页面与要嵌入的文件合并成一个新的 JSP 页面，然后再由 JSP 引擎将新页面转化为 Java 文件处理并运行。

例3-5 编写两个 JSP 页面 example3_5.jsp 和 example3_5_1.jsp，在 example3_5.jsp 页面中使用 include 指令标记静态嵌入 example3_5_1.jsp 页面，访问 example3_5.jsp 页面，运行效果如图 3.5 所示。

图 3.5　include 指令标记的使用

页面文件 example3_5.jsp 的代码如下：

```
<%@ page language="java" contentType="text/html; charset=UTF-8" pageEncoding="UTF-8"%>
<!DOCTYPE html>
<html>
<head>
<meta charset="UTF-8">
<title>example3_5.jsp</title>
</head>
<body>
静态嵌入 example3_5_1.jsp 之前
<br>
<%@include file="example3_5_1.jsp" %>
<br>
静态嵌入 example3_5_1.jsp 之后
</body>
</html>
```

页面文件 example3_5_1.jsp 的代码如下：

```
<%@ page language="java" contentType="text/html; charset=UTF-8" pageEncoding="UTF-8"%>
<!DOCTYPE html>
<html>
<head>
<meta charset="UTF-8">
<title>example3_5_1.jsp</title>
</head>
<body>
    <font color="red" size=4>example3_5_1.jsp 文件的内容</font>
</body>
</html>
```

example3_5.jsp 页面静态嵌入 example3_5_1.jsp 页面，此时需要先将 example3_5_1.jsp 中的所有代码全部嵌入到 example3_5.jsp 指定位置，形成一个新的 JSP 文件，然后再将新文件提交给 JSP 引擎处理，如图 3.6 所示。

图 3.6　静态嵌入的原理

在使用 include 指令标记时，需要注意嵌入文件后必须保证新合成的 JSP 页面符合 JSP 的语法规则，比如例 3-5 的 example3_5.jsp 和 example3_5_1.jsp 两个页面的 page 指令就不能指定不同的 contentType 值，否则合并后的 JSP 页面就两次使用 page 指令为 contentType 属性设置了不同的属性值，导致语法错误。

3.3.3　实践环节——制作导航栏

编写 3 个 JSP 页面 index.jsp、main.jsp 和 head.jsp，在 index.jsp 和 main.jsp 页面中分别使用 include 指令静态嵌入 head.jsp（导航栏）。导航栏 head.jsp 的运行效果如图 3.7 所示。

图 3.7　导航栏的运行效果

3.4 JSP 动作标记

视频讲解

常用的 JSP 动作标记有 include、forward、param、useBean、getProperty 和 setProperty。useBean、getProperty 和 setProperty 将在本书的第 5 章介绍。

3.4.1 include 动作标记

动作标记 include 的作用是将 JSP 文件、HTML 网页文件或其他文本文件动态嵌入到当前的 JSP 网页中，该指令的语法有以下两种格式：

```
<jsp:include page="文件的URL"/>
```

或

```
<jsp:include page="文件的URL">
    子标记
</jsp:include>
```

当动作标记 include 不需要子标记时使用上述第一种形式。

所谓动态嵌入就是"先处理后包含"，在运行阶段完成对文件的嵌入。也就是在将 JSP 页面转译成 Java 文件时并不合并两个页面，而是在 Java 文件的字节码文件被加载并执行时才去处理 include 动作标记中引入的文件。与静态嵌入方式相比，动态嵌入的执行速度稍慢，但是灵活性较高。

【例】3-6 编写两个 JSP 页面 example3_6.jsp 和 example3_6_1.jsp，在 example3_6.jsp 页面中使用 include 标记动态嵌入 example3_6_1.jsp 页面。运行 example3_6.jsp 页面。

页面文件 example3_6.jsp 的代码如下：

```
<%@ page language="java" contentType="text/html; charset=UTF-8" pageEncoding="UTF-8"%>
<!DOCTYPE html>
<html>
<head>
<meta charset="UTF-8">
<title>example3_6.jsp</title>
</head>
<body>
动态嵌入 example3_6_1.jsp 之前
<br>
<jsp:include page="example3_6_1.jsp" />
<br>
动态嵌入 example3_6_1.jsp 之后
</body>
</html>
```

页面文件 example3_6_1.jsp 的代码如下：

```
<%@ page language="java" contentType="text/html; charset=UTF-8" pageEncoding=
"UTF-8"%>
<!DOCTYPE html>
<html>
<head>
<meta charset="UTF-8">
<title>example3_6_1.jsp</title>
</head>
<body>
<font color="red" size=4>example3_6_1.jsp 文件的内容</font>
</body>
</html>
```

页面文件 example3_6.jsp 通过动作标记 include 动态嵌入了 example3_6_1.jsp，此时 JSP 引擎不会将两个文件合并成一个 JSP 页面，而是分别将文件 example3_6.jsp 和 example3_6_1.jsp 转化成对应的 Java 文件和字节码文件。当 JSP 解释器解释执行 example3_6.jsp 页面时会遇到动作指令<jsp:include page="example3_6_1.jsp"/>对应的代码，此时才会执行 example3_6_1.jsp 页面对应的字节码文件，然后将执行的结果发送到客户端，并由客户端负责显示这些结果，所以 example3_6.jsp 和 example3_6_1.jsp 页面中 page 指令的 contentType 属性值可以不同。

3.4.2 forward 动作标记

动作标记 forward 的作用是从该标记出现处停止当前 JSP 页面的继续执行，转向执行 forward 动作标记中 page 属性值指定的 JSP 页面。该标记有以下两种格式：

```
<jsp: forward page="文件的 URL"/>
```

或

```
<jsp: forward page="文件的 URL">
    子标记
</jsp: forward >
```

当动作标记 forward 不需要子标记时使用上述第一种形式。

例 3-7 编写 JSP 页面 example3_7.jsp、oddNumber.jsp 和 evenNumbers.jsp，在 example3_7.jsp 页面中随机获取 0~10 的整数，当该数为偶数时转向页面 evenNumbers.jsp，否则转向页面 oddNumber.jsp。访问 example3_7.jsp 页面。

页面文件 example3_7.jsp 的代码如下：

```
<%@ page language="java" contentType="text/html; charset=UTF-8" pageEncoding=
"UTF-8"%>
<!DOCTYPE html>
<html>
<head>
```

```
<meta charset="UTF-8">
<title>example3_7.jsp</title>
</head>
<body>
<%
long i = Math.round(Math.random() * 10);
if(i%2 == 0){
    System.out.println("获得的整数是偶数，即将跳转到偶数页面evenNumbers.jsp。");
%>
    <jsp:forward page="evenNumbers.jsp"/>
<%
    System.out.println("我是偶数尝试一下能看到我吗？");
}else{
    System.out.println("获得的整数是奇数，即将跳转到奇数页面oddNumber.jsp。");
%>
<jsp:forward page="oddNumber.jsp"/>
<%
    System.out.println("我是奇数尝试一下能看到我吗？");
}
%>
</body>
</html>
```

页面文件 evenNumbers.jsp 的代码如下：

```
<%@ page language="java" contentType="text/html; charset=UTF-8" pageEncoding="UTF-8"%>
<!DOCTYPE html>
<html>
<head>
<meta charset="UTF-8">
<title>evenNumbers.jsp</title>
</head>
<body>
    我是偶数页。
</body>
</html>
```

页面文件 oddNumber.jsp 的代码如下：

```
<%@ page language="java" contentType="text/html; charset=UTF-8" pageEncoding="UTF-8"%>
<!DOCTYPE html>
<html>
<head>
<meta charset="UTF-8">
<title> oddNumber.jsp</title>
</head>
<body>
```

```
    我是奇数页。
</body>
</html>
```

3.4.3　param 动作标记

动作标记 param 不能独立使用，但可以作为 include、forward 动作标记的子标记来使用，该标记以"名字-值"对的形式为对应页面传递参数。该标记的格式如下：

```
<jsp:父标记 page="接收参数页面的 URL">
    <jsp:param name="参数名" value="参数值"/>
</jsp:父标记>
```

接收参数的页面可以使用内置对象 request 调用 getParameter("参数名")方法获取动作标记 param 传递过来的参数值，内置对象将在本书的第 4 章介绍。

用户可以使用 param 子标记向页面传递多个参数，格式如下：

```
<jsp:父标记  page="接收参数页面的 URL">
    <jsp:param name="参数名 1" value="参数值 1"/>
    <jsp:param name="参数名 2" value="参数值 2"/>
    <jsp:param name="参数名 3" value="参数值 3"/>
    …
</jsp:父标记>
```

例 3-8　编写两个页面 example3_8.jsp 和 computer.jsp，在页面 example3_8.jsp 中使用 include 动作标记动态包含文件 computer.jsp，并向它传递一个矩形的长和宽；computer.jsp 收到参数后计算矩形的面积，并显示结果。运行 example3_8.jsp 页面。

页面文件 example3_8.jsp 的代码如下：

```
<%@ page language="java" contentType="text/html; charset=UTF-8" pageEncoding="UTF-8"%>
<!DOCTYPE html>
<html>
<head>
<meta charset="UTF-8">
<title>example3_8.jsp</title>
</head>
<body>
    加载 computer.jsp 页面计算矩形的面积<br><br>
    <jsp:include page="computer.jsp">
        <jsp:param value="10" name="length"/>
        <jsp:param value="6" name="width"/>
    </jsp:include>
</body>
</html>
```

页面文件 computer.jsp 的代码如下:

```jsp
<%@ page language="java" contentType="text/html; charset=UTF-8" pageEncoding="UTF-8"%>
<!DOCTYPE html>
<html>
<head>
<meta charset="UTF-8">
<title>computer.jsp</title>
</head>
<body>
    <%
        String m = request.getParameter("length");
        String n = request.getParameter("width");
        double a = Double.parseDouble(m);
        double b = Double.parseDouble(n);
        out.print("我是被加载的页面，负责计算矩形的面积" + "<br>");
        out.print("给我传递的矩形的长度是："+a+",宽度是:" + b + "<br>");
        out.print("矩形的面积是:" + a * b);
    %>
</body>
</html>
```

3.4.4 实践环节——include 和 param 动作标记的应用

编写 JSP 页面 input.jsp、first.jsp 和 second.jsp，将 3 个 JSP 文件保存在同一个 Web 服务目录中，input.jsp 使用 include 动作标记加载 first.jsp 和 second.jsp 页面。first.jsp 页面可以画出一个表格，second.jsp 页面可以计算出两个正整数的最大公约数。当 first.jsp 被加载时获取 input.jsp 页面中 include 动作标记的 param 子标记提供的表格的行数和列数，当 second.jsp 被加载时获取 input.jsp 页面中 include 动作标记的 param 子标记提供的两个正整数的值。

3.4.5 实践环节——登录验证

编写 JSP 页面 login.jsp、validate.jsp 和 success.jsp，在 login.jsp 页面中输入用户名和密码信息，提交给 validate.jsp 页面进行用户验证，如果验证为合法用户（用户名为 tom，密码为 jenny），则转到（forward 动作标记）success.jsp 页面，否则转到 login.jsp 页面重新登录。

3.5 本章小结

本章主要介绍了 JSP 页面的组成、JSP 脚本元素和 JSP 标记。一个 JSP 页面通常由普通的 HTML 标记、JSP 注释、Java 脚本元素以及 JSP 标记组成。JSP 脚本元素包括 Java 程序片、JSP 页面成员变量与方法的声明、Java 表达式。JSP 标记包括指令标记和动作标记。

习题 3

1. JSP 页面由哪几种主要元素组成？
2. 如果有 3 个用户访问一个 JSP 页面，则该页面中的 Java 程序片将被执行几次？
3. "<%!"和"%>"之间声明的变量与"<%"和"%>"之间声明的变量有何不同？
4. 动作标记 include 和指令标记 include 的区别是什么？
5. 在一个 JSP 页面中是否允许使用 page 指令为 contentType 属性设置多个值？是否允许使用 page 指令为 import 属性设置多个值？
6. 用户可以在以下（   ）标记之间插入 Java 程序片。
   A．"<%"和"%>"　　　　　　　　B．"<%"和"/>"
   C．"</"和"%>"　　　　　　　　D．"<%"和"!>"
7. JSP 的 page 指令的 language 属性的默认值是（   ）。
   A．Java　　　B．C　　　C．C#　　　D．SQL
8. 用户可以在以下（   ）标记之间插入变量与方法声明。
   A．"<%"和"%>"　　　　　　　　B．"<%!"和"%>"
   C．"</"和"%>"　　　　　　　　D．"<%"和"!>"
9. 下列变量声明在（   ）范围内有效。

```
<%! Date dateTime;
    int countNum;
%>
```

   A．从定义开始处有效，客户之间不共享
   B．在整个页面内有效，客户之间不共享
   C．在整个页面内有效，被多个客户共享
   D．从定义开始处有效，被多个客户共享

10. 在"<%!"和"%>"标记之间声明的 Java 方法称为页面的成员方法，其（   ）有效。
    A．从定义处之后　　　　　　　B．在整个页面内
    C．从定义处之前　　　　　　　D．不确定

11. 在"<%="和"%>"标记之间放置（   ）可以直接输出其值。
    A．变量　　　B．Java 表达式　　　C．字符串　　　D．数字

12. include 指令用于在 JSP 页面中静态插入一个文件，所插入文件可以是 JSP 页面、HTML 网页、文本文件或一段 Java 代码，但必须保证插入后形成的文件是（   ）。
    A．一个完整的 HTML 文件　　　B．一个完整的 JSP 文件
    C．一个完整的 TXT 文件　　　　D．一个完整的 Java 源文件

13. JSP 页面可以在"<%="和"%>"标记之间放置 Java 表达式，直接输出 Java 表达式的值，组成"<%="标记的各字符之间（   ）。
    A．可以有空格　　　　　　　　B．不可以有空格
    C．必须有空格　　　　　　　　D．不确定

第4章 JSP 内置对象

学习目的与要求

本章主要介绍 request、response、out、session、application、pageContext、page、config 和 exception 等内置对象。通过本章的学习，读者能够理解 JSP 内置对象的含义，掌握 JSP 内置对象的使用方法。

本章主要内容

- request 对象
- response 对象
- out 对象
- session 对象
- application 对象
- pageContext 对象
- page 对象
- config 对象
- exception 对象
- JSP 的 4 种作用域

有些对象在 JSP 页面中不需要声明和实例化，可以直接在 Java 程序片和 Java 表达式部分使用，通常称这样的对象为 JSP 内置对象。JSP 内置对象由 Web 服务器负责实现和管理，JSP 自带了 9 个功能强大的内置对象，共分为四大类。

❶ 与 Input/Output 有关的内置对象

与 Input/Output 有关的内置对象包括 request、response 和 out，该类对象主要用来作为客户端和服务器间通信的桥梁。request 对象表示客户端对服务器端发送的请求；response 对象表示服务器对客户端的响应；而 out 对象负责把处理结果输出到客户端。

❷ 与 Context 有关的内置对象

与 Context（上下文）有关的内置对象包括 session、application 和 pageContext。其中，session 对象表示浏览器与服务器会话的上下文环境；application 对象表示应用程序（Web

应用）的上下文环境；pageContext 对象表示当前 JSP 页面的上下文环境。

❸ **与 Servlet 有关的内置对象**

与 Servlet 有关的内置对象包括 page 和 config。page 对象表示 JSP 文件转换为 Java 文件后的 Servlet 对象；config 对象表示 JSP 文件转换为 Java 文件后的 Servlet 的 ServletConfig 对象。

❹ **与 Error 有关的内置对象**

与 Error 有关的内置对象只有 exception 对象。当 JSP 网页有错误时将产生异常，该对象就是用来处理这个异常。

本章涉及的 JSP 页面保存在工程 ch4 的 WebContent 中。

4.1 request 对象

视频讲解

request 对象的类型为 javax.servlet.http.HttpServletRequest。当客户端请求一个 JSP 页面时，JSP 页面所在的服务器将客户端发出的所有请求信息封装在内置对象 request 中，因此使用该对象就可以获取客户端提交的信息。

4.1.1 常用方法

request 内置对象的常用方法如表 4.1 所示。

表 4.1 request 内置对象的常用方法

序号	方法	功能说明
1	Object getAttribute(String name)	返回指定属性的属性值
2	Enumeration getAttributeNames()	返回所有可用属性名的枚举
3	String getCharacterEncoding()	返回字符编码方式
4	int getContentLength()	返回请求体的字节数
5	String getContentType()	返回请求体的 MIME 类型
6	ServletInputStream getInputStream()	返回请求体中一行的二进制流
7	String getParameter(String name)	返回 name 指定参数的参数值
8	Enumeration getParameterNames()	返回可用参数名的枚举
9	String[] getParameterValues(String name)	返回包含参数 name 的所有值的数组
10	String getProtocol()	返回请求用的协议类型及版本号
11	String getServerName()	返回接受请求的服务器主机名
12	int getServerPort()	返回服务器接受此请求所用的端口号
13	String getRemoteAddr()	返回发送此请求的客户端 IP 地址
14	String getRemoteHost()	返回发送此请求的客户端主机名
15	void setAttribute(String key, Object obj)	设置属性的属性值
16	String getRealPath(String path)	返回一虚拟路径的真实路径

❶ **使用 request 对象获取客户端提交的信息**

1）String getParameter(String name)

该方法以字符串的形式返回客户端传来的某个参数的值，该参数名由 name 指定。

例 4-1 调用方法 getParameter(String name)获取表单信息。

页面文件 example4_1.jsp 的代码如下：

```
<%@ page language="java" contentType="text/html; charset=UTF-8"
pageEncoding="UTF-8"%>
<!DOCTYPE html>
<html>
<head>
<meta charset="UTF-8">
<title>example4_1.jsp</title>
</head>
<body>
<form action="getValue.jsp" >
    <input type="text" name="userName"/>
    <input type="submit" value="提交"/>
</form>
</body>
</html>
```

页面文件 getValue.jsp 的代码如下：

```
<%@ page language="java" contentType="text/html; charset=UTF-8"
pageEncoding="UTF-8"%>
<!DOCTYPE html>
<html>
<head>
<meta charset="UTF-8">
<title>getValue.jsp</title>
</head>
<body>
<%
    String name = request.getParameter("userName");
    //userName 为 example4_1.jsp 页面中的表单参数名
    out.println(name);
%>
</body>
</html>
```

2）String[] getParameterValues(String name)

该方法以字符串数组的形式返回客户端向服务器端传递的指定参数名的所有值。

例 4-2 调用方法 getParameterValues(String name)获取表单信息。

页面文件 example4_2.jsp 的代码如下：

```
<%@ page language="java" contentType="text/html; charset=UTF-8"
pageEncoding="UTF-8"%>
<!DOCTYPE html>
<html>
```

```html
<head>
<meta charset="UTF-8">
<title>example4_2.jsp</title>
</head>
<body>
<form action="getValues.jsp">
    选择你去过的城市：<br/>
    <input type="checkbox" name="cities" value="北京"/>北京
    <input type="checkbox" name="cities" value="上海"/>上海
    <input type="checkbox" name="cities" value="香港"/>香港
    <input type="submit" value="提交"/>
</form>
</body>
</html>
```

页面文件 getValues.jsp 的代码如下：

```jsp
<%@ page language="java" contentType="text/html; charset=UTF-8"
    pageEncoding="UTF-8"%>
<!DOCTYPE html>
<html>
<head>
<meta charset="UTF-8">
<title>getValues.jsp</title>
</head>
<body>
您去过的城市：<br>
<%
    String yourCities[] = request.getParameterValues("cities");
    for(int i = 0; i < yourCities.length; i ++){
        out.println(yourCities[i] + "<br>");
    }
%>
</body>
</html>
```

❷ **NullPointerException 异常**

如果不选择 example4_2.jsp 页面中的城市直接单击"提交"按钮，那么 getValues.jsp 页面就会提示出现 NullPointerException 异常。为了避免在运行时出现 NullPointerException 异常，在 getValues.jsp 页面中使用如下代码：

```java
if(yourCities != null){
      for(int i = 0;i < yourCities.length; i ++){
          out.print(yourCities[i] + "<br>");
      }
}
```

4.1.2 使用 request 对象存取数据

request 对象可以通过 setAttribute(String key, Object obj)方法将参数 obj 指定的对象保存到 request 对象中，其中 key 为所保存的对象指定一个关键字。若保存的两个对象的关键字相同，则先保存的对象将被清除。

request 对象可以通过 Object getAttribute(String key)方法获取请求域（例如 forward 转发）中的关键字为 key 的对象（属性值）。

在实际工程中，request 对象经常用于存储、传递本次请求的处理结果。

例 4-3 编写两个 JSP 页面 example4_3.jsp 和 example4_3_1.jsp，在 example4_3.jsp 页面中输入一个整数提交给 example4_3_1.jsp 页面求平方。当输入的值非整数时，在 example4_3_1.jsp 页面中使用 request 对象的 setAttribute(String key, Object obj)方法将错误消息存储到 request 对象中，同时使用 forward 标记转发到 example4_3.jsp 页面并显示错误消息。

页面文件 example4_3.jsp 的代码如下：

```jsp
<%@ page language="java" contentType="text/html; charset=UTF-8"
    pageEncoding="UTF-8"%>
<!DOCTYPE html>
<html>
<head>
<meta charset="UTF-8">
<title>example4_3.jsp</title>
</head>
<body>
    <%
        //从请求域中获取属性 errorMsg 的错误消息
        String error = (String)request.getAttribute("errorMsg");
        if(error != null){
            out.print("<font color='red'>" + error + "</font>");
        }
    %>
    <form action="example4_3_1.jsp">
        输入一个整数求平方：
        <input type="text" name="mynumber">
        <br>
        <input type="submit" value="提交">
    </form>
</body>
</html>
```

页面文件 example4_3_1.jsp 的代码如下：

```jsp
<%@ page language="java" contentType="text/html; charset=UTF-8"
```

```jsp
pageEncoding="UTF-8"%>
<!DOCTYPE html>
<html>
<head>
<meta charset="UTF-8">
<title>example4_3_1.jsp</title>
</head>
<body>
    <%
        String number = request.getParameter("mynumber");
        int toNumber = 0;
        try{
            toNumber = Integer.parseInt(number);
            out.print(toNumber*toNumber);
        }catch(Exception e){
            request.setAttribute("errorMsg", "请输入一个整数！");
    %>
            <jsp:forward page="example4_3.jsp"></jsp:forward>
    <%
        }
    %>
</body>
</html>
```

在例 4-3 中，错误消息被以请求域属性的形式保存到 request 对象中，并通过请求转发的方式将请求对象再转发给 example4_3.jsp，这样在 example4_3.jsp 页面中便可以从 request 对象中获取到属性值，从而实现了错误消息在一次 request 请求范围内传递。

4.1.3 中文乱码问题

Java 的内核和 class 文件是基于 unicode 的，这使得 Java 程序具有良好的跨平台性，但也产生了一些中文乱码问题。

如果在例 4-1 的 example4_1.jsp 页面的文本框中输入中文姓名，那么 getValue.jsp 页面获得的姓名可能是乱码。如果出现中文乱码，可以使用 request 对象的 setCharacterEncoding(String code)方法设置统一字符编码，其中参数 code 以字符串形式传入要设置的编码格式，但这种方法仅对于提交方式是 post 的表单（表单默认的提交方式是 get）有效。例如，使用该方法解决例 4-1 中的 getValue.jsp 页面出现的中文乱码问题需要完成下面两件事。

首先将 example4_1.jsp 中的表单提交方式改为 "post"，具体代码如下：

```html
<form action="getValue.jsp" method="post">
```

其次在 getValue.jsp 中获取表单信息之前设置统一编码，具体代码如下：

```jsp
request.setCharacterEncoding("UTF-8");
```

在使用该方法解决中文乱码问题时，接受参数的每个页面都需要执行 request.setCharacterEncoding("UTF-8")。为了避免每个页面都编写 request.setCharacterEncoding("UTF-8") 语句，可以使用过滤器对所有 JSP 页面进行编码处理。过滤器将在本书的第 8 章讲解。

4.1.4 实践环节——获取客户端的基本信息

编写一个 JSP 页面 practice4_1.jsp，在该页面中使用 request 对象的方法获取客户端的 IP 地址、客户机名称、服务器名称以及服务器端口号。

4.2 response 对象

视频讲解

当客户端请求服务器的一个页面时会提交一个 HTTP 请求，服务器收到请求后返回 HTTP 响应。request 对象对请求信息进行封装，与 request 对象对应的对象是 response 对象。response 对象的类型为 javax.servlet.http.HttpServletResponse，对客户端的请求做出动态响应。动态响应通常有动态改变 contentType 属性值、设置响应表头和 response 重定向。

4.2.1 动态改变 contentType 属性值

JSP 页面用 page 指令标记设置了页面的 contentType 属性值，response 对象按照此属性值的方式对客户端做出响应。在 page 指令中只能为 contentType 属性指定一个值。如果想动态改变 contentType 属性值，换一种方式来响应客户端，可以让 response 对象调用 setContentType(String s)方法来重新设置 contentType 的属性值。

例4-4 编写一个 JSP 页面 example4_4.jsp，客户端通过单击页面上的不同按钮可以改变页面响应的 MIME 类型。当单击 word 按钮时，JSP 页面动态改变 contentType 的属性值为 application/msword，浏览器启用本地的 Word 软件来显示当前页面内容；当单击 excel 按钮时，JSP 页面动态改变 contentType 的属性值为 application/vnd.ms-excel，浏览器启用本地的 Excel 软件来显示当前页面内容。页面效果如图 4.1 所示。

图 4.1 例 4-4 的效果图

(c) application/vnd.ms-excel 响应方式

图 4.1（续）

页面文件 example4_4.jsp 的代码如下：

```jsp
<%@ page language="java" contentType="text/html; charset=UTF-8"
    pageEncoding="UTF-8"%>
<!DOCTYPE html>
<html>
<head>
<meta charset="UTF-8">
<title>example4_4.jsp</title>
</head>
<body>
    <form action="" method="post">
    <p>我们在学习使用 response 动态改变 contentType 属性值
    <p>
        <input type="submit" value="word" name="submit">
        <input type="submit" value="excel" name="submit">
        <%
            String str = request.getParameter("submit");
            if ("word".equals(str)) {
                //response 调用 setContentType()方法设置 MIME 类型为
                //application/msword
                response.setContentType("application/msword");
            } else if ("excel".equals(str)) {
                //response 调用 setContentType()方法设置 MIME 类型为
                //application/vnd.ms-excel
                response.setContentType("application/vnd.ms-excel");
            }
        %>
    </form>
</body>
</html>
```

4.2.2 设置响应表头

response 对象可以通过方法 setHeader(String name,String value)设置指定名字的 HTTP 文件头值，以此来操作 HTTP 文件头。如果希望某页面每 3 秒钟刷新一次，那么在该页面中添加如下代码：

```
response.setHeader("refresh","3");
```

有时候希望从当前页面几秒钟后自动跳转到另一个页面，比如打开 one.jsp 页面 3 秒钟后自动跳转到 another.jsp 页面（one.jsp 与 another.jsp 在同一个 Web 服务目录下）。这该如何实现呢？只需要为 one.jsp 设置一个响应表头即可，也就是在 one.jsp 页面中添加如下代码：

```
response.setHeader("refresh","3;url=another.jsp");
```

例 4-5 编写一个 JSP 页面 example4_5.jsp，在该页面中使用 response 对象设置一个响应表头 "refresh"，其值是 "3"。那么在用户收到这个响应表头之后，该页面会每 3 秒钟刷新一次。

页面文件 example4_5.jsp 的代码如下：

```jsp
<%@ page language="java" contentType="text/html; charset=UTF-8"
    pageEncoding="UTF-8"%>
<%@ page import="java.util.*" %>
<!DOCTYPE html>
<html>
<head>
<meta charset="UTF-8">
<title>example4_5.jsp</title>
</head>
<body>
    <h2>该页面每 3 秒钟刷新 1 次</h2>
    <p>现在的秒钟时间是:
    <%
        Date d = new Date();
        out.print("" + d.getSeconds());
        response.setHeader("refresh","3");
    %>
</body>
</html>
```

4.2.3 使用 response 对象重定向

当需要将客户端引导至另一个页面时，可以使用 reponse 对象的 sendRedirect(String url) 方法实现客户端的重定向。例如客户端输入的表单信息不完整或有误，应该再次被重定向到输入页面。

例 4-6 编写两个 JSP 页面 login.jsp 和 validate.jsp，如果在页面 login.jsp 中输入正确的密码 "2019manpinde"，单击 "提交" 按钮后将提交给页面 validate.jsp，如果输入不正确，重新定向到 login.jsp 页面。运行 login.jsp 页面，页面效果如图 4.2 所示。

（a）login.jsp 页面

（b）validate.jsp 页面

图 4.2　例 4-6 的效果图

页面文件 login.jsp 的代码如下：

```
<%@ page language="java" contentType="text/html; charset=UTF-8"
    pageEncoding="UTF-8"%>
<!DOCTYPE html>
<html>
<head>
<meta charset="UTF-8">
<title>login.jsp</title>
</head>
<body>
<form action="validate.jsp" method="post" name=form>
<p>
    输入密码：
<br>
<input type="password" name="pwd"/>
<input type="submit" value="提交">
</form>
</body>
</html>
```

页面文件 validate.jsp 的代码如下：

```
<%@ page language="java" contentType="text/html; charset=UTF-8"
    pageEncoding="UTF-8"%>
<!DOCTYPE html>
<html>
<head>
<meta charset="UTF-8">
<title>validate.jsp</title>
</head>
<body>
<%
String str = request.getParameter("pwd");
```

```
if (!"2019manpinde".equals(str)) {
    response.sendRedirect("login.jsp");
} else {
    out.print("2019年是蛮拼的一年！");
}
%>
</body>
</html>
```

response 对象的 sendRedirect()方法是在客户端的浏览器中工作，Web 服务器要求浏览器重新发送一个到被定向页面的请求。在浏览器的地址栏上将出现重定向页面的 URL，且为绝对路径。

forward 动作标记也可以实现页面的跳转，例如<jsp:forward page="info.jsp"/>。但使用 forward 动作标记与使用 response 对象调用 sendRedirect()不同。二者的比较如下：

① forward 为服务器端跳转，浏览器的地址栏不变；sendRedirect()为客户端跳转，浏览器的地址栏改变为新页面的 URL。

② 执行到 forward 标记出现处停止当前 JSP 页面的继续执行，而转向标记中 page 属性指定的页面；sendRedirect()在所有代码执行完毕之后再跳转。

③ 使用 forward，request 请求信息能够保留到下一个页面；使用 sendRedirect()不能保留 request 请求信息。

forward 传递参数的格式如下：

```
<jsp:forward page="info.jsp">
    <jsp:param name="no" value="001"/>
    <jsp:param name="age" value="18"/>
</jsp:forward>
```

response 对象的 sendRedirect()传递参数的格式如下：

```
response.sendRedirect("info.jsp?sno=001&sage=18") ;
```

4.2.4 实践环节——登录验证

编写 JSP 页面 login_1.jsp、server.jsp 和 loginSuccess.jsp。在页面 login_1.jsp 中输入用户名和密码，单击"提交"按钮将输入的信息提交给页面 server.jsp。在 server.jsp 页面中进行登录验证：如果输入正确（用户名为"zhangsan"，密码为"123"），提示"成功登录，3 秒钟后进入 loginSuccess.jsp 页面"；如果输入不正确，重新定向到 login_1.jsp 页面。运行 login_1.jsp 页面，页面运行效果如图 4.3 所示。

（a）login_1.jsp 页面

图 4.3 4.2.4 实践环节的效果图

成功登录，3秒钟后进入loginSuccess.jsp页面

（b）server.jsp 页面

欢迎张三登录成功！

（c）loginSuccess.jsp 页面

图 4.3（续）

4.3 out 对象

out 对象的类型为 javax.servlet.jsp.JspWriter，是一个输出流，用来向客户端浏览器输出数据。其常用方法如表 4.2 所示。

表 4.2 out 对象的常用方法

序 号	方 法	功 能 说 明
1	void clear()	清除缓冲区的内容
2	void clearBuffer()	清除缓冲区的当前内容
3	void flush()	清空流
4	int getBufferSize()	返回缓冲区字节数，如不设缓冲区将返回 0
5	int getRemaining()	返回缓冲区剩余大小
6	boolean isAutoFlush()	返回缓冲区满时是自动清空还是抛出异常
7	void close()	关闭输出流
8	void print()	输出各种数据类型
9	void newLine()	输出一个换行符

例 4-7 编写一个页面 example4_7.jsp，在该页面中使用 out 对象输出信息。

页面文件 example4_7.jsp 的代码如下：

```
<%@ page language="java" contentType="text/html; charset=UTF-8"
pageEncoding="UTF-8"%>
<%@ page import="java.util.Date"%>
<!DOCTYPE html>
<html>
<head>
<meta charset="UTF-8">
<title>example4_7.jsp</title>
</head>
<body>
    <%
        int myNumber = 1000;
        Date myDate = new Date();
```

```
            out.print(myNumber);
            out.print("<br>");
            out.print(myDate);
        %>
    </body>
</html>
```

4.4　session 对象

视频讲解

浏览器与 Web 服务器之间使用 HTTP 协议进行通信。HTTP 是一种无状态协议，客户端向服务器发出请求（request），服务器返回响应（response），连接就被关闭了，在服务器端不保留连接的相关信息，所以服务器必须采取某种手段来记录每个客户端的连接信息。Web 服务器可以使用内置对象 session 来存放有关连接的信息，session 对象的类型为 javax.servlet.http.HttpSession。session 对象指的是客户端与服务器端的一次会话，从客户端连到服务器端的一个 Web 应用程序开始，直到客户端与服务器端断开为止。

4.4.1　session 对象的 ID

Web 服务器会给每一个用户自动创建一个 session 对象，为每个 session 对象分配一个唯一标识的 String 类型的 session ID，这个 ID 用于区分其他用户。这样每个用户都对应着一个 session 对象，不同用户的 session 对象互不相同。session 对象调用 getId()方法就可以获取当前 session 对象的 ID。

例4-8　编写 JSP 页面 example4_8_1.jsp、example4_8_2.jsp 和 example4_8_3.jsp，其中，example4_8_2.jsp 存放在目录 tom 中，example4_8_3.jsp 存放在目录 cat 中。用户首先访问 example4_8_1.jsp 页面，从该页面链接到 example4_8_2.jsp 页面，然后再从 example4_8_2.jsp 页面链接到 example4_8_3.jsp，效果如图 4.4 所示。

> http://localhost:8080/ch4/example4_8_1.jsp
>
> 年轻人如何养生呢？
>
> 先看看Web服务器给我分配的session对象的ID：
> BD6235DC13B7BE2C5EBFB35D42E57B23
>
> 点击链接去吃睡篇看看吧？
>
> （a）example4_8_1.jsp 页面效果

> http://localhost:8080/ch4/tom/example4_8_2.jsp
>
> 欢迎您进入养生之吃睡篇！
>
> 先看看Web服务器给我分配的session对象的ID：
> BD6235DC13B7BE2C5EBFB35D42E57B23
>
> 吃，不忌嘴，五谷杂粮、蔬菜水果通吃不挑食
> 睡，早睡早起不熬夜
>
> 点击链接去运动篇看看吧？
>
> （b）example4_8_2.jsp 页面效果

图 4.4　获取 session 对象的 ID

欢迎您进入养生之**运动篇**!

先看看Web服务器给我分配的session对象的ID:
BD6235DC13B7BE2C5EBFB35D42E57B23

动,坚持运动——这一点年轻人很多都做不好,
高兴起来就拼命打球,懒起来拼命睡觉,不好!
总之,生活规律化,坚持长期运动

点击链接去首页看看吧?

(c) example4_8_3.jsp 页面效果

图 4.4(续)

页面文件 example4_8_1.jsp 的代码如下:

```
<%@ page language="java" contentType="text/html; charset=UTF-8"
pageEncoding="UTF-8"%>
<!DOCTYPE html>
<html>
<head>
<meta charset="UTF-8">
<title>example4_8_1.jsp</title>
</head>
<body>
年轻人如何养生呢?<br><br>
先看看Web服务器给我分配的session对象的ID:
<%
    String id = session.getId();
%>
<br>
<%=id %>
<br><br>
点击链接去<a href="tom/example4_8_2.jsp">吃睡篇</a>看看吧?
</body>
</html>
```

页面文件 example4_8_2.jsp 的代码如下:

```
<%@ page language="java" contentType="text/html; charset=UTF-8"
pageEncoding="UTF-8"%>
<!DOCTYPE html>
<html>
<head>
<meta charset="UTF-8">
<title>example4_8_2.jsp</title>
</head>
<body>
欢迎您进入养生之<font size=5>吃睡篇</font>! <br><br>
```

先看看Web服务器给我分配的session对象的ID：
```
<%
    String id = session.getId();
%>
<br>
<%=id %>
<br><br>
吃，不忌嘴，五谷杂粮、蔬菜水果通吃不挑食<br>
睡，早睡早起不熬夜<br><br>
点击链接去<a href="../cat/example4_8_3.jsp">运动篇</a>看看吧？
</body>
</html>
```

页面文件example4_8_3.jsp的代码如下：

```
<%@ page language="java" contentType="text/html; charset=UTF-8"
pageEncoding="UTF-8"%>
<!DOCTYPE html>
<html>
<head>
<meta charset="UTF-8">
<title>example4_8_3.jsp</title>
</head>
<body>
欢迎您进入养生之<font size=5>运动篇</font>！<br><br>
先看看Web服务器给我分配的session对象的ID：
<%
    String id = session.getId();
%>
<br>
<%=id %>
<br><br>
动，坚持运动——这一点年轻人很多都做不好，<br>高兴起来就拼命打球，懒起来拼命睡觉，不好！<br>
总之，生活规律化，坚持长期运动<br><br>
点击链接去<a href="../example4_8_1.jsp">首页</a>看看吧？
</body>
</html>
```

从例4-8各个页面的运行结果看，一个用户在同一个Web服务中只有一个session对象，当用户访问相同Web服务的其他页面时，Web服务器不会再重新分配session对象，直到用户关闭浏览器或这个session对象达到了它的生存期限。当用户重新打开浏览器再访问该Web服务时，Web服务器为该用户再创建一个新的session对象。

需要注意的是，同一用户在多个不同的Web服务中所对应的session对象是不同的，一个Web服务对应一个session对象。

4.4.2 使用 session 对象存取数据

使用 session 对象可以保存用户在访问某个 Web 服务期间的有关数据，有关处理数据的方法如下。

① public void setAttribute(String key, Object obj)：将参数 obj 指定的对象保存到 session 对象中，key 为所保存的对象指定一个关键字。若保存的两个对象的关键字相同，则先保存的对象被清除。

② public Object getAttribute(String key)：获取 session 中关键字是 key 的对象。

③ public void removeAttribute(String key)：从 session 中删除关键字 key 所对应的对象。

④ public Enumeration getAttributeNames()：产生一个枚举对象，该枚举对象可以使用方法 nextElements()遍历 session 中各个对象所对应的关键字。

例 4-9 使用 session 对象模拟在线考试系统。编写 JSP 页面 example4_9_1.jsp、example4_9_2.jsp 和 example4_9_3.jsp，在 example4_9_1.jsp 页面中考试，在 example4_9_2.jsp 页面中显示答题结果，在 example4_9_3.jsp 页面中计算并公布考试成绩。运行 example4_9_1.jsp 页面，效果如图 4.5 所示。

（a）试卷页面

（b）确认页面

（c）成绩公布页面

图 4.5　session 对象模拟考试系统

页面文件 example4_9_1.jsp 的代码如下：

```jsp
<%@ page language="java" contentType="text/html; charset=UTF-8"
pageEncoding="UTF-8"%>
<!DOCTYPE html>
<html>
<head>
<meta charset="UTF-8">
<title>example4_9_1.jsp</title>
</head>
<body>
<form action="example4_9_2.jsp" method="post">
        考号：
        <input type="text" name="id"/>
        <p>
            一、单项选择题（每题 2 分）
            <br/><br/>
            1.下列哪个方法是获取 session 中关键字是 key 的对象（  ）。
            <br/>
            <input type="radio" name="one" value="A"/>
            A. public void setAttribute(String key, Object obj)<br/>
            <input type="radio" name="one" value="B"/>
            B. public void removeAttribute(String key)<br/>
            <input type="radio" name="one" value="C"/>
            C. public Enumeration getAttributeNames()<br/>
            <input type="radio" name="one" value="D"/>
            D. public Object getAttribute(String key)<br/>
        </p>
        <p>
            二、判断题（每题 2 分）
            <br/><br/>
            1.同一用户在多个 Web 服务中，所对应的 session 对象是互不相同的。
            <br/>
            <input type="radio" name="two" value="True"/>
            True
            <input type="radio" name="two" value="False"/>
            False
        </p><br/>
        <input type="submit" value="提交" name=submit>
        <input type="reset" value="重置" name=reset>
    </form>
</body>
</html>
```

页面文件 example4_9_2.jsp 的代码如下：

```jsp
<%@ page language="java" contentType="text/html; charset=UTF-8"
pageEncoding="UTF-8"%>
<!DOCTYPE html>
<html>
<head>
<meta charset="UTF-8">
<title>example4_9_2.jsp</title>
</head>
<body>
<form action="example4_9_3.jsp" method="post">
        <%
            //考号
            String id = request.getParameter("id");
            //把考号id以"id"为关键字存储到session对象中
            session.setAttribute("id", id);
            //单项选择第一题
            String first = request.getParameter("one");
            //把答案first以"one"为关键字存储到session对象中
            session.setAttribute("one", first);
            //判断第一题
            String second = request.getParameter("two");
            //把答案second以"two"为关键字存储到session对象中
            session.setAttribute("two", second);
        %>
        您的考号：<%=id%><br/>
        一、单项选择题（每题2分）
        <br/>
        1.<%=first%>
        <br/>
        二、判断题（每题2分）
        <br/>
        1.<%=second%><br/>
        <input type="submit" value="确认完毕"/>
        <a href="example4_9_1.jsp">重新答题</a>
    </form>
</body>
</html>
```

页面文件 example4_9_3.jsp 的代码如下：

```jsp
<%@ page language="java" contentType="text/html; charset=UTF-8"
pageEncoding="UTF-8"%>
<!DOCTYPE html>
<html>
<head>
<meta charset="UTF-8">
```

```jsp
<title>example4_9_3.jsp</title>
</head>
<body>
<%
    //获取考号
    //获取 session 中关键字是 id 的对象（考号）
    String id = (String) session.getAttribute("id");
    //计算成绩
    int sum = 0;
    //如果单项选择第一题选中 D 选项，得 2 分
    //获取 session 中关键字是 one 的对象（选择答案）
    String first = (String) session.getAttribute("one");
    if ("D".equals(first)) {
        sum += 2;
    }
    //如果判断第一题选中 True，得 2 分
    //获取 session 中关键字是 two 的对象（判断答案）
    String second = (String) session.getAttribute("two");
    if ("True".equals(second)) {
        sum += 2;
    }
%>
您的成绩公布如下：
<table border="1">
    <tr>
        <th width="50%">
            考号
        </th>
        <th width="50%">
            成绩
        </th>
    </tr>
    <tr>
        <td><%=id%></td>
        <td align="right"><%=sum%></td>
    </tr>
</table>
</body>
</html>
```

4.4.3　session 对象的生命周期

一个用户在某个 Web 服务中 session 对象的生命周期依赖于以下几个因素。
① 用户是否关闭浏览器。

② session 对象是否调用 invalidate()方法。
③ session 对象是否达到设置的最长"发呆"时间。
与 session 对象生命周期相关的方法如表 4.3 所示。

表 4.3 session 对象的方法

序号	方法	功能说明
1	long getCreationTime()	返回 session 的创建时间
2	long getLastAccessedTime ()	返回此 session 里客户端最近一次请求的时间
3	int getMaxInactiveInterval()	返回两次请求的间隔时间（单位是秒）
4	void invalidate()	使 session 失效
5	boolean isNew()	判断客户端是否已经加入服务器创建的 session
6	void setMaxInactiveInterval()	设置两次请求的间隔时间（单位是秒）

例 4-10 编写一个 JSP 页面 example4_10.jsp。如果用户是第一次访问该页面，会显示欢迎信息，并输出 session 对象允许的最长"发呆"时间、创建时间，以及 session 对象的 ID。在 example4_10.jsp 页面中，session 对象使用 setMaxInactiveInterval(int maxValue) 方法设置最长"发呆"时间为 10 秒。如果用户两次刷新的间隔时间超过 10 秒，用户先前的 session 被取消，用户将获得一个新的 session 对象。页面运行效果如图 4.6 所示。

http://localhost:8080/ch4/example4_10.jsp
欢迎您第一次访问当前Web服务。
session允许的最长发呆时间为：10秒。
session的创建时间为：北京时间：2018年10月13日07时04分09秒星期六。
session的id为：B91ACCD1FCD8DCC458C6C440D8D9C86B。

（a）第一次或 10 秒后访问该页面

http://localhost:8080/ch4/example4_10.jsp
session允许的最长发呆时间为：10秒。
session的创建时间为：北京时间：2018年10月13日07时04分09秒星期六。
session的id为：B91ACCD1FCD8DCC458C6C440D8D9C86B。

（b）10 秒之内访问该页面

图 4.6 session 对象的生命周期

页面文件 example4_10.jsp 的代码如下：

```
<%@ page language="java" contentType="text/html; charset=UTF-8"
pageEncoding="UTF-8"%>
<!DOCTYPE html>
<%@ page import="java.util.*"%>
<%@ page import="java.text.*"%>
<html>
<head>
<meta charset="UTF-8">
<title>example4_10.jsp</title>
</head>
<body>
```

```
<%
//session 调用 setMaxInactiveInterval(int n)方法设置最长"发呆"时间为 10 秒
session.setMaxInactiveInterval(10);
//session 调用 isNew()方法判断 session 是不是新创建的
    boolean flg = session.isNew();
    if (flg) {
        out.println("欢迎您第一次访问当前 Web 服务目录。");
        out.println("<hr/>");
    }
    out.println("session 允许的最长发呆时间为: " +
    session.getMaxInactiveInterval()+ "秒。");
    //获取 session 对象被创建的时间
    long num = session.getCreationTime();
    //将整数转换为 Date 对象
    Date time = new Date(num);
    //用给定的模式和默认语言环境的日期格式符号构造 SimpleDateFormat 对象
    SimpleDateFormat matter = new SimpleDateFormat(
            "北京时间: yyyy 年 MM 月 dd 日 HH 时 mm 分 ss 秒 E。");
    //得到格式化后的字符串
    String strTime = matter.format(time);
    out.println("<br/>session 的创建时间为: " + strTime);
    out.println("<br/>session 的 id 为: " + session.getId() + "。");
%>
</body>
</html>
```

从例 4-10 中可以看出，如果用户长时间不关闭浏览器，session 对象也没有调用 invalidate()方法，那么用户的 session 可能消失。例如该例中的 JSP 页面在 10 秒钟之内如果不被访问，它先前创建的 session 对象会消失，服务器又重新创建一个 session 对象。这是因为 session 对象达到了它的最长"发呆"时间。所谓"发呆"时间，是指用户对某个 Web 服务发出的两次请求之间的时间。

用户对某个 Web 服务的 JSP 页面发出请求并得到响应，如果用户不再对该 Web 服务发出请求，比如不再操作浏览器，那么用户对该 Web 服务进入"发呆"状态，直到用户再次请求该 Web 服务时"发呆"状态结束。

Tomcat 服务器允许用户最长的"发呆"时间为 30 分钟。用户可以通过修改 Tomcat 安装目录中 conf 文件夹下的配置文件 web.xml，找到下面的片段，修改其中的默认值"30"，这样就可以重新设置各个 Web 服务目录下的 session 对象的最长"发呆"时间。这里的时间单位为分钟。

```
<session-config>
    <session-timeout>30</session-timeout>
</session-config>
```

用户也可以通过 session 对象调用 setMaxInactiveInterval(int time)方法来设置最长"发

呆"时间，其参数的时间单位为秒。

4.4.4 实践环节——购物车

用户到便民超市采购商品，在购物前需要登录会员卡号，购物时先将选购的商品放入购物车，最后到柜台清点商品。请借助于 session 对象模拟购物车，并存储客户的会员卡号和购买的商品名称。会员卡号在输入后可以修改，购物车中的商品可以查看。编写程序模拟上述过程。loginID.jsp 实现会员卡号的输入，shop.jsp 实现商品导购，food.jsp 实现商品购物，count.jsp 实现清点商品。本节实践环节的 4 个 JSP 页面都保存在目录 practice4 中，运行 loginID.jsp 页面，运行效果如图 4.7 所示。

（a）会员卡号输入页面

（b）导购页面

（c）购物页面

（d）查看购物车商品页面

图 4.7 session 模拟购物车

4.5 application 对象

4.5.1 什么是 application

不同用户的 session 对象互不相同，但有时候用户之间可能需要共享一个对象，Web 服务器启动后就产生了这样一个唯一的内置对象 application。application 对象实现了 javax.servlet.ServletContext 接口。任何用户在访问同一个 Web 服务的各个页面时共享一个 application 对象，直到服务器关闭这个 application 对象才被取消。

4.5.2 使用 application 对象存取数据

application 对象和 session 对象一样也可以进行数据的存取，处理数据的方法如下。

① public void setAttribute(String key, Object obj)：将参数 obj 指定的对象保存到 application 对象中，key 为所保存的对象指定一个关键字。若保存的两个对象的关键字相同，则先保存的对象被清除。

② public Object getAttribute(String key)：获取 application 中关键字是 key 的对象。

③ public void removeAttribute(String key)：从 application 中删除关键字 key 所对应的对象。

④ public Enumeration getAttributeNames()：产生一个枚举对象，该枚举对象可以使用方法 nextElements()遍历 application 中各个对象所对应的关键字。

【例】4-11 用 application 对象模拟 "成语接龙"，用户通过 example4_11_1.jsp 向 example4_11_2.jsp 页面提交四字成语，example4_11_2.jsp 页面获取成语内容后，用同步方法将该成语内容和以前的成语内容进行连接，然后将这些四字成语内容添加到 application 对象中。页面运行效果如图 4.8 所示。

（a）成语提交页面

（b）接龙成功页面

图 4.8 成语接龙

页面文件 example4_11_1.jsp 的代码如下：

```jsp
<%@ page language="java" contentType="text/html; charset=UTF-8"
pageEncoding="UTF-8"%>
<!DOCTYPE html>
<html>
<head>
<meta charset="UTF-8">
<title>example4_11_1.jsp</title>
</head>
<body>
<h2>四字成语接龙</h2>
<%
//取出 application 中关键字是 message 的对象（成语内容）
StringBuffer s = (StringBuffer)application.getAttribute("message");
if(s != null){
    out.print(s.toString());
}
else{
    out.print("还没有词语，请您龙头开始！<br>");
}
%>
<form action="example4_11_2.jsp" method="post">
四字成语输入：<input type="text" name="mes"/><br>
<input type="submit" value="提交"/>
</form>
</body>
</html>
```

页面文件 example4_11_2.jsp 的代码如下：

```jsp
<%@ page language="java" contentType="text/html; charset=UTF-8"
pageEncoding="UTF-8"%>
<!DOCTYPE html>
<html>
<head>
<meta charset="UTF-8">
<title>example4_11_2.jsp</title>
</head>
<body>
<%!
    StringBuffer message = new StringBuffer("");
    ServletContext application;
    synchronized void sendMessage(String s){
        application = getServletContext();
        message.append(s + "->") ;
        //把成语内容 message 以 "message" 为关键字存储到 application 对象中
        application.setAttribute("message", message);
    }
%>
<%
```

```
        request.setCharacterEncoding("utf-8");
        String content = request.getParameter("mes");
        sendMessage(content);
        out.print("您的四字成语已经提交！3 秒钟后回到成语页面，继续接龙！");
        response.setHeader("refresh", "3;url=example4_11_1.jsp");
    %>
    </body>
</html>
```

例 4-11 中的成语接龙方法 sendMessage()为什么定义为同步方法呢？这是因为 application 对象对所有用户都是相同的，任何用户对该对象中存储的数据进行操作都会影响到其他用户。

如果客户端浏览不同的 Web 服务，将产生不同的 application 对象。同一个 Web 服务的所有 JSP 页面共享一个 application 对象，即使浏览这些 JSP 页面的是不同的客户端也是如此。因此，保存在 application 对象中的数据不仅可以跨页面分享，还可以由所有用户共享。

有些 Web 服务器不能直接使用 application 对象，必须使用父类 ServletContext 声明这个对象，然后使用 getServletContext()方法为 application 对象进行实例化。例如，例 4-11 中 example4_11_2.jsp 页面中的代码。

4.5.3 实践环节——网站访客计数器

使用 application 对象实现网站访客计数器的功能。

4.6 pageContext 对象

视频讲解

pageContext 对象即页面上下文对象，表示当前页面运行环境，用于获取当前 JSP 页面的相关信息，它的作用范围为当前 JSP 页面。pageContext 对象的类型为 javax.servlet.jsp.HttpJspPage.PageContext。pageContext 对象可以访问当前 JSP 页面的所有内置对象，如表 4.4 所示。另外，pageContext 对象提供了存取属性的方法，如表 4.5 所示。

表 4.4 pageContext 对象获取内置对象的方法

序 号	方 法	功 能 说 明
1	ServletRequest getRequest()	获取当前 JSP 页面的请求对象
2	ServletResponse getResponse()	获取当前 JSP 页面的响应对象
3	HttpSession getSession()	获取和当前 JSP 页面有关的会话对象
4	ServletConfig getServletConfig()	获取当前 JSP 页面的 ServletConfig 对象
5	ServletContext getServletContext()	获取当前 JSP 页面的运行环境的 application 对象
6	Object getPage()	获取当前 JSP 页面的 Servlet 实体的 page 对象
7	Exception getException()	获取当前 JSP 页面的异常对象 exception，这时此页面的 page 指令的 isErrorPage 属性要设置为 true
8	JspWriter getOut()	获取当前 JSP 页面的输出流对象 out

表 4.5 pageContext 对象存取属性的方法

序号	方　　法	功　能　说　明
1	Object getAttribute(String key, int scope)	获取范围为 scope、关键字为 key 的属性对象
2	void setAttribute(String key, Object value, int scope)	以 K/V 对的方式存储 scope 范围域属性对象
3	void removeAttribute(String key, int scope)	从 scope 范围移除关键字为 key 的属性对象
4	Enumeration getAttributeNamesInScope(int scope)	从 scope 范围中获取所有属性对象对应的关键字

例 4-12 编写一个 JSP 页面 example4_12.jsp，在该页面中使用 pageContext 对象添加和获取请求域属性值。

页面文件 example4_12.jsp 的代码如下：

```jsp
<%@ page language="java" contentType="text/html; charset=UTF-8"
pageEncoding="UTF-8"%>
<!DOCTYPE html>
<html>
<head>
<meta charset="UTF-8">
<title>Insert title here</title>
</head>
<body>
    <%
        //添加页面域属性值
        pageContext.setAttribute("pageKey", "页面域属性");
        //获取页面域属性值
        String pageValue = (String)pageContext.getAttribute("pageKey");
        //添加请求域属性值
        pageContext.getRequest().setAttribute("requestKey", "请求域属性");
        //获取请求域属性值
        String requestValue=(String)pageContext.getAttribute("requestKey", 2);
        out.print(pageValue + "<br>");
        out.print(requestValue);
    %>
</body>
</html>
```

4.7 page 对象

视频讲解

　　page 对象是一个与 Servlet 有关的内置对象，它表示 JSP 文件转译后的 Servlet 对象，代表 JSP 页面本身，即 this。因此它可以调用 Servlet 类所定义的方法。page 对象的类型为 javax.servlet.jsp.HttpJspPage，在实际应用中很少在 JSP 页面使用 page 对象。

　　例 4-13 编写一个 JSP 页面 example4_13.jsp，在该页面中使用 page 指令的 info 属性设置页面的说明信息，并分别使用 this 和 page 对象获取页面的说明信息。

页面文件 example4_13.jsp 的代码如下：

```jsp
<%@ page language="java" contentType="text/html; charset=UTF-8"
"pageEncoding="UTF-8" info="page 内置对象测试"%>
<!DOCTYPE html>
<html>
<head>
<meta charset="UTF-8">
<title>example4_13.jsp</title>
</head>
<body>
    使用 this 获取本页面的说明信息：<%=this.getServletInfo() %><br>
    使用 page 对象获取本页面的说明信息:<%=((HttpJspPage)page).getServletInfo() %>
</body>
</html>
```

4.8　config 对象

　　config 对象即页面配置对象，表示当前 JSP 页面转译后的 Servlet 的 ServletConfig 对象，存储一些初始数据。config 对象实现了 java.servlet.ServletConfig 接口。config 对象和 page 对象一样，也很少被用到。

　　例 4-14　编写一个 JSP 页面 example4_14.jsp，在该页面中使用 config 对象获得当前 JSP 页面转译 Java 后的 Servlet 对象名。

　　页面文件 example4_14.jsp 的代码如下：

```jsp
<%@ page language="java" contentType="text/html; charset=UTF-8"
pageEncoding="UTF-8"%>
<!DOCTYPE html>
<html>
<head>
<meta charset="UTF-8">
<title>example4_14.jsp</title>
</head>
<body>
<!--获得当前 JSP 页面转译 Java 后的 Servlet 对象名-->
<%=config.getServletName()%>
</body>
</html>
```

4.9　exception 对象

　　exception 对象是一个与 Error 有关的内置对象，表示 JSP 页面产生的异常。如果一个 JSP 页面需要使用此对象，必须将页面中 page 指令的 isErrorPage 属性设置为 true，否则无法编译。

例 4-15 编写两个 JSP 页面 example4_15.jsp 和 example4_15_1.jsp。在页面 example4_15.jsp 中使用语句"exception.printStackTrace(response.getWriter());"输出 JSP 页面产生的异常信息；在 example4_15_1.jsp 页面中产生数组越界异常，并设置该页面 page 指令的 errorPage 属性值为 example4_15.jsp。

页面文件 example4_15.jsp 的代码如下：

```
<%@ page language="java" contentType="text/html; charset=UTF-8"pageEncoding="UTF-8" isErrorPage="true"%>
<!DOCTYPE html>
<html>
<head>
<meta charset="UTF-8">
<title>example4_15.jsp</title>
</head>
<body>
    <%
        exception.printStackTrace(response.getWriter());
    %>
</body>
</html>
```

页面文件 example4_15_1.jsp 的代码如下：

```
<%@ page language="java" contentType="text/html; charset=UTF-8"pageEncoding="UTF-8" errorPage="example4_15.jsp"%>
<!DOCTYPE html>
<html>
<head>
<meta charset="UTF-8">
<title>example4_15_1.jsp</title>
</head>
<body>
<%
    int a[] = {1,2,3,4,5};
    for(int i = 0; i <= 5; i++){
        out.print(a[i]);
    }
%>
</body>
</html>
```

运行页面 example4_15_1.jsp，结果如图 4.9 所示。

> http://localhost:8080/ch4/example4_15_1.jsp
>
> java.lang.ArrayIndexOutOfBoundsException: 5 at org.apache.jsp.example4_005f15_005f
> org.apache.jasper.runtime.HttpJspBase.service(HttpJspBase.java:70) at javax.servlet.http.

图 4.9　页面 example4_15_1.jsp 的运行结果

4.10　JSP 的 4 种作用域

对象的作用域就是对象的生命周期和可访问性，在 JSP 中有 4 种作用域，即页面域、请求域、会话域和应用域。

❶ 页面域

页面域（page scope）的生命周期是指页面执行期间，存储在页面域中的对象只能在它所在的页面被访问。

❷ 请求域

请求域（request scope）的生命周期是指一次请求过程，包括请求被转发（forward）或者被包含（include）的情况，存储在请求域中的对象只有在此次请求过程中才可以被访问。

❸ 会话域

会话域（session scope）的生命周期是指某个客户端与服务器所连接的时间，客户端在第 1 次访问服务器时创建会话，在会话过期或用户主动退出后会话结束，存储在会话域中的对象在整个会话期间（可以包含多次请求）都可以被访问。

❹ 应用域

应用域（application scope）的生命周期是指从服务器开始执行服务到服务器关闭为止，其生命周期是 4 个作用域中时间最长的，存储在应用域中的对象在整个应用程序运行期间可以被所有 JSP 和 Servlet 共享访问。

JSP 的 4 种作用域分别对应 pageContext、request、session 和 application 内置对象，这 4 个内置对象都可以通过 setAttribute(String key, Object value) 方法存储数据，通过 getAttribute(String key) 获取数据。

例 4-16　编写一个 JSP 页面 example4_16.jsp，在该页面中使用 pageContext、session 和 application 对象分别实现页面域、会话域和应用域的页面访问统计情况。

页面文件 example4_16.jsp 的代码如下：

```
<%@ page language="java" contentType="text/html; charset=UTF-8"
pageEncoding="UTF-8"%>
<!DOCTYPE html>
<html>
<head>
<meta charset="UTF-8">
<title>example4_16.jsp</title>
</head>
<body>
```

```
<%
    int pageSum = 1;
    int sessionSum = 1;
    int applicationSum = 1;
    //页面域计数
    if(pageContext.getAttribute("pageCount") != null){
        pageSum = Integer.parseInt(pageContext.getAttribute("pageCount").
        toString());
        pageSum++;
    }
    pageContext.setAttribute("pageCount", pageSum);
    //会话域计数
    if(session.getAttribute("sessionCount") != null){
        sessionSum = Integer.parseInt(session.getAttribute("sessionCount").
        toString());
        sessionSum++;
    }
    session.setAttribute("sessionCount", sessionSum);
    //应用域计数
    if(application.getAttribute("applicationCount") != null){
        applicationSum = Integer.parseInt(application.getAttribute
        ("applicationCount").toString());
        applicationSum++;
    }
    application.setAttribute("applicationCount", applicationSum);
%>
<p>页面域访问统计：<%=pageSum%></p>
<p>会话域访问统计：<%=sessionSum%></p>
<p>应用域访问统计：<%=applicationSum%></p>
</body>
</html>
```

第 1 次访问 example4_16.jsp 页面，运行结果如图 4.10 所示。

多次刷新浏览器窗口，运行结果如图 4.11 所示。

图 4.10 第 1 次访问 example4_16.jsp 页面

图 4.11 多次刷新 example4_16.jsp 页面

打开另一个浏览器窗口，再次访问 example4_16.jsp 页面，运行结果如图 4.12 所示。

```
← → C  ⓘ localhost:8080/ch4/example4_16.jsp
```

页面域访问统计：1

会话域访问统计：1

应用域访问统计：5

图 4.12　新打开浏览器窗口访问 example4_16.jsp 页面

从图 4.10 到图 4.12 的运行结果可以看出，pageContext 域的访问范围为当前 JSP 页面，因此访问计数始终为 1；session 域的访问范围为当前浏览器与服务器的会话，因此刷新页面访问计数会累加，但新开启浏览器窗口时会新建一个会话，计数又从 1 开始；application 域的访问范围为整个应用，所以只要服务器不停止运行，计数就会不断累加。

4.11　本章小结

本章重点介绍了 request、session 和 application 对象，三者的区别如下：

request 对象内数据的存活范围在 request 对象的存活范围内，当客户端向服务器端发送一个请求，服务器向客户端返回一个响应后，该请求对象就被销毁了；之后再向服务器端发送新的请求时，服务器会创建新的 request 对象，该 request 对象与之前的 request 对象没有任何关系，因此也无法获得在之前 request 对象中所存放的任何数据。

session 是服务器端的行为，用于跟踪客户端的状态，当客户端去访问某个站点时，服务器端就会为客户产生一个 sessionID，以 cookie 的方式返回给客户端，当客户端去访问该站点的其他服务时，就会带着当前 sessionID 一起发出请求，以识别是哪个客户端，一个客户端就好比一个 session 对象，互不干扰。

session 对象内数据的生命周期也就是 session 对象的生命周期（只要浏览器不关闭，session 对象就会一直存在），因此在同一个浏览器窗口中无论向服务器端发送多少个请求，session 对象只有一个。session 对象经常用在登录和购物车场景。

application 对象是生命周期最长的对象，只要服务器没有关闭，application 对象中的数据就会一直存在。在整个服务器运行过程当中，application 对象只有一个。所有客户端共享这个 application 对象。该对象经常用在统计网站访问次数的场景。

request、session 以及 application 这 3 个对象的范围是逐个增加的：request 只在一个请求范围内；session 是在客户端与服务器会话范围内；application 则是在整个服务器的运行过程中。

习题 4

1. 下面（　　）操作不能关闭 session 对象。

    A．用户刷新当前页面调用

B．用户关闭浏览器
C．session 达到设置的最长"发呆"时间
D．session 对象的 invalidate()方法

2．有如下程序片段：

```
<form>
    <input type = "text" name = "id">
    <input type= "submit" value ="提交">
</form>
```

下面（　　）语句可以获取用户输入的信息。

　　A．request.getParameter("id");

　　B．request.getAttribute("submit");

　　C．session.getParameter(key, "id");

　　D．session.getAttribute(key, "id");

3．下面的（　　）内置对象是对客户的请求做出响应，向客户端发送数据的。

　　A．request　　　B．session　　　C．response　　　D．application

4．可以使用（　　）方法实现客户的重定向。

　　A．response.setStatus();　　　　B．response.setHeader();

　　C．response.setContentType();　　D．response.sendRedirect();

5．下面不属于 JSP 内置对象的是（　　）。

　　A．out 对象　　B．response 对象　　C．application 对象　　D．cookie 对象

6．下面对象中作用域最大的是（　　）。

　　A．page　　　B．request　　　C．application　　　D．session

7．在 JSP 中保存用户会话信息使用（　　）对象。

　　A．page　　　B．request　　　C．application　　　D．session

8．如果选择一个对象保存聊天室信息，则选择（　　）。

　　A．page　　　B．request　　　C．application　　　D．session

9．什么对象是 JSP 的内置对象？常见的内置对象有哪些？

10．request、session 和 application 对象的作用域是什么？它们之间有什么区别？

11．一个用户在不同 Web 服务中的 session 对象相同吗？一个用户在同一 Web 服务的不同子目录中的 session 对象相同吗？

12．session 对象的生命周期依赖于哪些因素？

13．简述 forward 动作标记与 response.sendRedirect()两种跳转的区别。

第 5 章 JSP 与 JavaBean

学习目的与要求

本章主要介绍在 JSP 程序中如何使用 JavaBean 组件。通过本章的学习，读者能够熟练掌握<jsp:useBean>、<jsp:setProperty>、<jsp:getProperty>等 JSP 的操作指令。

本章主要内容

- 编写 JavaBean
- 在 JSP 中使用 JavaBean

视频讲解

一个 JSP 页面通过使用 HTML 标记为用户显示数据（静态部分），页面中变量的声明、程序片以及表达式为动态部分，对数据进行处理。如果 Java 程序片和 HTML 标记大量掺杂在一起使用，将不利于 JSP 页面的扩展和维护。JSP 和 JavaBean 技术的结合不仅可以实现数据的表示和处理分离，而且可以提高 JSP 程序代码重用的程度，是 JSP 编程中常用的技术。

本章涉及的 Java 源文件保存在工程 ch5 的 src 中，涉及的 JSP 页面保存在工程 ch5 的 WebContent 中。

5.1 编写 JavaBean

JavaBean 是一个可重复使用的软件组件，是遵循一定标准、用 Java 语言编写的一个类，该类的一个实例称为一个 JavaBean，简称 bean。JavaBean 具有可重用、升级方便、不依赖于平台等特点。JavaBean 又可以分为业务 bean 和数据 bean。业务 bean 用于封装业务逻辑、数据库操作等；数据 bean 用于封装数据。

编写一个 JavaBean 就是编写一个 Java 类（该类必须带有包名），这个类创建的一个对象称为一个 bean，为了让 JSP 引擎（比如 Tomcat）知道这个 bean 的属性和方法，JavaBean 类必须遵守以下规则。

① 如果类的成员变量的名字是 name，那么为了获取或更改成员变量的值，在类中必须提供以下两个方法。
- getName()：用来获取属性 name。
- setName()：用来修改属性 name。

即方法的名字用 get 或 set 作前缀，后缀是首字母大写的成员变量的名字。

② 对于 boolean 类型的成员变量，允许使用"is"代替上面的"get"和"set"。

③ 类中方法的访问权限必须是 public。

④ 如果有构造方法，必须提供一个无参数的构造方法。

【例】5-1 创建 bean 的源文件 Rectangle.java（在包 com.bean 中）。该 bean 的作用是封装矩形的长和宽，并计算矩形的面积和周长。

Rectangle.java 的代码如下：

```java
package com.bean;
public class Rectangle{
    private double length;
    private double width;
    public Rectangle(){
        length = 20;
        width = 10;
    }
    public double getLength(){
        return length;
    }
    public void setLength(double length){
        this.length = length;
    }
    public double getWidth(){
        return width;
    }
    public void setWidth(double width){
        this.width = width;
    }
    public double computerArea(){
        return length*width;
    }
    public double computerLength(){
        return (length+width)*2;
    }
}
```

JavaBean 可以在任何 Java 程序编写环境下完成编写，再通过编译成为一个字节码文件，为了让 JSP 引擎（比如 Tomcat）找到这个字节码，必须把字节码文件放在特定的位置。

本书使用 Eclipse 集成环境开发 JSP 程序，Java 类的字节码文件由 Eclipse 自动保存到 Web 工程的 build\classes 中。例如，例 5-1 中的 Rectangle.class 文件保存在 ch5\build\classes\com\bean 目录中。

5.2 在 JSP 中使用 JavaBean

在 JSP 页面中使用 bean 时，首先使用 page 指令的 import 属性导入创建 bean 的类，例如：

```
<%@ page import="com.bean.*"%>
```

5.2.1 动作标记 useBean

useBean 动作标记用来查找或者实例化一个 JavaBean。useBean 标记的格式如下：

```
<jsp:useBean id="bean 的名字" class="创建 bean 的类" scope="bean 的有效范围" />
```

或

```
<jsp:useBean id="bean 的名字" type="创建 bean 的类" scope="bean 的有效范围" />
```

例如：

```
<jsp:useBean id= "rectangle" class= "com.bean.Rectangle" scope= "page"/>
```

useBean 标记中各属性的含义如表 5.1 所示。

表 5.1 useBean 标记中各属性的含义

属 性 名	描 述
id	指定该 JavaBean 实例的变量名，通过 id 可以访问这个实例
class	指定 JavaBean 的类名。如果需要创建一个新的实例，Web 容器会使用 class 指定的类，并调用无参数的构造方法来完成实例化
scope	指定 JavaBean 的作用范围，包括 page、request、session 和 application。其默认值为 page，表明此 JavaBean 只能应用于当前页；值为 request 表明此 JavaBean 只能应用于当前的请求；值为 session 表明此 JavaBean 能应用于当前会话；值为 application 表明此 JavaBean 能应用于整个应用程序内
type	指定 JavaBean 对象的类型，通常在查找已存在的 JavaBean 时使用，这时使用 type 将不会产生新的对象。如果是查找已存在的 JavaBean 对象，type 属性的值可以是此对象的准确类名、其父类或者其实现的接口；如果是新建实例，则只能是准确类名或者父类。另外，如果能够确定此 JavaBean 的对象存在，则指定 type 属性后可以省略 class 属性

当含有 useBean 动作标记的 JSP 页面被 Web 容器加载执行时，Web 容器首先根据 id 的名字在 pageContext 内置对象中查看是否含有名字为 id 和作用域为 scope 的对象。如果该对象存在，Web 容器就将这个对象的副本（bean）分配给 JSP 页面使用；如果没有找到，就根据 class 指定的类创建一个名字是 id 的 bean，并添加到 pageContext 对象中，同时将这个 bean 分配给 JSP 页面使用。useBean 动作标记的执行流程如图 5.1 所示。

图 5.1 useBean 动作标记的执行流程图

【例】5-2 编写一个 JSP 页面 example5_2.jsp，在 JSP 页面中使用 useBean 标记获得一个 bean，负责创建 bean 的类是例 5-1 中的 Rectangle 类，bean 的名字是 rectangle，rectangle 的 scope 取值为 page。JSP 页面的运行效果如图 5.2 所示。

矩形的长是：20.0

矩形的宽是：10.0

矩形的面积是：200.0

矩形的周长是：60.0

图 5.2 使用 bean 的 JSP 页面

页面文件 example5_2.jsp 的代码如下：

```
<%@ page language="java" contentType="text/html; charset=UTF-8"
pageEncoding="UTF-8"%>
<!DOCTYPE html>
<html>
<head>
<meta charset="UTF-8">
<title>example5_2.jsp</title>
```

```
</head>
<body>
<jsp:useBean id="rectangle" class="com.bean.Rectangle" scope="page"/>
<p>矩形的长是：<%=rectangle.getLength()%>
<p>矩形的宽是：<%=rectangle.getWidth() %>
<p>矩形的面积是：<%=rectangle.computerArea() %>
<p>矩形的周长是：<%=rectangle.computerLength() %>
</body>
</html>
```

5.2.2 动作标记 getProperty

JavaBean 的实质是遵守一定规范的类所创建的对象，用户可以通过两种方式获取 bean 的属性：①在 Java 程序片或表达式中使用 bean 对象调用 getXxx()方法获取 bean 的属性值，比如例 5-2 中的语句<%=rectangle.getLength()%>；②先通过<jsp:useBean>标记获得一个 bean，再通过<jsp:getProperty>标记获取 bean 的属性值。

使用 getProperty 动作标记可以获得 bean 的属性值。注意，在使用该动作标记之前必须事先使用 useBean 动作标记获得一个相应的 bean。getProperty 动作标记的语法格式如下：

```
<jsp:getProperty name= "bean 的名字" property= "bean 的属性" />
```

或

```
<jsp:getProperty name= "bean 的名字" property= "bean 的属性"/>
</jsp:getProperty>
```

其中，name 的取值是 bean 的名字，与 useBean 标记中的 id 对应；property 的取值是 bean 的一个属性名字，和创建该 bean 的类的成员变量名对应。这条指令相当于在 Java 表达式中使用 bean 的名字调用 getXxx()方法。

例 5-3 创建 bean 的源文件 NewRectangle.java，该 bean 的作用是计算矩形的面积和周长。编写一个 JSP 页面 useGetProperty.jsp，在该 JSP 页面中使用 useBean 标记创建一个名字是 pig 的 bean，并使用 getProperty 动作标记获得 pig 的每个属性值，负责创建 pig 的类是 NewRectangle。JSP 页面的运行效果如图 5.3 所示。

http://localhost:8080/ch5/useGetProperty.jsp

矩形的长是：30.0

矩形的宽是：20.0

矩形的面积是：600.0

矩形的周长是：100.0

图 5.3 使用 getProperty 标记获得 bean 的属性值

NewRectangle.java 的代码如下：

```
package com.bean;
```

```java
public class NewRectangle{
    double length;
    double width;
    double rectangleArea;
    double rectangleLength;
    public NewRectangle(){
        length = 20;
        width = 10;
    }
    public double getLength(){
        return length;
    }
    public void setLength(double length){
        this.length = length;
    }
    public double getWidth(){
        return width;
    }
    public void setWidth(double width){
        this.width = width;
    }
    public double getRectangleArea(){
        return length * width;
    }
    public double getRectangleLength(){
        return 2 * (width + length);
    }
}
```

页面文件 useGetProperty.jsp 的代码如下：

```jsp
<%@ page language="java" contentType="text/html; charset=UTF-8"
pageEncoding="UTF-8"%>
<!DOCTYPE html>
<html>
<head>
<meta charset="UTF-8">
<title>useGetProperty.jsp</title>
</head>
<body>
<jsp:useBean id="pig" class="com.bean.NewRectangle" scope="page"/>
<%pig.setLength(30);%>
<%pig.setWidth(20);%>
<p>矩形的长是：<jsp:getProperty property="length" name="pig"/>
<p>矩形的宽是：<jsp:getProperty property="width" name="pig"/>
<p>矩形的面积是：<jsp:getProperty property="rectangleArea" name="pig"/>
<p>矩形的周长是：<jsp:getProperty property="rectangleLength" name="pig"/>
```

```
        </body>
        </html>
```

5.2.3 动作标记 setProperty

除了可以在 Java 程序片中使用 bean 对象调用 setXxx()方法修改 bean 的属性值外，还可以使用 setProperty 动作标记修改 bean 的属性值。注意，在使用该动作标记之前必须事先使用 useBean 动作标记获得一个相应的 bean。使用 setProperty 动作标记进行 bean 属性值的设置有以下 3 种方式。

❶ **用表达式或字符串设置 bean 的属性**

（1）用表达式设置 bean 的属性：

```
<jsp:setProperty name="bean 的名字" property= "bean 的属性" value= "<%=
expression%>" />
```

（2）用字符串设置 bean 的属性：

```
<jsp:setProperty name="bean 的名字" property= "bean 的属性" value="字符串" />
```

在用表达式修改 bean 属性值时，表达式值的类型必须与 bean 的属性类型一致。在用字符串修改 bean 属性值时，字符串会被转换为 bean 的属性类型，不能成功转换的可能会抛出 NumberFormatException 异常。

❷ **通过 HTTP 表单参数值设置 bean 的属性**

通过 HTTP 表单参数值设置 bean 的属性：

```
<jsp:setProperty name="bean 的名字" property="*" />
```

这时，表单参数的名字必须与 bean 属性的名字相同，服务器会根据名字自动匹配，类型会自动转换。

❸ **任意指定请求参数设置 bean 的属性**

任意指定请求参数设置 bean 的属性：

```
<jsp:setProperty name="bean 的名字" property="属性名" param="参数名"/>
```

用户可以根据自己的需要任意选择传递的参数，请求参数名无须与 bean 属性名相同。

例 5-4 用表达式或字符串修改 bean 的属性。具体要求如下：

（1）创建 bean 的源文件 Car.java，该 bean 的作用是描述小汽车的一些属性。

（2）编写 JSP 页面 car.jsp，在该 JSP 页面中使用 useBean 标记创建一个名字是 smallCar 的 bean，其有效范围是 page，并使用动作标记修改、获取该 bean 的属性值，负责创建 smallCar 的类是 Car。JSP 页面的运行效果如图 5.4 所示。

```
http://localhost:8080/ch5/car.jsp
汽车的品牌是：宝马X6
汽车的牌号是：京A8888
```

图 5.4 使用字符串或表达式修改 bean 的属性

Car.java 的代码如下：

```java
package com.bean;
public class Car {
    String tradeMark;
    String number;
    public String getTradeMark() {
        return tradeMark;
    }
    public void setTradeMark(String tradeMark) {
        this.tradeMark = tradeMark;
    }
    public String getNumber() {
        return number;
    }
    public void setNumber(String number) {
        this.number = number;
    }
}
```

页面文件 car.jsp 的代码如下：

```jsp
<%@ page language="java" contentType="text/html; charset=UTF-8"
    pageEncoding="UTF-8"%>
<!DOCTYPE html>
<html>
<head>
<meta charset="UTF-8">
<title>car.jsp</title>
</head>
<body>
<jsp:useBean id="smallCar" class="com.bean.Car" scope="page"/>
<%
    String carNo = "京A8888";
%>
<%--使用 setProperty 标记设置 smallCar 的 tradeMark 属性值为"宝马 X6"--%>
<jsp:setProperty property="tradeMark" name="smallCar" value="宝马 X6"/>
<%--使用 setProperty 标记设置 smallCar 的 number 属性值为 carNo--%>
<jsp:setProperty property="number" name="smallCar" value="<%=carNo %>"/>
汽车的品牌是：<jsp:getProperty property="tradeMark" name="smallCar"/>
<br>汽车的牌号是：<jsp:getProperty property="number" name="smallCar"/>
</body>
</html>
```

例 5-5 通过 HTTP 表单参数值设置 bean 的属性。具体要求如下：

（1）编写 JSP 页面 inputCar.jsp 和 showCar.jsp。

(2) 在 inputCar.jsp 页面中输入信息后提交给 showCar.jsp 页面显示信息。
(3) JSP 页面中用到的 bean 是例 5-4 中的 Car 类创建的。JSP 页面的运行效果如图 5.5 所示。

(a) 信息输入页面

(b) 信息显示页面

图 5.5 JSP 页面运行效果

页面文件 inputCar.jsp 的代码如下：

```
<%@ page language="java" contentType="text/html; charset=UTF-8"
pageEncoding="UTF-8"%>
<!DOCTYPE html>
<html>
<head>
<meta charset="UTF-8">
<title>inputCar.jsp</title>
</head>
<body>
<form action="showCar.jsp" method="post" >
    请输入汽车品牌：
    <input type="text" name="tradeMark"/>
    <br>
    请输入汽车牌号：
    <input type="text" name="number"/>
    <br>
    <input type="submit" value="提交"/>
</form>
</body>
</html>
```

页面文件 showCar.jsp 的代码如下：

```
<%@ page language="java" contentType="text/html; charset=UTF-8"
pageEncoding="UTF-8"%>
<!DOCTYPE html>
<html>
```

```
<head>
<meta charset="UTF-8">
<title>showCar.jsp</title>
</head>
<body>
<%
request.setCharacterEncoding("utf-8");
%>
<jsp:useBean id="smallCar" class="com.bean.Car" scope="page"/>
<%--通过HTTP表单的参数值设置bean的属性（表单参数与属性自动匹配）--%>
<jsp:setProperty property="*" name="smallCar"/>
汽车的品牌是：<jsp:getProperty property="tradeMark" name="smallCar"/>
<br>汽车的牌号是：<jsp:getProperty property="number" name="smallCar"/>
</body>
</html>
```

5.2.4 实践环节——使用 JavaBean 计算三角形的面积和周长

编写两个 JSP 页面 inputTriangle.jsp 和 showTriangle.jsp。inputTriangle.jsp 提供一个表单，用户可以通过表单输入三角形的三条边提交给 showTriangle.jsp。用户提交表单后，JSP 页面将计算三角形面积和周长的任务交给一个 bean 去完成，创建 bean 的源文件是 Triangle.java。页面运行效果如图 5.6 所示。

输入三角形三边：
边A: 3
边B: 4
边C: 5
提交

（a）三角形边长输入页面

三角形的三边是：
边A: 3.0 边B: 4.0 边C: 5.0

这三个边能构成一个三角形吗？ true

面积是： 6.0

（b）信息显示页面

图 5.6 页面运行效果

5.3 本章小结

本章重点介绍了在 JSP 中如何使用动作标记 useBean、getProperty 和 setProperty。

习题 5

1. 下面方式中正确使用 JavaBean 的是（   ）。
   A. <jsp:useBean id="address" class="tom.AddressBean" scope="page"/>
   B. <jsp:useBean name="address" class="tom.AddressBean" scope="page"/>
   C. <jsp:useBean bean="address" class="tom.AddressBean" scope="page" />
   D. <jsp:useBean beanName="address" class="AddressBean" scope="page" />
2. JavaBean 中方法的访问属性必须是（   ）。
   A. private   B. public   C. protected   D. friendly
3. 在 JSP 中调用 JavaBean 时不会用到的标记是（   ）。
   A. <javabean>   B. <jsp:useBean>
   C. <jsp:setProperty>   D. <jsp:getProperty>
4. JavaBean 的作用域可以是（   ）、page、session 和 application。
   A. request   B. response   C. out   D. 以上都不对
5. 在 test.jsp 文件中有如下一行代码：

   `<jsp:useBean class="tom.jiafei.Test" id="user" scope="____" />`

   要使 user 对象一直存在于会话中，直到终止或被删除为止，在下画线处应填入（   ）。
   A. page   B. request   C. session   D. application
6. 在 JSP 中使用<jsp:useBean>动作可以将 JavaBean 引入 JSP 页面，对 JavaBean 的访问范围不能是（   ）。
   A. page   B. request   C. response   D. application
7. 下面语句与<jsp:getProperty name="aBean" property="jsp"/>等价的是（   ）。
   A. <%=jsp()%>   B. <%out.print(aBean.getJsp());%>
   C. <%=aBean.setJsp()%>   D. <%aBean.setJsp();%>
8. 以下是有关 jsp:setProperty 和 jsp:getProperty 标记的描述，正确的是（   ）。
   A. <jsp:setProperty>和<jsp:getProperty>标记都必须在<jsp:useBean>的开始标记和结束标记之间
   B. 这两个标记的 name 属性值必须与<jsp:useBean>标记的 id 属性值相对应
   C. 这两个标记的 name 属性值可以与<jsp:useBean>标记的 id 属性值不同
   D. 以上均不对
9. 在 JSP 中使用<jsp:getProperty>标记时不会出现的属性是（   ）。
   A. name   B. property   C. value   D. 以上皆不会出现

第 6 章

Java Servlet 基础

学习目的与要求

本章主要介绍 Java Servlet 的基础知识,包括部署 Servlet、Servlet 的生命周期、JSP 页面如何访问 Servlet、重定向与转发等。通过本章的学习,读者能够熟练掌握 Servlet 对象的创建与运行,理解 Servlet 的生命周期与工作原理。

本章主要内容
- Servlet 类与 Servlet 对象
- Servlet 对象的创建与运行
- Servlet 的生命周期
- 通过 JSP 页面访问 Servlet
- doGet 和 doPost 方法
- 重定向与转发
- 在 Java Servlet 中使用 session
- 基于 Servlet 的 MVC 模式

Java Servlet 的核心思想是在 Web 服务器端创建用来响应客户端请求的对象,该对象被称为一个 Servlet 对象。JSP 技术以 Java Servlet 为基础,当客户端请求一个 JSP 页面时,Web 服务器(如 Tomcat 服务器)会自动生成一个对应的 Java 文件,编译该 Java 文件,并用编译得到的字节码文件在服务器端创建一个 Servlet 对象。但实际的 Web 应用需要 Servlet 对象具有特定的功能,这时就需要 Web 开发人员编写创建 Servlet 对象的类。如何编写 Servlet 类,又如何使用 Servlet 类,将在本章中重点介绍。

本章涉及的 Java 源文件保存在工程 ch6 的 src 中,涉及的 JSP 页面保存在工程 ch6 的 WebContent 中。

6.1 Servlet 类与 Servlet 对象

编写一个 Servlet 类很简单,只要继承 javax.servlet.http 包中的 HttpServlet 类,并重写响应 HTTP 请求的方法即可。HttpServlet 类实现了 Servlet 接口,实现了响应用户请求的接口方法。HttpServlet 类的一个子类习惯地称为一个 Servlet 类，这样的子类创建的对象又习惯地称为 Servlet 对象。

例 6-1 一个简单的 Servlet 类。

FirstServlet.java 的代码如下：

```java
package servlet;
import java.io.IOException;
import java.io.PrintWriter;
import javax.servlet.ServletConfig;
import javax.servlet.ServletException;
import javax.servlet.http.HttpServlet;
import javax.servlet.http.HttpServletRequest;
import javax.servlet.http.HttpServletResponse;
public class FirstServlet extends HttpServlet{
    private static final long serialVersionUID = 1L;
    public void init(ServletConfig config) throws ServletException{
        super.init(config);
    }
    public void service(HttpServletRequest request,HttpServletResponse response)
            throws IOException{
        //设置响应的内容类型
        response.setContentType("text/html;charset=utf-8");
        //取得输出对象
        PrintWriter out = response.getWriter();
        out.println("<html><body>");
        //在浏览器中显示"第一个 Servlet 类"
        out.println("第一个 Servlet 类");
        out.println("</body></html>");
    }
}
```

编写 Servlet 类时，必须有包名。也就是说，必须在包中编写 Servlet 类。在本章中新建一个 Web 工程 ch6，所有的 Servlet 类放在 src 目录下的 servlet 包中。

例 6-1 中 Servlet 类的源文件 FirstServlet.java 保存在 Eclipse 的 Web 工程 ch6 的 src 目录下的 servlet 包中。FirstServlet.java 源文件由 Eclipse 自动编译生成字节码文件 FirstServlet.class，保存在 build\classes\servlet 中。Servlet 类的源文件与字节码文件保存目录如图 6.1 所示。

编写完 Servlet 类的源文件，并编译了源文件，是不是就可以运行 Servlet 对象呢？不可以，需要部署 Servlet 后，才可以运行 Servlet 对象。

图 6.1　Servlet 类的源文件与字节码文件保存目录

6.2　Servlet 对象的创建与运行

要想让 Web 服务器使用 Servlet 类编译后的字节码文件创建 Servlet 对象处理用户请求，必须先为 Web 服务器部署 Servlet。部署 Servlet 目前有两种方式：一是在 web.xml 中部署 Servlet；二是基于注解的方式部署 Servlet。

6.2.1　在 web.xml 中部署 Servlet

web.xml 文件由 Web 服务器负责管理，该文件是 Web 应用的部署描述文件，包含如何将用户请求 URL 映射到 Servlet。在 Web 工程的 WebContent\WEB-INF 目录下创建 web.xml 文件（可以在创建 Java Web 应用 ch6 时，首先在对话框的 Project Name 文本框中输入项目名称 "ch6"，然后两次单击 Next 按钮后，选中 Generate web.xml deployment descriptor 复选框，自动生成 web.xml），并部署自己的 Servlet。

❶ 部署 Servlet

为了在 web.xml 文件中部署 6.1 节中的 FirstServlet，需要在 web.xml 文件中找到 <web-app></web-app> 标记，然后在 <web-app></web-app> 标记中添加如下内容：

```xml
<servlet>
   <servlet-name>firstServlet</servlet-name>
   <servlet-class>servlet.FirstServlet</servlet-class>
</servlet>
<servlet-mapping>
   <servlet-name>firstServlet</servlet-name>
   <url-pattern>/firstServlet</url-pattern>
</servlet-mapping>
```

❷ 运行 Servlet

在访问 Servlet 之前，需要将它发布到 Web 服务器（右击 Servlet 类的源文件，选择 Run As/Run on Server 命令）。这时，在 Eclipse 内嵌的浏览器中可看到如图 6.2 所示的运行效果。

图 6.2　第一个 Servlet 类的运行效果

将 Servlet 发布到 Web 服务器后，也可以在浏览器的地址栏中输入"http://localhost:8080/ch6/firstServlet"来运行 Servlet。

❸ web.xml 文件中与 Servlet 部署有关的标记及其说明

1）根标记<web-app>

XML 文件中必须有一个根标记，web.xml 的根标记是<web-app>。

2）<servlet>标记及其子标记

web.xml 文件中可以有若干个<servlet>标记，该标记的内容由 Web 服务器负责处理。<servlet>标记中有两个子标记：<servlet-name>和<servlet-class>。其中，<servlet-name>子标记的内容是 Web 服务器创建的 Servlet 对象的名字。web.xml 文件中可以有若干个<servlet>标记，但要求它们的<servlet-name>子标记的内容互不相同。<servlet-class>子标记的内容指定 Web 服务器用哪个类来创建 Servlet 对象；如果 Servlet 对象已经创建，那么 Web 服务器就不再使用指定的类创建。

3）<servlet-mapping>标记及其子标记

web.xml 文件中出现一个<servlet>标记就会对应地出现一个<servlet-mapping>标记。<servlet-mapping>标记中也有两个子标记：<servlet-name>和<url-pattern>。其中，<servlet-name>子标记的内容是 Web 服务器创建的 Servlet 对象的名字（该名字必须和<servlet>标记的子标记<servlet-name>的内容相同）；<url-pattern>子标记用来指定用户用怎样的模式请求 Servlet 对象，例如，<url-pattern>子标记的内容是"/firstServlet"，用户需要请求服务器运行 Servlet 对象 firstServlet 为其服务，那么可在浏览器的地址栏中输入"http://localhost:8080/ch6/firstServlet"。

一个 Web 工程的 web.xml 文件负责管理该 Web 工程的所有 Servlet 对象，当 Web 工程需要提供更多的 Servlet 对象时，只要在 web.xml 文件中添加<servlet>和<servlet-mapping>标记即可。

6.2.2 基于注解的方式部署 Servlet

从 6.2.1 节可知，每开发一个 Servlet，都要在 web.xml 文件中部署 Servlet 才能使用。这样会给 Web 工程的维护带来非常大的麻烦。在 Servlet 3.0 中提供了注解@WebServlet，使得不再需要在 web.xml 文件中进行 Servlet 的部署描述，简化开发流程。本书中后续的 Servlet 都是基于注解的方式部署。

注解虽然方便了开发人员，但在后期会让维护和调试成本增加。为了方便后期维护，建议开发人员部署 Servlet 时把@WebServlet 的属性 urlPatterns 的值设置为 Servlet 类的名字。例如：

```
@WebServlet(name = "secondServlet", urlPatterns = { "/secondServlet" })
public class SecondServlet extends HttpServlet {
}
```

❶ @WebServlet

@WebServlet 用于将一个类声明为 Servlet 对象，该注解将会在部署时被 Web 容器处理，Web 容器根据具体属性将相应的类部署为 Servlet 对象。该注解具有一些常用属性，如表 6.1 所示。

表 6.1 @WebServlet 的常用属性列表

属性名	类型	描述
name	String	指定 Servlet 的 name 属性，等价于<servlet-name>。如果没有显式指定，则该 Servlet 的取值即为类的全名
value	String[]	该属性等价于 urlPatterns 属性。两个属性不能同时使用
urlPatterns	String[]	指定一组 Servlet 的 URL 匹配模式。等价于<url-pattern>标签
loadOnStartup	int	指定 Servlet 的加载顺序，等价于<load-on-startup>标签
initParams	WebInitParam[]	指定一组 Servlet 初始化参数，等价于<init-param>标签

表 6.1 中的所有属性均为可选属性，但是 value 或者 urlPatterns 通常是必需的，且二者不能共存，如果同时指定，通常是忽略 value 的取值。可使用简化注解，如@WebServlet("/secondServlet")，其中"/secondServlet"为请求的 URL，即 urlPatterns 属性值。

例 6-2 基于注解的 Servlet 类——SecondServlet。

SecondServlet.java 的代码如下：

```
package servlet;
import java.io.IOException;
import java.io.PrintWriter;
import javax.servlet.ServletConfig;
import javax.servlet.ServletException;
import javax.servlet.annotation.WebServlet;
import javax.servlet.http.HttpServlet;
import javax.servlet.http.HttpServletRequest;
import javax.servlet.http.HttpServletResponse;
```

```java
//建议urlPatterns的值和类名一样,方便维护。可使用@WebServlet("/secondServlet")
//简化注解
@WebServlet(name = "secondServlet", urlPatterns = { "/secondServlet" })
public class SecondServlet extends HttpServlet {
    private static final long serialVersionUID = 1L;
    public void init(ServletConfig config) throws ServletException {
    }
    protected void service(HttpServletRequest request, HttpServletResponse
response) throws ServletException, IOException {
        //设置响应的内容类型
        response.setContentType("text/html;charset=utf-8");
        //取得输出对象
        PrintWriter out = response.getWriter();
        out.println("<html><body>");
        //在浏览器中显示"第二个Servlet类"
        out.println("第二个Servlet类");
        out.println("</body></html>");
    }
}
```

SecondServlet.java 的代码中使用 "@WebServlet(name="secondServlet", urlPatterns={ "/secondServlet" })" 部署之后,就不必在 web.xml 中部署相应的<servlet>和<servlet-mapping>元素了,Web 容器会在部署时根据指定的属性将该类发布为 Servlet 对象。"@WebServlet(name = "secondServlet", urlPatterns = { "/secondServlet" })" 等价的 web.xml 部署形式如下:

```xml
<servlet>
    <servlet-name>secondServlet</servlet-name>
    <servlet-class>servlet.SecondServlet</servlet-class>
</servlet>
<servlet-mapping>
    <servlet-name>secondServlet</servlet-name>
    <url-pattern>/secondServlet</url-pattern>
</servlet-mapping>
```

❷ @WebInitParam

@WebInitParam 注解通常不单独使用,而是配合@WebServlet 和@WebFilter(第 8 章讲解)使用。它的作用是为 Servlet 或 Filter 指定初始化参数,这等价于 web.xml 中<servlet>的<init-param>子标签。@WebInitParam 具有一些常用属性,如表 6.2 所示。

表 6.2 @WebInitParam 常用属性列表

属性名	类型	是否可选	描述
name	String	否	指定参数的名字,等价于<param-name>
value	String	否	指定参数的值,等价于 <param-value>

@WebInitParam 注解的示例代码如下:

```java
@WebServlet(name = "thirdServlet", urlPatterns = { "/thirdServlet" },
```

```
initParams={@WebInitParam(name = "firstParam", value = "one"),@WebInitParam
(name = "secondParam", value = "two")})
```

6.2.3 实践环节——@WebServlet 的应用

首先，将 web.xml 文件中有关 Servlet 的部署代码删除，然后使用注解的方式部署例 6-1 的 Servlet，并运行部署后的 Servlet。

6.3 Servlet 的生命周期

视频讲解

一个 Servlet 对象的生命周期主要由下列 3 个过程组成。

❶ 初始化 Servlet 对象

当 Servlet 对象第一次被请求加载时，服务器会创建一个 Servlet 对象，该 Servlet 对象调用 init()方法完成必要的初始化工作。

❷ service()方法响应请求

创建的 Servlet 对象再调用 service()方法响应客户的请求。

❸ Servlet 对象消亡

当服务器关闭时，Servlet 对象调用 destroy()方法使自己消亡。

从上面 3 个过程来看，init()方法只能被调用一次，即在 Servlet 第一次被请求加载时调用。当客户端请求 Servlet 服务时，服务器将启动一个新的线程，在该线程中，Servlet 对象调用 service()方法响应客户端的请求。那么多客户端请求 Servlet 服务时，服务器会怎么办呢？服务器会为每个客户端启动一个新的线程，在每个线程中，Servlet 对象调用 service()方法响应客户端的请求。也就是说，每个客户端请求都会导致 service()方法被调用执行，分别运行在不同的线程中。

【例】6-3 Servlet 接口的 init()、service()和 destroy()方法。

ThirdServlet.java 的代码如下：

```
package servlet;
import java.io.IOException;
import javax.servlet.ServletConfig;
import javax.servlet.ServletException;
import javax.servlet.annotation.WebInitParam;
import javax.servlet.annotation.WebServlet;
import javax.servlet.http.HttpServlet;
import javax.servlet.http.HttpServletRequest;
import javax.servlet.http.HttpServletResponse;
@WebServlet(name = "thirdServlet", urlPatterns = { "/thirdServlet" },
initParams={@WebInitParam(name = "firstParam", value = "one"),
        @WebInitParam(name = "secondParam", value = "two")})
public class ThirdServlet extends HttpServlet {
    private static final long serialVersionUID = 1L;
```

```java
    private String first = null;
    private String second = null;
    private static int count = 0;
    public void init(ServletConfig config) throws ServletException {
        //获得参数 firstParam 的值
        first = config.getInitParameter("firstParam");
        second = config.getInitParameter("secondParam");
        System.out.println("第一个参数值: " + first);
        System.out.println("第二个参数值: " + second);
    }
    protected void service(HttpServletRequest request, HttpServletResponse
    response) throws ServletException, IOException {
        count ++ ;
        System.out.println("您是第" + count + "个客户端请求该Servlet!");
    }
    public void destroy() {
    }
}
```

在 ThirdServlet 的 init()方法中,通过 ServletConfig 的对象 config 调用 getInitParameter() 方法来获取参数的值。请求该 Servlet 三次后，在 Eclipse 的控制台中打印出如图 6.3 所示的结果。

图 6.3 请求三次 ThirdServlet 的结果

从图 6.3 看出，不管请求几次 ThirdServlet，其中的 init()方法只执行一次。而 service() 方法每请求一次，就执行一次。

6.4 通过 JSP 页面访问 Servlet

视频讲解

可以通过 JSP 页面的表单或超链接请求某个 Servlet。通过 JSP 页面访问 Servlet 的好处 是：JSP 页面负责页面的静态信息处理，而动态信息处理由 Servlet 完成。

❶ 通过表单访问 Servlet

假设在 JSP 页面中，有如下表单：

```
<form action="isLogin" method="post">
    ...
```

```
</form>
```

那么该表单的处理程序（action）就是一个 Servlet，为该 Servlet 部署时，@WebServlet 的 urlPatterns 属性值为"{ "/ isLogin " }"。

❷ **通过超链接访问 Servlet**

在 JSP 页面中，可以单击超链接访问 Servlet 对象，也可以通过超链接向 Servlet 提交信息，例如，<a href="loginServlet?user=taipingle&&pwd=zhenzuile">查看用户名和密码</a>，"查看用户名和密码"这个超链接就将 user=taipingle 和 pwd=zhenzuile 两个信息提交给 Servlet 处理。

例 6-4 编写 JSP 页面 login.jsp，在该页面中通过表单向 urlPatterns 为 "{ "/ loginServlet" }" 的 Servlet（由 LoginServlet 类负责创建）提交用户名和密码，Servlet 负责判断输入的用户名和密码是否正确，并把判断结果返回。运行效果如图 6.4 所示。

图 6.4 例 6-4 的运行效果

页面文件 login.jsp 的代码如下：

```
<%@ page language="java" contentType="text/html; charset=UTF-8"
pageEncoding="UTF-8"%>
<!DOCTYPE html>
<html>
<head>
<meta charset="UTF-8">
<title>login.jsp</title>
</head>
  <body>
    <form action="loginServlet" method="post">
    <table>
```

```html
        <tr>
            <td>用户名：</td>
            <td><input type="text" name="user"/></td>
        </tr>
        <tr>
            <td>密    码：</td>
            <td><input type="password" name="pwd"/></td>
        </tr>
        <tr>
            <td><input type="submit" value="提交"/></td>
            <td><input type="reset" value="重置"/></td>
        </tr>
    </table>
  </form>
 </body>
</html>
```

LoginServlet.java 的代码如下：

```java
package servlet;
import java.io.IOException;
import java.io.PrintWriter;
import javax.servlet.ServletException;
import javax.servlet.annotation.WebServlet;
import javax.servlet.http.HttpServlet;
import javax.servlet.http.HttpServletRequest;
import javax.servlet.http.HttpServletResponse;
@WebServlet(name = "loginServlet", urlPatterns = { "/loginServlet" })
public class LoginServlet extends HttpServlet {
    private static final long serialVersionUID = 1L;
    protected void service(HttpServletRequest request, HttpServletResponse response) throws ServletException, IOException {
    response.setContentType("text/html;charset=UTF-8");
    PrintWriter out = response.getWriter();
    request.setCharacterEncoding("UTF-8");          //设置编码，防止中文乱码
    String name = request.getParameter("user");     //获取客户提交的信息
    String password = request.getParameter("pwd");  //获取客户提交的信息
    out.println("<html><body>");
    if(name == null || name.length() == 0){
        out.println("请输入用户名");
    }
    else if(password == null || password.length() == 0){
        out.println("请输入密码");
    }
    else if(name.length() > 0 && password.length() > 0){
        if(name.equals("zhangsan") && password.equals("lisi")){
```

```
            out.println("信息输入正确");
        }else{
            out.println("信息输入错误");
        }
    }
    out.println("</body></html>");
}
```

6.5　doGet()和doPost()方法

视频讲解

　　当服务器接收到一个 Servlet 请求时，就会产生一个新线程，在这个线程中 Servlet 对象调用 service()方法为请求做出响应。service()方法首先检查 HTTP 请求类型（get 或 post），然后在 service()方法中根据用户的请求方式，对应地调用 doGet()或 doPost()方法。

　　HTTP 请求类型为 get 方式时，service()方法调用 doGet()方法响应用户请求；HTTP 请求类型为 post 方式时，service()方法调用 doPost()方法响应用户请求。因此，在 Servlet 类中，没有必要重写 service()方法，直接继承即可。

　　在 Servlet 类中重写 doGet()或 doPost()方法来响应用户的请求，这样可以增加响应的灵活性，同时减轻服务器的负担。

　　一般情况下，如果不论用户请求类型是 get 还是 post，服务器的处理过程完全相同，那么可以只在 doPost()方法中编写处理过程，而在 doGet()方法中再调用 doPost()方法；或只在 doGet()方法中编写处理过程，而在 doPost()方法中再调用 doGet()方法。

　　【例】6-5　编写 JSP 页面 inputLader.jsp，在该页面中使用表单向 urlPatterns 为 "{ "/getLengthOrAreaServlet" }"的 Servlet 提交矩形的长与宽。Servlet（由 GetLengthOrAreaServlet 负责创建）处理手段依赖表单提交数据的方式。当提交方式为 get 时，Servlet 计算矩形的周长；当提交方式为 post 时，Servlet 计算矩形的面积。页面运行效果如图 6.5 所示。

图 6.5　例 6-5 的运行效果

页面文件 inputLader.jsp 的代码如下：

```jsp
<%@ page language="java" contentType="text/html; charset=UTF-8"
pageEncoding="UTF-8"%>
<!DOCTYPE html>
<html>
<head>
<meta charset="UTF-8">
<title>inputLader.jsp</title>
</head>
<body>
    <h2>输入矩形的长和宽，提交给 servlet（post 方式）求面积：</h2>
    <form action="getLengthOrAreaServlet" method="post">
     长：<input type="text" name="length"/><br/>
     宽：<input type="text" name="width"/><br/>
     <input type="submit" value="提交"/>
    </form>
     <br/>
    <h2>输入矩形的长和宽，提交给 servlet（get 方式）求周长：</h2>
    <form action="getLengthOrAreaServlet" method="get">
     长：<input type="text" name="length"/><br/>
     宽：<input type="text" name="width"/><br/>
     <input type="submit" value="提交"/>
    </form>
  </body>
</html>
```

GetLengthOrAreaServlet.java 的代码如下：

```java
package servlet;
import java.io.IOException;
import java.io.PrintWriter;
import javax.servlet.ServletException;
import javax.servlet.annotation.WebServlet;
import javax.servlet.http.HttpServlet;
import javax.servlet.http.HttpServletRequest;
import javax.servlet.http.HttpServletResponse;
@WebServlet(name = "getLengthOrAreaServlet", urlPatterns =
{ "/getLengthOrAreaServlet" })
public class GetLengthOrAreaServlet extends HttpServlet {
    private static final long serialVersionUID = 1L;
    protected void doGet(HttpServletRequest request, HttpServletResponse
    response) throws ServletException, IOException {
        response.setContentType("text/html;charset=utf-8");
        PrintWriter out = response.getWriter();
        String l = request.getParameter("length");
```

```
        String w = request.getParameter("width");
        out.println("<html><body>");
        double m = 0,n = 0;
        try{
            m = Double.parseDouble(l);
            n = Double.parseDouble(w);
            out.println("矩形的周长是:" + ( m + n ) * 2 );
        }catch(NumberFormatException e){
            out.println("请输入数字字符!");
        }
        out.println("</body></html>");
    }
    protected void doPost(HttpServletRequest request, HttpServletResponse response) throws ServletException, IOException {
        response.setContentType("text/html;charset=utf-8");
        PrintWriter out = response.getWriter();
        String l = request.getParameter("length");
        String w = request.getParameter("width");
        out.println("<html><body>");
        double m = 0, n = 0;
        try{
            m = Double.parseDouble(l);
            n = Double.parseDouble(w);
            out.println("矩形的面积是:" + m * n);
        }catch(NumberFormatException e){
            out.println("请输入数字字符!");
        }
        out.println("</body></html>");
    }
}
```

6.6 重定向与转发

视频讲解

重定向是将用户从当前 JSP 页面或 Servlet 定向到另一个 JSP 页面或 Servlet，以前的 request 中存放的信息全部失效，并进入一个新的 request 作用域；转发是将用户对当前 JSP 页面或 Servlet 的请求转发给另一个 JSP 页面或 Servlet，以前的 request 中存放的信息不会失效。

6.6.1 重定向

在 Servlet 中通过调用 HttpServletResponse 类中的方法 sendRedirect(String location)来实现重定向，重定向的目标页面或 Servlet（由参数 location 指定），无法从以前的 request 对

象中获取用户提交的数据。

6.6.2 转发

javax.servlet.RequestDispatcher 对象可以将用户对当前 JSP 页面或 Servlet 的请求转发给另一个 JSP 页面或 Servlet。实现转发需要两个步骤。

❶ 获得 RequestDispatcher 对象

在当前 JSP 页面或 Servlet 中，使用 request 对象调用 public RequestDispatcher getRequestDispatcher(String url)方法返回一个 RequestDispatcher 对象，其中参数 url 就是要转发的 JSP 页面或 Servlet 的地址。例如：

```
RequestDispatcher dis=request.getRequestDispatcher("dologin");
```

❷ 使用 RequestDispatcher 对象调用 forward()方法实现转发

获得 RequestDispatcher 对象之后，就可以使用该对象调用 public void forward(ServletRequest request,ServletResponse response)方法将用户对当前 JSP 页面或 Servlet 的请求转发给 RequestDispatcher 对象所指定的 JSP 页面或 Servlet。例如：

```
dis.forward(request,response);
```

转发是服务器行为，重定向是客户端行为。具体工作流程如下：

转发过程：客户浏览器发送 http 请求，Web 服务器接收此请求，调用内部的一个方法在容器内部完成请求处理和转发动作，将目标资源发送给客户；在这里，转发的路径必须是同一个 Web 容器下的 URL，不能转向到其他的 Web 路径上去，中间传递的是自己的容器内的 request。在客户浏览器的地址栏中显示的仍然是其第一次访问的路径，也就是说客户是感觉不到服务器做了转发的。转发行为是浏览器只做了一次访问请求。

重定向过程：客户浏览器发送 http 请求，Web 服务器接收后发送 302 状态码响应及对应新的 location 给客户浏览器，客户浏览器发现是 302 响应，则自动再发送一个新的 http 请求，请求 URL 是新的 location 地址，服务器根据此请求寻找资源并发送给客户。在这里，location 可以重定向到任意 URL，既然是浏览器重新发出了请求，就没有什么 request 传递的概念了。在客户浏览器的地址栏中显示的是其重定向的路径，客户可以观察到地址的变化。重定向行为是浏览器做了至少两次的访问请求。

例6-6 编写 JSP 页面 redirectForward.jsp，在该 JSP 页面中通过表单向 urlPatterns 为"{ "/ redirectForwardServlet" }"的 Servlet（由 RedirectForwardServlet 负责创建）提交用户名和密码。如果用户输入的数据不完整，redirectForwardServlet 将用户重定向到 redirectForward.jsp 页面；如果用户输入的数据完整，redirectForwardServlet 将用户对 redirectForward.jsp 页面的请求转发给 urlPatterns 为"{ "/ showServlet" }"的 Servlet（由 ShowServlet 负责创建），showServlet 显示用户输入的信息。

页面文件 redirectForward.jsp 的代码如下：

```
<%@ page language="java" contentType="text/html; charset=UTF-8"
pageEncoding="UTF-8"%>
<!DOCTYPE html>
```

```html
<html>
<head>
<meta charset="UTF-8">
<title>redirectForward.jsp</title>
</head>
  <body>
    <form action="redirectForwardServlet" method="post">
     <table>
        <tr>
            <td>用户名：</td>
            <td><input type="text" name="user"/></td>
        </tr>
        <tr>
            <td>密　码：</td>
            <td><input type="password" name="pwd"/></td>
        </tr>
        <tr>
            <td><input type="submit" value="提交"/></td>
            <td><input type="reset" value="重置"/></td>
        </tr>
     </table>
    </form>
  </body>
</html>
```

RedirectForwardServlet.java 的代码如下：

```java
package servlet;
import java.io.IOException;
import javax.servlet.RequestDispatcher;
import javax.servlet.ServletException;
import javax.servlet.annotation.WebServlet;
import javax.servlet.http.HttpServlet;
import javax.servlet.http.HttpServletRequest;
import javax.servlet.http.HttpServletResponse;
@WebServlet(name = "redirectForwardServlet", urlPatterns =
{ "/redirectForwardServlet" })
public class RedirectForwardServlet extends HttpServlet {
private static final long serialVersionUID = 1L;
    protected void doGet(HttpServletRequest request,
            HttpServletResponse response) throws ServletException, IOException {
        doPost(request, response);
    }
    protected void doPost(HttpServletRequest request,
            HttpServletResponse response) throws ServletException, IOException {
        request.setCharacterEncoding("utf-8");
        String name = request.getParameter("user");
```

```
            String password = request.getParameter("pwd");
            if (name == null || name.length() == 0) {
                //使用 response 调用 sendRedirect 方法重定向到 redirectForward.jsp
                response.sendRedirect("redirectForward.jsp");
            } else if (password == null || password.length() == 0) {
                response.sendRedirect("redirectForward.jsp");
            } else if (name.length() > 0 && password.length() > 0) {
                //转发
                RequestDispatcher dis = request.getRequestDispatcher("showServlet");
                dis.forward(request, response);
            }
        }
    }
}
```

ShowServlet.java 的代码如下：

```
package servlet;
import java.io.IOException;
import java.io.PrintWriter;
import javax.servlet.ServletException;
import javax.servlet.annotation.WebServlet;
import javax.servlet.http.HttpServlet;
import javax.servlet.http.HttpServletRequest;
import javax.servlet.http.HttpServletResponse;
@WebServlet(name = "showServlet", urlPatterns = { "/showServlet" })
public class ShowServlet extends HttpServlet {
private static final long serialVersionUID = 1L;
    protected void doGet(HttpServletRequest request,
            HttpServletResponse response) throws ServletException, IOException {
        doPost(request, response);
    }
    protected void doPost(HttpServletRequest request,
            HttpServletResponse response) throws ServletException, IOException {
        response.setContentType("text/html;charset=utf-8");
        PrintWriter out = response.getWriter();
        request.setCharacterEncoding("utf-8");
        String name = request.getParameter("user");
        String password = request.getParameter("pwd");
        out.println("您的用户名是：" + name);
        out.println("<br>您的密码是：" + password);
    }
}
```

6.6.3　实践环节——登录验证

编写登录页面 login_1.jsp，在该 JSP 页面中通过表单向 urlPatterns 为"{ "/ loginServlet_1" }"

的 Servlet（由 LoginServlet_1 类负责创建）提交用户名和密码。如果用户输入的数据不完整，loginServlet_1 将用户重定向到 login_1.jsp 页面；如果用户输入的数据完整且正确（用户名"zhangsan"，密码"123"），loginServlet_1 将用户的请求转发给 loginSuccess_1.jsp 页面，loginSuccess_1.jsp 页面显示用户输入的信息。

6.7 在 Java Servlet 中使用 session

视频讲解

在 Servlet 中获得当前请求的会话对象可通过调用 HttpServletRequest 的 getSession()方法实现，例如：

```
HttpSession session = request.getSession(true);//若存在会话则返回该会话，否则
                                               //新建一个会话
```

或

```
HttpSession session = request.getSession(false);//若存在会话则返回该会话，否
                                                //则返回 null
```

通常情况下，通过第一种方式获得 session，即指定 getSession()的参数为 true。默认参数为 true，即 request.getSession(true)等同于 request.getSession()。

【例】6-7　编写 JSP 页面 useSession.jsp，在该页面中通过表单向名为 useSession 的 Servlet 对象（由 UseSessionServlet 类负责创建）提交用户名，useSession 将用户名存入用户的 session 对象中，然后用户请求另一个 Servlet 对象 showName（由 ShowNameServlet 类负责创建），showName 从用户的 session 对象中取出存储的用户名，并显示在浏览器中。程序运行效果如图 6.6 所示。

图 6.6　在 Servlet 中使用 session

页面文件 useSession.jsp 的代码如下：

```html
<%@ page language="java" contentType="text/html; charset=UTF-8"
pageEncoding="UTF-8"%>
<!DOCTYPE html>
<html>
<head>
<meta charset="UTF-8">
<title>useSession.jsp</title>
 </head>
 <body>
  <form action="sendMyName" method="post">
   <table>
      <tr>
         <td>用户名：</td>
         <td><input type="text" name="user"/></td>
      </tr>
      <tr>
         <td><input type="submit" value="提交"/></td>
      </tr>
   </table>
  </form>
 </body>
</html>
```

UseSessionServlet.java 的代码如下：

```java
package servlet;
import java.io.*;
import javax.servlet.*;
import javax.servlet.annotation.WebServlet;
import javax.servlet.http.*;
@WebServlet(name = "useSession", urlPatterns = { "/sendMyName" })
public class UseSessionServlet extends HttpServlet {
    private static final long serialVersionUID = 1L;
    public void init(ServletConfig config) throws ServletException {
        super.init(config);
    }
    public void doPost(HttpServletRequest request, HttpServletResponse response)
          throws ServletException, IOException {
        response.setContentType("text/html;charset=utf-8");
        PrintWriter out = response.getWriter();
        request.setCharacterEncoding("utf-8");
        String name = request.getParameter("user");
        if (null == name || name.trim().length() == 0) {
            response.sendRedirect("useSession.jsp");
```

```java
        } else {
            HttpSession session = request.getSession(true);
            session.setAttribute("myName", name);
            out.println("<htm><body>");
            out.println("您请求的 servlet 对象是: " + getServletName());
            out.println("<br>您的会话 ID 是: " + session.getId());
            out.println("<br>请单击请求另一个 servlet: ");
            out.println("<br><a href='showMyName'>请求另一个 servlet</a>");
            out.println("</body></htm>");
        }
    }
    public void doGet(HttpServletRequest request, HttpServletResponse response)
            throws ServletException, IOException {
        doPost(request, response);
    }
}
```

ShowNameServlet.java 的代码如下:

```java
package servlet;
import java.io.*;
import javax.servlet.*;
import javax.servlet.annotation.WebServlet;
import javax.servlet.http.*;
@WebServlet(name = "showName", urlPatterns = { "/showMyName" })
public class ShowNameServlet extends HttpServlet {
    private static final long serialVersionUID = 1L;
    public void init(ServletConfig config) throws ServletException {
        super.init(config);
    }
    public void doPost(HttpServletRequest request, HttpServletResponse response)
            throws ServletException, IOException {
        response.setContentType("text/html;charset=utf-8");
        PrintWriter out = response.getWriter();
        HttpSession session = request.getSession(true);
        String name = (String) session.getAttribute("myName");
        out.println("<htm><body>");
        out.println("您请求的 servlet 对象是: " + getServletName());
        out.println("<br>您的会话 ID 是: " + session.getId());
        out.println("<br>您的会话中存储的用户名是: " + name);
        out.println("</body></htm>");

    }
    public void doGet(HttpServletRequest request, HttpServletResponse response)
            throws ServletException, IOException {
        doPost(request, response);
```

        }
    }

6.8 基于 Servlet 的 MVC 模式

视频讲解

本节将重点介绍基于 Servlet 的 MVC 模式。

6.8.1 MVC 模式

❶ MVC 的概念

MVC 是 Model、View、Controller 的缩写，分别代表 Web 应用程序中的 3 种职责。

① 模型：用于存储数据以及处理用户请求的业务逻辑。

② 视图：向控制器提交数据，显示模型中的数据。

③ 控制器：根据视图提出的请求，判断将请求和数据交给哪个模型处理，处理后的有关结果交给哪个视图更新显示。

❷ 基于 Servlet 的 MVC 模式

基于 Servlet 的 MVC 模式的具体实现如下。

① 模型：一个或多个 JavaBean 对象，用于存储数据（实体模型，由 JavaBean 类创建）和处理业务逻辑（业务模型，由一般的 Java 类创建）。

② 视图：一个或多个 JSP 页面，向控制器提交数据和为模型提供数据显示，JSP 页面主要使用 HTML 标记和 JavaBean 标记来显示数据。

③ 控制器：一个或多个 Servlet 对象，根据视图提交的请求进行控制，即将请求转发给处理业务逻辑的 JavaBean，并将处理结果存放到实体模型 JavaBean 中，输出给视图显示。

基于 Servlet 的 MVC 模式的流程如图 6.7 所示。

图 6.7　JSP 中的 MVC 模式

6.8.2 使用 JSP、Servlet、JavaBean 实现 MVC

例 6-8 使用 MVC 模式实现简单的用户登录验证程序，其中包括实体模型 User、业务模型 UserCheck、控制器 LoginCheckServlet 和两个视图页面，即登录页面和登录成功页面。

❶ 定义实体模型来表示数据

User 类用于创建实体模型存储用户信息，代码如下：

```java
package dto;
public class User {
    private String name;
    private String pwd;
    public String getName() {
        return name;
    }
    public void setName(String name) {
        this.name = name;
    }
    public String getPwd() {
        return pwd;
    }
    public void setPwd(String pwd) {
        this.pwd = pwd;
    }
}
```

❷ 定义业务模型来处理业务

UserCheck 类用于判断用户名和密码是否正确，代码如下：

```java
package service;
import dto.User;
public class UserCheck {
    //验证登录
    public boolean validate(User user) {
        if (user != null && user.getName().equals("JSPMVC")) {
            if (user.getPwd().equals("MVC")) {
                return true;
            }
            return false;
        }
        return false;
    }
}
```

❸ 编写 Servlet 处理请求

LoginCheckServlet 完成请求控制，代码如下：

```java
package servlet;
import java.io.IOException;
import javax.servlet.RequestDispatcher;
import javax.servlet.ServletException;
import javax.servlet.annotation.WebServlet;
import javax.servlet.http.HttpServlet;
import javax.servlet.http.HttpServletRequest;
import javax.servlet.http.HttpServletResponse;
import service.UserCheck;
import dto.User;
@WebServlet(name = "loginCheckServlet", urlPatterns = { "/loginCheckServlet" })
public class LoginCheckServlet extends HttpServlet {
    private static final long serialVersionUID = 1L;
    protected void doGet(HttpServletRequest request,
            HttpServletResponse response) throws ServletException, IOException {
        doPost(request, response);
    }
    protected void doPost(HttpServletRequest request,
            HttpServletResponse response) throws ServletException, IOException {
        request.setCharacterEncoding("utf-8");
        String name = request.getParameter("name");
        String pwd = request.getParameter("pwd");
        User user = new User();              //实例化实体模型 user
        user.setName(name);                  //将数据存在模型 user 中
        user.setPwd(pwd);                    //将数据存在模型 user 中
        UserCheck uc = new UserCheck();      //实例化业务模型 userCheck
        if (uc.validate(user)) {
            // 将装有数据的实体模型 user 存储到 request 范围内
            request.setAttribute("user", user);
            RequestDispatcher dis = request
                    .getRequestDispatcher("loginSuccess.jsp");
            dis.forward(request, response);
        } else {
            response.sendRedirect("loginCheck.jsp");
        }
    }
}
```

❹ 编写视图

登录页面 loginCheck.jsp 的代码如下：

```jsp
<%@ page language="java" contentType="text/html; charset=UTF-8"
```

```
pageEncoding="UTF-8"%>
<!DOCTYPE html>
<html>
<head>
<meta charset="UTF-8">
<title>loginCheck.jsp</title>
  </head>
  <body>
    <form action="loginCheckServlet" method="post">
    <table>
        <tr>
            <td>用户名：</td>
            <td><input type="text" name="name"/></td>
        </tr>
        <tr>
            <td>密　码：</td>
            <td><input type="password" name="pwd"/></td>
        </tr>
        <tr>
            <td><input type="submit" value="提交"/></td>
            <td><input type="reset" value="重置"/></td>
        </tr>
    </table>
    </form>
  </body>
</html>
```

登录成功页面 loginSuccess.jsp 的代码如下：

```
<%@ page language="java" contentType="text/html; charset=UTF-8"
pageEncoding="UTF-8"%>
<!DOCTYPE html>
<html>
<head>
<meta charset="UTF-8">
<title>loginSuccess.jsp</title>
</head>
<body>
<jsp:useBean id="user" type="dto.User" scope="request"/>
恭喜<jsp:getProperty property="name" name="user"/>登录成功！
</body>
</html>
```

6.8.3　模型周期

在基于 Servlet 的 MVC 模式中，控制器 Servlet 创建的实体模型 JavaBean 也涉及生命周期，生命周期分别为 request、session 和 application。下面以例 6-8 中的实体模型 user 来讨论这 3 种生命周期的用法。

❶ request 周期的模型

使用 request 周期的模型一般需要以下几个步骤：

1）创建模型并将数据保存到模型中

在 Servlet 中需要这样的代码：

```
User user = new User ();              //实例化模型 user
user.setName(name);                   //将数据存在模型 user 中
user.setPwd(pwd);                     //将数据存在模型 user 中
```

2）将模型保存到 request 对象中并转发给视图 JSP

在 Servlet 中需要这样的代码：

```
request.setAttribute("user", user);   //将装有数据的模型 user 输出给视图
                                      //loginSuccess.jsp 页面
    RequestDispatcher dis=request.getRequestDispatcher("loginSuccess.jsp");
dis.forward(request, response);
```

request.setAttribute("user", user)这句代码指定了查找 JavaBean 的关键字，并决定了 JavaBean 的生命周期为 request。

3）视图更新

Servlet 所转发的页面，如 loginSuccess.jsp 页面，必须使用 useBean 标记获得 Servlet 所创建的 JavaBean 对象（视图不负责创建 JavaBean）。在 JSP 页面中需要使用这样的代码：

```
<jsp:useBean id="user" type="dto.User" scope="request"/>
<jsp:getProperty property="name" name="user"/>
```

标记中的 id 就是 Servlet 所创建的模型 JavaBean，它和 request 对象中的关键字对应。因为在视图中不创建 JavaBean 对象，所以在 useBean 标记中使用 type 属性，而不使用 class 属性。useBean 标记中的 scope 必须和存储模型时的范围（request）一致。

❷ session 周期的模型

使用 session 周期的模型一般需要以下几个步骤。

1）创建模型并将数据保存到模型中

在 Servlet 中需要这样的代码：

```
User user = new User ();              //实例化模型 user
user.setName(name);                   //将数据存在模型 user 中
user.setPwd(pwd);                     //将数据存在模型 user 中
```

2）将模型保存到 session 对象中并转发给视图 JSP

在 Servlet 中需要这样的代码：

```
session.setAttribute("user", user);   //将装有数据的模型 user 输出给视图
                                      //loginSuccess.jsp 页面
RequestDispatcher dis=request.getRequestDispatcher("loginSuccess.jsp");
dis.forward(request, response);
```

session.setAttribute("user", user)这句代码指定了查找 JavaBean 的关键字，并决定了 JavaBean 的生命周期为 session。

3）视图更新

Servlet 所转发的页面，如 loginSuccess.jsp 页面，必须使用 useBean 标记获得 Servlet 所创建的 JavaBean 对象（视图不负责创建 JavaBean）。在 JSP 页面中需要使用这样的代码：

```
<jsp:useBean id="user" type="dto.User" scope="session"/>
<jsp:getProperty property="name" name="user"/>
```

标记中的 id 就是 Servlet 所创建的模型 JavaBean，它和 session 对象中的关键字对应。因为在视图中不创建 JavaBean 对象，所以在 useBean 标记中使用 type 属性，而不使用 class 属性。useBean 标记中的 scope 必须和存储模型时的范围（session）一致。

注：对于生命周期为 session 的模型，Servlet 不仅可以使用 RequestDispatcher 对象转发给 JSP 页面，还可以使用 response 的重定向方法（sendRedirect()）定向到 JSP 页面。

❸ **application 周期的模型**

使用 application 周期的模型一般需要以下几个步骤：

1）创建模型并将数据保存到模型中

在 Servlet 中需要这样的代码：

```
User user=new User ();                //实例化模型 user
user.setName(name);                   //将数据存在模型 user 中
user.setPwd(pwd);                     //将数据存在模型 user 中
```

2）将模型保存到 application 对象中并转发给视图 JSP

在 Servlet 中需要这样的代码：

```
application.setAttribute("user", user);      //将装有数据的模型 user 输出给视图
                                             //loginSuccess.jsp 页面
RequestDispatcher dis=request.getRequestDispatcher("loginSuccess.jsp");
dis.forward(request, response);
```

"application.setAttribute("user", user);"这句代码指定了查找 JavaBean 的关键字，并决定了 JavaBean 的生命周期为 application。

3）视图更新

Servlet 所转发的页面，例如 loginSuccess.jsp 页面，必须使用 useBean 标记获得 Servlet 所创建的 JavaBean 对象（视图不负责创建 JavaBean）。在 JSP 页面中需要使用这样的代码：

```
<jsp:useBean id="user" type="dto.User" scope="application"/>
<jsp:getProperty property="name" name="user"/>
```

标记中的 id 就是 Servlet 所创建的模型 JavaBean，它和 application 对象中的关键字对应。因为在视图中不创建 JavaBean 对象，所以在 useBean 标记中使用 type 属性，而不使用 class 属性。useBean 标记中的 scope 必须和存储模型时的范围（application）一致。

注：对于生命周期为 session 或 application 的模型，Servlet 不仅可以使用 RequestDispatcher 对象转发给 JSP 页面，还可以使用 response 的重定向方法（sendRedirect()）

定向到 JSP 页面。

6.8.4 实践环节——四则运算

模仿例 6-8，使用基于 Servlet 的 MVC 模式设计一个 Web 应用，要求如下：
用户通过 JSP 页面 inputNumber.jsp 输入两个操作数，并选择一种运算符，单击"提交"按钮后，调用 HandleComputer.java 这个 Servlet。在 HandleComputer 中首先获取用户输入的数字和运算符并将这些内容存入实体模型（由 Computer.java 创建）中，然后调用业务模型（由 CalculateBean.java 创建）进行计算并将结果存入实体模型中，在 showResult.jsp 中调用 JavaBean 显示计算的结果。

6.9 本章小结

本章使用了 Servlet 的注解机制部署 Servlet，简化了 Servlet 的开发流程，使得 web.xml 部署描述文件从 Servlet 3.0 开始不再是必选的了。

习题 6

1. Servlet 对象是在服务器端被创建的，还是在用户端被创建的？
2. 什么是转发？什么是重定向？它们有什么区别？
3. 简述 Servlet 的生命周期与运行原理。
4. Servlet 对象初始化时是调用 init 方法还是 service 方法？
5. 下列选项中，（　　）不是 Servlet 中使用的方法。
   A．doGet()　　　　B．doPost()　　　　C．service()　　　　D．close()
6. 下列选项中，对 Servlet 描述错误的是（　　）。
   A．Servlet 是一个特殊的 Java 类，它必须直接或间接实现 Servlet 接口
   B．Servlet 接口定义了 Servlet 的生命周期方法
   C．当多个客户请求一个 Servlet 时，服务器为每一个客户启动一个进程
   D．Servlet 客户线程调用 service 方法响应客户的请求
7. 下列 Servlet 的方法中，（　　）方法载入时被执行，且只执行一次，负责对 Servlet 进行初始化。
   A．service()　　　　B．init()　　　　C．doPost()　　　　D．destroy()
8. 下列 Servlet 的方法中，（　　）方法用来为请求服务，在 Servlet 生命周期中，Servlet 每被请求一次它就会被调用一次。
   A．service()　　　　B．init()　　　　C．doPost()　　　　D．destroy()
9. 下列方法中，（　　）方法当服务器关闭时被调用，用来释放 Servlet 所占的资源。
   A．service()　　　　B．init()　　　　C．doPost()　　　　D．destroy()

10. 阅读下面代码片段:

```
JavaBeanClass bean=new JavaBeanClass(parameter);
HttpSession session=request.getSession(true);
session.setAttribute("keyword",bean);
```

关于该段代码创建的 bean，下列叙述正确的是（　　）。

    A. request 周期的 bean　　B. application 周期的 bean

    C. session 周期的 bean　　D. Page 周期的 bean

11. 阅读下面代码片段:

```
RequestDispatcher dispatcher=request.getRequestDispatcher("a.jsp");
dispatcher.forward(request,response);
```

关于该段代码的作用，下列叙述正确的是（　　）。

    A. 页面重定向到 a.jsp 页面　　B. 将请求转发到 a.jsp 页面

    C. 从 a.jsp 定向到当前页面　　D. 从 a.jsp 转发到当前页面

第7章 会话跟踪技术

学习目的与要求

本章重点对会话跟踪技术进行详细的介绍。通过本章的学习，读者能够理解会话跟踪的含义，掌握会话跟踪技术的应用。

本章主要内容

- Cookie 技术
- session 技术
- URL 重写技术
- 隐藏表单域技术

客户端与服务器进行通信的协议是 HTTP 协议，该协议本身是基于请求/响应模式的、无状态的协议，服务器不会记录客户端的任何信息，这样客户端每次发送的请求都是独立的，这样的方式在工程实践中是不可用的。而会话（session）正是一种能将客户端信息保存在服务器端的技术，它可以记录客户端到服务器的一系列请求。

会话跟踪技术作为在客户端与服务器之间保持 HTTP 状态的解决方案，主要包括 Cookie 技术、session 技术、URL 重写技术以及隐藏表单域技术。

本章涉及的 Java 源文件保存在工程 ch7 的 src 中，涉及的 JSP 页面保存在工程 ch7 的 WebContent 中。

7.1 Cookie 技术

视频讲解

Cookie 技术是一种在客户端保持会话跟踪的解决方案。通过 Cookie，服务器在接收来自客户端浏览器的请求时，能够通过分析请求头的内容得到客户端特有的信息，从而动态生成与该客户端相对应的内容。例如，在登录页面中记住密码。另外，一些商业网站根据用户的偏好，进行风格设置、广告投放等个性化推荐，这些功能都可以通过存储在客户端的 Cookie 实现。

Cookie 可以通过 javax.servlet.http.Cookie 类的构造方法 Cookie(String name, String value) 创建，示例代码如下：

```
Cookie vistedCountC = new Cookie("vistedCount", "1000");
```

其中，第 1 个参数 name 用于指定 Cookie 的属性名；第 2 个参数 value 用于指定属性值。

创建完成的 Cookie 对象，可以使用 HttpServletResponse 对象的 addCookie()方法添加到响应对象中，进而让 Cookie 对象存储在客户端机器上。存储在客户端的 Cookie，可以通过 HttpServletRequest 对象的 getCookies()方法获取。

Cookie 有一定的存活时间，不会在客户端一直保存。默认情况下，在浏览器关闭时失效。若要让其长时间保存，可以通过 Cookie 对象的 setMaxAge(int time)方法设置其存活时间（以秒为单位），若为正整数，表示其存活的秒数；若为负数，表示其为临时 Cookie（在浏览器关闭时失效）；若为 0，表示通知浏览器删除相应的 Cookie 对象。

需要注意的是，在使用 Cookie 时，要保证浏览器接受 Cookie。IE 浏览器的设置方法是：选择"Internet 选项"|"隐私"|"高级"选项，单击"接受"单选按钮，如图 7.1 所示。

图 7.1 接受 Cookie

例 7-1 编写一个 Servlet，在该 Servlet 中测试 Cookie 对象的常用方法。

CookieTest.java 的代码如下：

```
package servlet;
import java.io.IOException;
import java.io.PrintWriter;
import java.text.SimpleDateFormat;
import java.util.Date;
import javax.servlet.ServletException;
import javax.servlet.annotation.WebServlet;
import javax.servlet.http.Cookie;
import javax.servlet.http.HttpServlet;
```

```java
import javax.servlet.http.HttpServletRequest;
import javax.servlet.http.HttpServletResponse;
@WebServlet("/CookieTest")
public class CookieTest extends HttpServlet {
    private static final long serialVersionUID = 1L;
    protected void doGet(HttpServletRequest request, HttpServletResponse response) throws ServletException, IOException {
        response.setContentType("text/html;charset=UTF-8");
        PrintWriter out = response.getWriter();
        //时间不能有空格，不然后续Cookie创建后运行报错
        // (An invalid character [32] was present in the Cookie value)
        SimpleDateFormat sdf = new SimpleDateFormat("yyyy年MM月dd日 HH:mm:ss");
        String nowTime = sdf.format(new Date());
        String lastVistTime = "";//访问时间
        int vistedCount = 0;//访问次数
        //获取客户端浏览器保存的所有Cookie
        Cookie[] myCookies = request.getCookies();
        if(myCookies != null) {
            for (Cookie cookie : myCookies) {
                //判断是否为最近访问过的Cookie，Cookie中属性名为lastVist,
                //属性值为上次访问的时间
                if("lastVist".equals(cookie.getName())) {
                    lastVistTime = cookie.getValue();
                }
                //判断是否为记录访问次数的Cookie,Cookie中属性名为vistedCount,
                //属性值为上次访问的次数
                if("vistedCount".equals(cookie.getName())) {
                    vistedCount = Integer.valueOf(cookie.getValue());
                }
            }
        }
        //不是第一次访问，输出上次访问时间
        if(!"".equals(lastVistTime)){
            out.println("您上次访问时间为：" + lastVistTime);
        }
        //输出访问次数
        out.println("您是第：" + (vistedCount + 1) + "次访问该网站。");
        //以本次访问时间创建同名Cookie
        Cookie lastVistTimeC = new Cookie("lastVist", nowTime);
        //设置最大存活时间为一年
        lastVistTimeC.setMaxAge(365 * 24 * 60 * 60);
        //以访问次数创建同名Cookie
        Cookie vistedCountC = new Cookie("vistedCount", (vistedCount + 1) + "");
        //设置最大存活时间为一年
        vistedCountC.setMaxAge(365 * 24 * 60 * 60);
        //将两个Cookie对象响应到客户端
        response.addCookie(lastVistTimeC);
```

```
            response.addCookie(vistedCountC);
    }
    protected void doPost(HttpServletRequest request, HttpServletResponse
response) throws ServletException, IOException {
        doGet(request, response);
    }
}
```

启动服务器，第 1 次请求 CookieTest 后（http://localhost:8080/ch7/CookieTest）的显示结果如图 7.2 所示。

图 7.2　第 1 次请求后的显示结果

第 2 次请求 CookieTest 后（http://localhost:8080/ch7/CookieTest）的显示结果如图 7.3 所示。

图 7.3　第 2 次请求后的显示结果

例 7-1 实例效果只限于使用同一个浏览器并且允许使用 Cookie 的情况下访问，这是由 Cookie 本身的局限性决定的。Cookie 的缺点主要集中在其安全性和隐私保护上，主要包括以下几种。

① Cookie 可能被禁用。有的用户担心隐私安全问题，很可能会禁用浏览器的 Cookie 功能。

② Cookie 是与浏览器相关的。即使访问同一个应用程序，不同浏览器之间所保存的 Cookie 也是不能互相访问的。

③ Cookie 可能被删除。因为每个 Cookie 都是硬盘上的一个文件，因此很有可能被用户删除。

④ Cookie 的大小和个数受限。单个 Cookie 保存的数据一般不能超过 4KB，很多浏览器都限制一个域名最大保存数量。

⑤ Cookie 安全性不够高。所有的 Cookie 都是以纯文本的形式记录于文件中，因此，如果要保存用户名和密码等信息时，最好事先做加密处理。

7.2　session 技术

session 技术是指使用 HttpSession 对象实现会话跟踪技术，是一种在服务器端保持会话跟踪的解决方案。HttpSession 对象将在用户第 1 次访问服务器时由容器创建（只有访问 JSP、

Servlet 等程序时才会创建，只访问 HTML、image 等静态资源并不创建），当用户调用其失效方法（invalidate()方法）或超过其最大不活动时间时会失效。在会话有效期间，用户与服务器之间的多次请求都属于同一个会话。

服务器在创建 session 对象时，会为其分配一个唯一的会话标识 sessionId，以 JSESSIONID 的属性名保存在客户端 Cookie 中。在用户后续的请求中，服务器通过读取 Cookie 中的 JSESSIONID 属性值来识别不同的用户，从而实现对每个用户的会话跟踪。有关 session 技术的使用参见 4.4 节和 6.7 节。

7.3 URL 重写技术

URL 重写是指服务器程序对接收的 URL 请求重新写成网站可以处理的另一个 URL 的过程。URL 重写技术是实现动态网站会话跟踪的重要保障。在实际应用中，当不能确定客户端浏览器是否支持 Cookie 的情况下，使用 URL 重写技术可以对请求的 URL 地址追加会话标识，从而实现用户的会话跟踪功能。URL 重写实现会话跟踪，要保证每个页面都必须使用 Servlet 或 JSP 动态页面。因为附加在 URL 上的 SessionID 是动态产生的，所以对于静态页面的跳转，URL 重写机制无能为力。

可以通过 HttpServletResponse 的 encodeURL()方法和 encodeRedirectURL()方法实现 URL 重写，其中，encodeRedirectURL()方法主要对使用 sendRedirect()方法的 URL 进行重写。URL 重写方法根据请求信息中是否包含 Set-Cookie 请求头来决定是否进行 URL 重写。若包含了该请求头，将 URL 原样输出；若不包含，则将会话标识 JSESSIONID 重写到 URL 中。URL 重写示例代码如下：

```
//对重定向的 URL 进行重写
String encodeURL = response.encodeRedirectURL("toRedirectServlet");
response.sendRedirect(encodeURL);
```

下面首先演示对谷歌浏览器 Cookie 的禁用设置，如图 7.4 所示。然后，使用禁用 Cookie 的谷歌浏览器运行例 6-7，程序运行结果如图 7.5～图 7.7 所示。

图 7.4　谷歌浏览器 Cookie 的禁用设置

图 7.5 信息输入页面

图 7.6 获取会话并存储数据

您请求的servlet对象是：showMyName
您的会话ID是：C10338663FDB40C017646108C18AB37F
您的会话中存储的用户名是：null

图 7.7 获取会话中的数据并显示

从图 7.6 和图 7.7 中可以看出 session 对象发生了改变，也就是说在 Cookie 禁用的情况下，无法使用 session 技术跟踪客户端的用户状态。幸运的是，我们可以使用 URL 重写技术解决这样的问题。

【例】7-2 使用 URL 重写技术跟踪例 6-7 的用户状态。

useSession.jsp 的代码如下：

```
<%@ page language="java" contentType="text/html; charset=UTF-8"
pageEncoding="UTF-8"%>
<!DOCTYPE html>
<html>
<head>
<meta charset="UTF-8">
<title>useSession.jsp</title>
</head>
  <body>
    <form action="<%=response.encodeURL("sendMyName")%>" method="post">
      <table>
        <tr>
          <td>用户名：</td>
          <td><input type="text" name="user"/></td>
        </tr>
        <tr>
          <td><input type="submit" value="提交"/></td>
        </tr>
```

```
        </table>
    </form>
  </body>
</html>
```

UseSessionServlet.java 的代码如下:

```java
package servlet;
import java.io.*;
import javax.servlet.*;
import javax.servlet.annotation.WebServlet;
import javax.servlet.http.*;
@WebServlet(name = "useSession", urlPatterns = { "/sendMyName" })
public class UseSessionServlet extends HttpServlet {
    private static final long serialVersionUID = 1L;
    public void init(ServletConfig config) throws ServletException {
        super.init(config);
    }
    public void doPost(HttpServletRequest request, HttpServletResponse response)
            throws ServletException, IOException {
        response.setContentType("text/html;charset=UTF-8");
        PrintWriter out = response.getWriter();
        request.setCharacterEncoding("UTF-8");
        String name = request.getParameter("user");
        if (null == name || name.trim().length() == 0) {
            response.sendRedirect(response.encodeRedirectURL("useSession.
            jsp"));
        } else {
            HttpSession session = request.getSession(true);
            session.setAttribute("myName", name);
            out.println("<htm><body>");
            out.println("您请求的 servlet 对象是: " + getServletName());
            out.println("<br>您的会话 ID 是: " + session.getId());
            out.println("<br>请单击请求另一个 servlet: ");
            out.println("<br><a href=" + response.encodeURL("showMyName")
            + ">请求另一个 servlet</a>");
            out.println("</body></htm>");
        }
    }
    public void doGet(HttpServletRequest request, HttpServletResponse response)
            throws ServletException, IOException {
        doPost(request, response);
    }
}
```

ShowNameServlet.java 的代码如下：

```java
package servlet;
import java.io.*;
import javax.servlet.*;
import javax.servlet.annotation.WebServlet;
import javax.servlet.http.*;
@WebServlet(name = "showMyName", urlPatterns = { "/showMyName" })
public class ShowNameServlet extends HttpServlet {
    private static final long serialVersionUID = 1L;
    public void init(ServletConfig config) throws ServletException {
        super.init(config);
    }
    public void doPost(HttpServletRequest request, HttpServletResponse response)
            throws ServletException, IOException {
        response.setContentType("text/html;charset=UTF-8");
        PrintWriter out = response.getWriter();
        HttpSession session = request.getSession(true);
        String name = (String) session.getAttribute("myName");
        if (null == name || name.trim().length() == 0) {
            response.sendRedirect(response.encodeRedirectURL("useSession.jsp"));
        }else{
            out.println("<htm><body>");
            out.println("您请求的servlet对象是：" + getServletName());
            out.println("<br>您的会话ID是：" + session.getId());
            out.println("<br>您的会话中存储的用户名是：" + name);
            out.println("<br><a href=" + response.encodeRedirectURL
            ("useSession.jsp")
                + ">重新登录</a>");
            out.println("</body></htm>");
        }
    }
    public void doGet(HttpServletRequest request, HttpServletResponse response)
            throws ServletException, IOException {
        doPost(request, response);
    }
}
```

例 7-2 运行结果如图 7.8~图 7.10 所示。

图 7.8　信息输入页面

第 7 章　会话跟踪技术

```
 ← → C  ⓘ localhost:8080/ch7/sendMyName;jsessionid=7F3C11EA46E9F386F289AB9B3812F003
您请求的servlet对象是：useSession
您的会话ID是：7F3C11EA46E9F386F289AB9B3812F003
请单击请求另一个servlet：
请求另一个servlet
```

图 7.9　获取会话并存储数据

```
 ← → C  ⓘ localhost:8080/ch7/showMyName;jsessionid=7F3C11EA46E9F386F289AB9B3812F003
您请求的servlet对象是：showMyName
您的会话ID是：7F3C11EA46E9F386F289AB9B3812F003
您的会话中存储的用户名是：陈恒
重新登录
```

图 7.10　获取会话中的数据并显示

从图 7.9 和图 7.10 运行结果可以看出，两个 Servlet 请求地址经 URL 重写后，都被附加了 JSESSIONID 标识。

7.4　隐藏表单域技术

利用 HTML 的 hidden 属性，将客户端的信息，在用户不察觉的情形下，随着请求一起发送给服务器处理。例如：

```
<input type="hidden" name="userID" value="15">
```

例7-3　编写 JSP 页面 login.jsp，在该页面中通过表单向名字为 login 的 Servlet 对象（由 LoginServlet 类负责创建）提交用户名，login 将用户名以隐藏表单域的形式提交给另一个 Servlet 对象 greet（由 GreetServlet 类负责创建），greet 将 login.jsp 提交的用户名显示在浏览器中。login 和 greet 是处理不同请求的 Servlet，但是也可以实现会话跟踪。

页面文件 login.jsp 的代码如下：

```jsp
<%@ page language="java" contentType="text/html; charset=GBK"
    pageEncoding="GBK"%>
<html>
  <head>
    <title>login.jsp</title>
  </head>
  <body>
    <form action="login" method="post">
      <table>
        <tr>
          <td>用户名：</td>
          <td><input type="text" name="user"/></td>
```

```html
            </tr>
            <tr>
                <td><input type="submit" value="提交"/></td>
            </tr>
        </table>
    </form>
  </body>
</html>
```

LoginServlet.java 的代码如下：

```java
package servlet;
import java.io.IOException;
import java.io.PrintWriter;
import javax.servlet.ServletConfig;
import javax.servlet.ServletException;
import javax.servlet.annotation.WebServlet;
import javax.servlet.http.HttpServlet;
import javax.servlet.http.HttpServletRequest;
import javax.servlet.http.HttpServletResponse;
@WebServlet(name = "login", urlPatterns = { "/login" })
public class LoginServlet extends HttpServlet {
    private static final long serialVersionUID = 1L;
    public void init(ServletConfig config) throws ServletException {
        super.init(config);
    }
    public void doPost(HttpServletRequest request, HttpServletResponse response)
            throws ServletException, IOException {
        response.setContentType("text/html;charset= UTF-8");
        PrintWriter out = response.getWriter();
        request.setCharacterEncoding("UTF-8");
        String name = request.getParameter("user");
        if (null == name || name.trim().length() == 0) {
            response.sendRedirect("login.jsp");
        }else{
        out.println("<htm><head><title>第二个页面</title></head><body>");
        out.println("<form action='greet' " + "method='post'>");
        out.println("<input type='hidden' name='user' value=" + name +">");//隐藏表单域
        out.println(name + "<br>");
        out.println("<input type='submit' value='转向下一个页面看一看' >");
        out.println("</form>");
        out.println("</body></htm>");
        }
        }
    public void doGet(HttpServletRequest request, HttpServletResponse response)
```

```
            throws ServletException, IOException {
        doPost(request, response);
    }
}
```

GreetServlet.java 的代码如下：

```java
package servlet;
import java.io.IOException;
import java.io.PrintWriter;
import javax.servlet.ServletConfig;
import javax.servlet.ServletException;
import javax.servlet.annotation.WebServlet;
import javax.servlet.http.HttpServlet;
import javax.servlet.http.HttpServletRequest;
import javax.servlet.http.HttpServletResponse;
@WebServlet(name = "greet", urlPatterns = { "/greet" })
public class GreetServlet extends HttpServlet {
    private static final long serialVersionUID = 1L;
    public void init(ServletConfig config) throws ServletException {
        super.init(config);
    }
    public void doPost(HttpServletRequest request, HttpServletResponse response)
            throws ServletException, IOException {
        response.setContentType("text/html;charset=UTF-8");
        PrintWriter out = response.getWriter();
        request.setCharacterEncoding("UTF-8");
        String name = request.getParameter("user");
        if (null == name || name.trim().length() == 0) {
            response.sendRedirect("login.jsp");
        }else{
            out.println("<htm><body>");
            out.println("<br>您的姓名是：" + name);
            out.println("<br><a href=login.jsp" + ">重新登录</a>");
            out.println("</body></htm>");
        }
    }
    public void doGet(HttpServletRequest request, HttpServletResponse response)
            throws ServletException, IOException {
        doPost(request, response);
    }
}
```

7.5 实践环节——登录注销系统

编写一个登录注销系统，在用户登录成功的情况下，将用户的信息保存在会话 HttpSession 中，用户信息可以在会话过程中的任意页面或 Servlet 中被获取并显示，使用 HttpSession

的 invalidate( )方法实现系统的注销功能。一旦注销，该用户信息丢失，用户需要重新登录。该系统需要考虑用户浏览器禁用 Cookie 的情况。

7.6 本章小结

　　HTTP 协议是一种无状态的协议，也就是说当用户请求一个资源，服务器根据客户端请求做出响应，回发到客户端后，就关闭了连接。但是，实际生活中可能有类似网上购物的场景，通过不同的请求购买到了不同的商品，并放在"购物车"中，那么对于服务器来说，它如何知道车中已经有了些什么物品？同样的，在支付费用时服务器又如何来确定以前所建立的购物车是属于哪一位客户的呢？很显然，HTTP 连接是一个请求，一个响应，独立的连接，无法解决。这个时候就出现了 4 种技术方法来满足这一实际需要——Cookie、session、URL 重写以及隐藏表单域。

习题 7

1. 会话跟踪技术有哪些？
2. 在工程实践中为什么要进行会话跟踪？
3. 浏览器禁用 Cookie 后，如何进行会话跟踪？

第 8 章
过滤器与监听器

学习目的与要求

本章主要介绍过滤器与监听器的概念、原理以及实际应用。通过本章的学习，读者能够理解过滤器与监听器的原理，掌握过滤器与监听器的实际应用。

本章主要内容

- Filter 类与 Filter 对象
- Filter 对象的部署与运行
- 过滤器的实际应用
- 监听器的分类与创建
- 监听器的实际应用

在开发一个网站时，可能有这样的需求：某些页面只希望几个特定的用户浏览。对于这样的访问权限的控制，该如何实现呢？Servlet 过滤器（Filter）就可以实现上述需求。过滤器位于服务器处理请求之前或服务器响应请求之前。也就是说，它可以过滤浏览器对服务器的请求，也可以过滤服务器对浏览器的响应，如图 8.1 所示。

图 8.1 过滤器

Servlet 监听器用于监听 Java Web 应用中某些对象信息的创建、销毁、增加、修改、删除等动作的发生，然后做出相应的响应处理。Servlet 监听器常用于统计在线人数和在线用户、系统加载时信息初始化、统计网站的访问量等。

本章涉及的 Java 源文件保存在工程 ch8 的 src 中，涉及的 JSP 页面保存在工程 ch8 的 WebContent 中。

8.1 Filter 类与 Filter 对象

编写一个过滤器类很简单，只要实现 javax.servlet 包中的 Filter 接口。实现 Filter 接口的类习惯地称为一个 Filter 类，这样的类创建的对象又习惯地称为 Filter 对象。

Filter 接口与 Servlet 接口很类似，同样都有 init()与 destroy()方法，还有一个 doFilter()方法类似于 Servlet 接口的 service()方法。下面分别介绍这 3 种方法的功能。

① public void init(FilterConfig fConfig) throws ServletException：该方法的功能是初始化过滤器对象。如果为过滤器设置了初始参数，则可以通过 FilterConfig 的 getInitParameter(String paramName)方法获得初始参数值。

② public void doFilter(ServletRequest request, ServletResponse response, FilterChain chain) throws IOException, ServletException：当 Web 服务器使用 Servlet 对象调用 service()方法处理请求前，发现应用了某个过滤器时，Web 服务器就会自动调用该过滤器的 doFilter()方法。在 doFilter()方法中有这样一条语句：

```
chain.doFilter(request, response);
```

如果执行了该语句，就会执行下一个过滤器；如果没有下一个过滤器，就返回请求目标程序。如果因为某个原因没有执行"chain.doFilter(request, response);"，则请求就不会继续交给以后的过滤器或请求目标程序，这就是所谓的拦截请求。

③ public void destroy()：当 Web 服务器终止服务时，destroy()方法会被执行，使 Filter 对象消亡。

例 8-1 在 Web 工程 ch8 中编写一个简单的 Filter 类 FirstFilter，Filter 类的功能要求：无论用户请求该 Web 工程的哪个页面或 Servlet，都会在浏览器中先出现"首先执行过滤器"这样的响应信息。

FirstFilter.java 的代码如下：

```java
package filter;
import java.io.IOException;
import java.io.PrintWriter;
import javax.servlet.Filter;
import javax.servlet.FilterChain;
import javax.servlet.FilterConfig;
import javax.servlet.ServletException;
import javax.servlet.ServletRequest;
import javax.servlet.ServletResponse;
public class FirstFilter implements Filter {
```

```java
        public void destroy() {
        }
        public void doFilter(ServletRequest request, ServletResponse response,
                FilterChain chain) throws IOException, ServletException {
            //设置响应类型
            response.setContentType("text/html;charset=utf-8");
            //获得输出对象 out
            PrintWriter out=response.getWriter();
            //在浏览器中输出
            out.print("首先执行过滤器<br>");
            //执行下一个过滤器
            chain.doFilter(request, response);
        }
        public void init(FilterConfig fConfig) throws ServletException {
        }
}
```

编写完 Filter 类的源文件，并编译了源文件，这时 Web 服务器是不是就可以运行 Filter 对象呢？不可以。需要部署 Filter 后，Web 服务器才可以运行 Filter 对象。

8.2　Filter 对象的部署与运行

与 Servlet 一样，部署过滤器有两种方式：一是在 web.xml 中部署 Filter；二是基于注解的方式部署 Filter。

8.2.1　在 web.xml 中部署过滤器

❶ 部署 Filter

为了在 web.xml 文件中部署 8.1 节中的 FirstFilter，需要在 web.xml 文件中找到 <web-app></web-app> 标记，然后在 <web-app></web-app> 标记中添加如下内容：

```xml
<filter>
    <filter-name>firstFilter</filter-name>
    <filter-class>filter.FirstFilter</filter-class>
</filter>
<filter-mapping>
    <filter-name>firstFilter</filter-name>
    <url-pattern>/*</url-pattern>
</filter-mapping>
```

❷ 运行 Filter

只要用户请求的 URL 和 <filter-mapping> 的子标记 <url-pattern> 指定的模式匹配，Web 服务器就会自动调用该 Filter 的 doFilter() 方法。如 8.1 节中的 FirstFilter 过滤器在 web.xml

中的<url-pattern>指定值为"/*","/*"代表任何页面或 Servlet 的请求。

为了测试过滤器,在 web 工程 ch8 中新建一个 JSP 页面 test.jsp,运行 test.jsp 显示如图 8.2 所示的结果。

图 8.2　首先执行过滤器

test.jsp 的代码如下:

```
<%@ page language="java" contentType="text/html; charset=UTF-8"
pageEncoding="UTF-8"%>
<!DOCTYPE html>
<html>
<head>
<meta charset="UTF-8">
<title>Insert title here</title>
</head>
<body>
    test.jsp
</body>
</html>
```

❸ 有关部署过滤器的标记

1)<filter>标记及其子标记

web.xml 文件中可以有若干个<filter>标记,该标记的内容由 Web 服务器负责处理。<filter>标记中有两个子标记:<filter-name>和<filter-class>,其中,<filter-name>子标记的内容是 Web 服务器创建的 Filter 对象的名字。web.xml 文件中可以有若干个<filter>标记,但要求它们的<filter-name>子标记的内容互不相同。<filter-class>子标记的内容指定 Web 服务器用哪个类来创建 Filter 对象,如果 Filter 对象已经创建,那么 Web 服务器就不再使用指定的类创建。

如果在过滤器初始化时需要读取一些参数的值,则可以在<filter>标记中使用<init-param>子标记设置。例如:

```
<filter>
        <filter-name>firstFilter</filter-name>
        <filter-class>filters.FirstFilter</filter-class>
        <init-param>
            <param-name>encoding</param-name>
            <param-value>GBK</param-value>
        </init-param>
    </filter>
```

那么就可以在 Filter 的 init()方法中,使用参数 fConfig(FilterConfig 的对象)调用

FilterConfig 的 getInitParameter(String paramName)方法获得参数值。例如：

```
public void init(FilterConfig fConfig) throws ServletException{
    String en = fConfig.getInitParameter("encoding");
}
```

2）<filter-mapping>标记及其子标记

web.xml 文件中出现一个<filter>标记就会对应地出现一个<filter-mapping>标记。<filter-mapping>标记中也有两个子标记：<filter-name>和<url-pattern>。其中，<filter-name>子标记的内容是 Web 服务器创建的 Filter 对象的名字（该名字必须和<filter>标记的子标记<filter-name>的内容相同）；<url-pattern>子标记用来指定用户用怎样的模式请求 Filter 对象。如果某个 URL 或 Servlet 需应用多个过滤器，则根据<filter-mapping>标记在 web.xml 中出现的先后顺序执行过滤器。

8.2.2 基于注解的方式部署过滤器

在 Servlet 3.0 中提供了注解@WebFilter，使得不再需要在 web.xml 文件中进行 Filter 的部署描述。但在实际的 Web 工程中不需要大量开发 Filter，因此，采用哪种方式部署过滤器都不太麻烦。

@WebFilter 用于将一个类声明为过滤器，该注解将会在部署时被容器处理，容器将根据具体的属性配置，将相应的类部署为过滤器。该注解的一些常用属性如表 8.1 所示。

表 8.1 @WebFilter 的常用属性列表

属 性 名	类 型	描 述
filterName	String	指定过滤器的 name 属性，等价于<filter-name>
value	String[]	该属性等价于 urlPatterns 属性，但两个属性不能同时使用
urlPatterns	String[]	指定一组过滤器的 URL 匹配模式，等价于<url-pattern>标签
servletNames	String[]	指定过滤器将应用于哪些 Servlet，取值是@WebServlet 中的 name 属性的取值，或者是 web.xml 中<servlet-name>的取值
initParams	WebInitParam[]	指定一组过滤器初始化参数，等价于<init-param>标签

表 8.1 中的所有属性均为可选属性，但是 value 或者 urlPatterns 通常是必需的，且二者不能共存，如果同时指定，通常是忽略 value 的取值。

例 8-2 基于注解的 Filter——SecondFilter。

SecondFilter.java 的代码如下：

```
package filter;
import java.io.IOException;
import java.io.PrintWriter;
import javax.servlet.Filter;
import javax.servlet.FilterChain;
import javax.servlet.FilterConfig;
import javax.servlet.ServletException;
import javax.servlet.ServletRequest;
```

```java
import javax.servlet.ServletResponse;
import javax.servlet.annotation.WebFilter;
@WebFilter(filterName = "secondFilter", urlPatterns = { "/*" })
public class SecondFilter implements Filter {
    public void destroy() {
    }
    public void doFilter(ServletRequest request, ServletResponse response,
            FilterChain chain) throws IOException, ServletException {
        //设置响应类型
        response.setContentType("text/html;charset=utf-8");
        //获得输出对象out
        PrintWriter out = response.getWriter();
        //在浏览器中输出
        out.print("执行第二个过滤器<br>");
        //执行下一个过滤器
        chain.doFilter(request, response);
    }
    public void init(FilterConfig fConfig) throws ServletException {
    }
}
```

SecondFilter.java 的代码中使用"@WebFilter(filterName = "secondFilter", urlPatterns = { "/*" })"这样的部署后,就不必在 web.xml 中部署相应的<filter>和<filter-mapping>元素了,Web 容器会在部署时根据指定的属性将该类发布为 Filter。与"@WebFilter(filterName = "secondFilter", urlPatterns = { "/*" })"等价的 web.xml 部署形式如下:

```xml
<filter>
    <filter-name>secondFilter</filter-name>
    <filter-class>filter.SecondFilter</filter-class>
</ filter>
<filter-mapping>
  <filter-name>secondFilter</filter-name>
    <url-pattern>/*</url-pattern>
</filter-mapping>
```

8.2.3 过滤器的执行顺序

为了验证过滤器的执行顺序,新建 SecondFilter、ThreeFilter、FourFilter 和 ZFilter。其中,SecondFilter、ThreeFilter 和 FourFilter 使用注解的方式部署,它们的 urlPatterns 都是"/*";而 ZFilter 在 web.xml 中部署,并部署在 FirstFilter 的前面。

web.xml 的代码如下:

```xml
<filter>
    <filter-name>zFilter</filter-name>
    <filter-class>filter.ZFilter</filter-class>
```

```xml
  </filter>
  <filter-mapping>
    <filter-name>zFilter</filter-name>
    <url-pattern>/*</url-pattern>
  </filter-mapping>
  <filter>
    <filter-name>firstFilter</filter-name>
    <filter-class>filter.FirstFilter</filter-class>
  </filter>
  <filter-mapping>
    <filter-name>firstFilter</filter-name>
    <url-pattern>/*</url-pattern>
  </filter-mapping>
```

SecondFilter.java 的 doFilter() 的代码如下：

```java
public void doFilter(ServletRequest request, ServletResponse response,
        FilterChain chain) throws IOException, ServletException {
    response.setContentType("text/html;charset=utf-8");
    PrintWriter out = response.getWriter();
    out.print("执行过滤器 SecondFilter<br>");
    chain.doFilter(request, response);
}
```

ThreeFilter.java 的 doFilter() 的代码如下：

```java
public void doFilter(ServletRequest request, ServletResponse response,
        FilterChain chain) throws IOException, ServletException {
    response.setContentType("text/html;charset=utf-8");
    PrintWriter out = response.getWriter();
    out.print("执行过滤器 ThreeFilter<br>");
    chain.doFilter(request, response);
}
```

FourFilter.java 的 doFilter() 的代码如下：

```java
public void doFilter(ServletRequest request, ServletResponse response,
        FilterChain chain) throws IOException, ServletException {
    response.setContentType("text/html;charset=utf-8");
    PrintWriter out = response.getWriter();
    out.print("执行过滤器 FourFilter<br>");
    chain.doFilter(request, response);
}
```

ZFilter.java 的 doFilter() 的代码如下：

```java
public void doFilter(ServletRequest request, ServletResponse response,
        FilterChain chain) throws IOException, ServletException {
    response.setContentType("text/html;charset=utf-8");
```

```
        PrintWriter out = response.getWriter();
        out.print("执行过滤器 ZFilter<br>");
        chain.doFilter(request, response);
    }
```

运行 test.jsp，结果如图 8.3 所示。

图 8.3　过滤器的执行顺序

从图 8.3 可以看出，过滤器拦截路径相同时，首先按照<filter-mapping>标记在 web.xml 中出现的先后顺序执行过滤器，然后按照过滤器类名的字典顺序执行注解的过滤器。

8.2.4　实践环节——过滤器的初始化参数

给 8.2.3 节中的 SecondFilter 添加初始化参数，并在该过滤器的 init()方法中获取这些参数。@WebInitParam 注解给过滤器添加参数的示例代码如下：

```
@WebFilter(filterName = "xxxFilter", urlPatterns = { "/*" }, initParams = {
        @WebInitParam(name = "firstParam", value = "one"),
        @WebInitParam(name = "secondParam", value = "two") })
```

8.3　过滤器的实际应用

视频讲解

过滤器是 Servlet 的一种特殊用法，主要用来完成一些通用的操作，例如编码的过滤，判断用户的登录状态等。

8.3.1　字符编码过滤器

在 4.1.3 节中介绍了中文乱码的解决方法，其中一种解决方法是：在获取表单信息之前，使用 request 对象调用 setCharacterEncoding(String code)方法设置统一字符编码。使用该方法解决中文乱码问题时，接收参数的每个页面或 Servlet 都需要执行 request.setCharacterEncoding("XXX")语句。为了避免每个页面或 Servlet 都编写 request.setCharacterEncoding("XXX")语句，可以使用过滤器进行字符编码处理。

【例】8-3　字符编码过滤器——SetCharacterEncodingFilter。

SetCharacterEncodingFilter.java 的代码如下：

```java
package filter;
import java.io.IOException;
import javax.servlet.Filter;
import javax.servlet.FilterChain;
import javax.servlet.FilterConfig;
import javax.servlet.ServletException;
import javax.servlet.ServletRequest;
import javax.servlet.ServletResponse;
import javax.servlet.annotation.WebFilter;
import javax.servlet.annotation.WebInitParam;
@WebFilter(filterName= "setCharacterEncodingFilter", urlPatterns = { "/*" },
initParams = { @WebInitParam(name = "encoding", value = "utf-8") })
public class SetCharacterEncodingFilter implements Filter {
    private static String encoding;
    public void destroy() {
    }
    public void doFilter(ServletRequest request, ServletResponse response,
            FilterChain chain) throws IOException, ServletException {
        request.setCharacterEncoding(encoding);
        chain.doFilter(request, response);
    }
    public void init(FilterConfig fConfig) throws ServletException {
        encoding = fConfig.getInitParameter("encoding");
    }
}
```

8.3.2 登录验证过滤器

在 Web 工程中，某些页面或 Servlet 只有用户登录成功才能访问。如果直接在应用程序每个相关的源代码中判断用户是否登录成功，并不是科学的做法。这时可以实现一个登录验证过滤器，不用在每个相关的源代码中验证用户是否登录成功。

例 8-4 新建一个 Web 工程 loginValidate，在该 Web 工程中至少编写两个 JSP 页面：login.jsp 与 loginSuccess.jsp，一个 Servlet（由 LoginServlet.java 负责创建）。用户在 login.jsp 页面中输入用户名和密码后，提交给 Servlet，在 Servlet 中判断用户名和密码是否正确，若正确跳转到 loginSuccess.jsp，若错误回到 login.jsp 页面。但该 Web 工程有另外一个要求：除了访问 login.jsp 页面和 LoginServlet 以外，别的页面或 Servlet 都不能直接访问，必须先登录成功才能访问。在设计这个 Web 工程时，编写了一个登录验证过滤器并在该 Web 工程中使用。页面运行效果如图 8.4 所示。

（a）登录画面

您没有登录，请先登录！3秒钟后回到登录页面。

（b）没有登录成功直接运行 loginSuccess.jsp

恭喜filter登录成功！

（c）登录成功页面

图 8.4　例 8-4 的运行效果

login.jsp 的代码如下：

```jsp
<%@ page language="java" contentType="text/html; charset=UTF-8"
    pageEncoding="UTF-8"%>
<!DOCTYPE html>
<html>
<head>
<meta charset="UTF-8">
<title>login.jsp</title>
 </head>
 <body bgcolor="lightPink">
   <form action="loginServlet" method="post">
   <table>
      <tr>
         <td>用户名：</td>
         <td><input type="text" name="name"/></td>
      </tr>
      <tr>
         <td>密　　码：</td>
         <td><input type="password" name="pwd"/></td>
      </tr>
      <tr>
         <td><input type="submit" value="提交"/></td>
         <td><input type="reset" value="重置"/></td>
      </tr>
   </table>
   </form>
 </body>
```

```
</html>
```

LoginServlet.java 的代码如下：

```java
package servlet;
import java.io.IOException;
import javax.servlet.ServletException;
import javax.servlet.annotation.WebServlet;
import javax.servlet.http.HttpServlet;
import javax.servlet.http.HttpServletRequest;
import javax.servlet.http.HttpServletResponse;
import javax.servlet.http.HttpSession;
@WebServlet(name = "loginServlet", urlPatterns = { "/loginServlet" })
public class LoginServlet extends HttpServlet {
    private static final long serialVersionUID = 1L;
    protected void doGet(HttpServletRequest request,
            HttpServletResponse response) throws ServletException, IOException {
        doPost(request, response);
    }
    protected void doPost(HttpServletRequest request,
            HttpServletResponse response) throws ServletException, IOException {
        String username = request.getParameter("name");
        String password = request.getParameter("pwd");
        if ("filter".equals(username) && "filter".equals(password)) {
            HttpSession session = request.getSession(true);
            session.setAttribute("user", username);
            response.sendRedirect("loginSuccess.jsp");
        } else {
            response.sendRedirect("login.jsp");
        }
    }
}
```

loginSuccess.jsp 的代码如下：

```jsp
<%@ page language="java" contentType="text/html; charset=UTF-8"
pageEncoding="UTF-8"%>
<!DOCTYPE html>
<html>
<head>
<meta charset="UTF-8">
<title>loginSuccess.jsp</title>
</head>
<body>
<%
String username=(String)session.getAttribute("user");
%>
```

恭喜<%=username %>登录成功！
</body>
</html>
```

LoginFilter.java 的代码如下：

```java
package filter;
import java.io.IOException;
import java.io.PrintWriter;
import javax.servlet.Filter;
import javax.servlet.FilterChain;
import javax.servlet.FilterConfig;
import javax.servlet.ServletException;
import javax.servlet.ServletRequest;
import javax.servlet.ServletResponse;
import javax.servlet.annotation.WebFilter;
import javax.servlet.http.HttpServletRequest;
import javax.servlet.http.HttpServletResponse;
import javax.servlet.http.HttpSession;
@WebFilter(filterName = "loginFilter", urlPatterns = { "/*" })
public class LoginFilter implements Filter {
    public void destroy() {
    }
    public void doFilter(ServletRequest request, ServletResponse response,
    FilterChain chain) throws IOException, ServletException {
        HttpServletRequest req = (HttpServletRequest) request;
        HttpServletResponse resp = (HttpServletResponse) response;
        HttpSession session = req.getSession(true);
        resp.setContentType("text/html;");
        resp.setCharacterEncoding("UTF-8");
        PrintWriter out = resp.getWriter();
        //得到用户请求的 URI
        String request_uri = req.getRequestURI();
        //得到 Web 应用程序的上下文路径
        String ctxPath = req.getContextPath();
        //去除上下文路径，得到剩余部分的路径
        String uri = request_uri.substring(ctxPath.length());
        //登录页面或 Servlet 不拦截
        if(uri.contains("login.jsp") || uri.contains("loginServlet")){
            chain.doFilter(request, response);
        }else{
            //判断用户是否已经登录
            if (null != session.getAttribute("user")) {
                //执行下一个过滤器
                chain.doFilter(request, response);
            } else {
```
```

```
                out.println("您没有登录，请先登录！3 秒钟后回到登录页面。");
                resp.setHeader("refresh", "3;url=" + ctxPath + "/login.jsp");
                return;
            }
        }
    }
    public void init(FilterConfig fConfig) throws ServletException {
    }
}
```

8.4 监听器的分类与创建

视频讲解

创建不同功能的监听器类，需要实现不同的监听接口。一个监听器类可以实现多个接口，即可以多种功能的监听器一起工作。根据监听的对象，可将监听器划分为 ServletContext 对象监听器、HttpSession 对象监听器和 ServletRequest 对象监听器；根据监听的事件，可将监听器划分为对象自身的创建和销毁的监听器，对象中属性的创建、修改和消除的监听器以及 session 中某个对象状态变化的监听器。下面根据监听的事件进行划分。

8.4.1 监听对象自身创建和销毁的监听器

HttpSessionListener、ServletContextListener 和 ServletRequestListener 监听接口分别监听 session、context 和 request 对象的创建与销毁。

① HttpSessionListener：监听 session 的创建与销毁。创建 session 时执行 sessionCreated (HttpSessionEvent se) 方法。超时或执行 session.invalidate() 时执行 sessionDestroyed (HttpSessionEvent se) 方法。该监听器可用于统计在线人数、记录访问日志等。

② ServletContextListener：监听 context 的创建与销毁。context 代表当前 Web 应用程序。服务器启动时执行 contextInitialized(ServletContextEvent sce)方法。服务器关闭时执行 contextDestroyed(ServletContextEvent sce)方法。该监听器可用于启动时获取 web.xml 中配置的初始化参数，可作为定时器、加载全局属性对象、创建全局数据库连接、加载缓存信息等。

③ ServletRequestListener：监听 request 的创建与销毁。用户每次请求 request 都将执行 requestInitialized(ServletRequestEvent sre)方法。request 处理完毕自动销毁前执行 requestDestroyed(ServletRequestEvent sre)方法。该监听器可用于读取 request 参数，记录访问历史。

【例】8-5 在 ch8 的 src 目录下创建一个名为 listener 的包，并在包中创建一个名为 MyObjectListener 的监听器类，该监听器类实现 HttpSessionListener、ServletContext Listener 和 ServletRequestListener 监听接口。具体代码如下：

```
package listener;
```

```java
import javax.servlet.ServletContext;
import javax.servlet.ServletContextEvent;
import javax.servlet.ServletContextListener;
import javax.servlet.ServletRequestEvent;
import javax.servlet.ServletRequestListener;
import javax.servlet.http.HttpServletRequest;
import javax.servlet.http.HttpSession;
import javax.servlet.http.HttpSessionEvent;
import javax.servlet.http.HttpSessionListener;
public class MyObjectListener implements HttpSessionListener,ServletContextListener,ServletRequestListener{
    @Override
    public void sessionCreated(HttpSessionEvent se) {
        HttpSession session = se.getSession();
        System.out.println("新创建一个session, ID为: " + session.getId());
    }
    @Override
    public void sessionDestroyed(HttpSessionEvent se) {
        HttpSession session = se.getSession();
        System.out.println("销毁一个session, ID为: " + session.getId());
    }
    @Override
    public void contextInitialized(ServletContextEvent sce) {
        ServletContext servletContext = sce.getServletContext();
        System.out.println("即将启动" + servletContext.getContextPath());
    }
    @Override
    public void contextDestroyed(ServletContextEvent sce) {
        ServletContext servletContext = sce.getServletContext();
        System.out.println("即将关闭" + servletContext.getContextPath());
    }
    @Override
    public void requestInitialized(ServletRequestEvent sre) {
        HttpServletRequest request = (HttpServletRequest) sre.getServletRequest();
        String uri = request.getRequestURI();
        uri = request.getQueryString() == null ? uri : (uri + "?" + request.getQueryString());
        request.setAttribute("dateCreated", System.currentTimeMillis());
        System.out.println("IP " + request.getRemoteAddr() + " 请求 " + uri);
    }
    @Override
    public void requestDestroyed(ServletRequestEvent sre) {
        HttpServletRequest request = (HttpServletRequest) sre.getServletRequest();
        long time = System.currentTimeMillis() - (Long) request.
```

```
                getAttribute("dateCreated");
            System.out.println(request.getRemoteAddr() + "请求处理结束,用时" +
            time + "毫秒。");
    }
}
```

编写完监听器类的源文件,并编译了源文件,这时 Web 服务器是不是就可以运行监听器对象呢?不可以。与过滤器一样,需要部署后,Web 服务器才可以运行监听器对象。在 web.xml 文件中部署监听器,示例代码如下:

```
<listener>
    <listener-class>完整类名</listener-class>
</listener>
```

因此,运行例 8-5 监听器前,需要在 ch8 的 web.xml 中部署该监听器,部署代码如下:

```
<listener>
        <listener-class>listener.MyObjectListener</listener-class>
</listener>
```

部署监听器后,通过 http://localhost:8080/ch8/test.jsp 运行 test.jsp 来测试监听器,测试结果如图 8.5 所示。

图 8.5 例 8-5 监听器的测试结果

注:监听器的启动顺序是 web.xml 的配置顺序;加载顺序是"监听器→过滤器→Servlet"。

8.4.2 监听对象属性变化的监听器

监听对象属性变化的监听器分别为 HttpSessionAttributeListener、ServletContextAttributeListener 和 ServletRequestAttributeListener。这 3 个监听接口分别用于监听 session、context 和 request 的属性变化。当在被监听对象中添加、更新、移除属性时，将分别执行 attributeAdded、attributeReplaced 和 attributeRemoved 方法。

例 8-6 在 ch8 的 listener 包中创建一个名为 MyAttributeListener 的监听器类，该监听器类实现 HttpSessionAttributeListener 监听接口。具体代码如下：

```java
package listener;
import javax.servlet.http.HttpSession;
import javax.servlet.http.HttpSessionAttributeListener;
import javax.servlet.http.HttpSessionBindingEvent;
public class MyAttributeListener implements HttpSessionAttributeListener{
    @Override
    public void attributeAdded(HttpSessionBindingEvent se) {
        String name = se.getName();
        System.out.println("新建 session 属性: " + name + ", 值为: " + se.getValue());
    }
    @Override
    public void attributeReplaced(HttpSessionBindingEvent se) {
        HttpSession session = se.getSession();
        String name = se.getName();
        Object oldValue = se.getValue();
        System.out.println("修改 session 属性: "+name + ", 原值: " + oldValue + ", 新值: " + session.getAttribute(name));
    }
    @Override
    public void attributeRemoved(HttpSessionBindingEvent se) {
        String name = se.getName();
        System.out.println("删除 session 属性: " + name + ", 值为: " + se.getValue());
    }
}
```

在 web.xml 文件中部署例 8-6 监听器，此监听器部署在例 8-5 监听器后，代码如下：

```xml
<listener>
    <listener-class>listener.MyAttributeListener</listener-class>
</listener>
```

在 test.jsp 页面中添加如下代码，测试例 8-6 监听器：

```
<%
    session.setAttribute("user", "陈恒");
```

```
        session.setAttribute("user", "陈恒恒");
        session.invalidate();
%>
```

启动服务器，通过 http://localhost:8080/ch8/test.jsp 运行 test.jsp 页面测试例 8-6 监听器，测试结果如图 8.6 所示。

图 8.6 例 8-6 监听器的测试结果

8.4.3 监听 session 中某个对象状态变化的监听器

监听绑定到 HttpSession 域中某个对象状态的事件监听器有 HttpSessionBindingListener 和 HttpSessionActivationListener。HttpSession 中对象的状态有：绑定→解除绑定，钝化→活化。

① 绑定：通过 setAttribute 保存到 session 对象当中。
② 解除绑定：通过 removeAttribue 去除绑定。
③ 钝化：将 session 对象持久化到存储设备上。
④ 活化：将 session 对象从存储设备上恢复。

实现钝化和活化的监听器对象必须实现 Serializable 接口，不需要在 web.xml 中部署。

- HttpSessionBindingListener：当对象被放到 session 中时，执行 valueBound(Http

SessionBindingEvent event)方法。当对象被从 session 中移除时，执行 valueUnbound (HttpSessionBindingEvent event)方法。对象必须实现该监听接口。
- HttpSessionActivationListener：当 session 中的对象被钝化时，执行 sessionWillPassivate(HttpSessionEvent se)方法。当对象被重新加载（活化）时，执行 sessionDidActivate(HttpSessionEvent se)方法。对象必须实现该监听接口。

例8-7 在 ch8 的 listener 包中创建一个名为 MySeesionListener 的监听器类，该监听器类实现 HttpSessionBindingListener、HttpSessionActivationListener 和 Serializable 接口。具体代码如下：

```java
package listener;
import java.io.Serializable;
import javax.servlet.http.HttpSession;
import javax.servlet.http.HttpSessionActivationListener;
import javax.servlet.http.HttpSessionBindingEvent;
import javax.servlet.http.HttpSessionBindingListener;
import javax.servlet.http.HttpSessionEvent;
public class MySeesionListener implements HttpSessionBindingListener,
HttpSessionActivationListener,
Serializable {
    private static final long serialVersionUID = -4780592776386225973L;
    private String name;
    /**
     * 活化
     */
    public void sessionDidActivate(HttpSessionEvent se) {
        HttpSession session = se.getSession();
        System.out.println(this + "已经成功从硬盘中加载。sessionId: " +
        session.getId());
    }
    /**
     * 即将被钝化到硬盘时
     */
    public void sessionWillPassivate(HttpSessionEvent se) {
        HttpSession session = se.getSession();
        System.out.println(this + "即将保存到硬盘。sessionId: " + session.getId());
    }

    /**
     * 绑定到 session
     */
    public void valueBound(HttpSessionBindingEvent event) {
        HttpSession session = event.getSession();
        String name = event.getName();
        System.out.println(this + "被绑定到 session \"" + session.getId() + "\
```

```java
        "的" + name + "属性上");
    }

    /**
     * 从 session 中移除后
     */
    public void valueUnbound(HttpSessionBindingEvent event) {
        HttpSession session = event.getSession();
        String name = event.getName();
        System.out.println(this + "被从 session \"" + session.getId() + "\
"的" + name + "属性上移除");
    }
    @Override
    public String toString() {
        return "MySessionListener(" + name + ")";
    }
    public String getName() {
        return name;
    }
    public void setName(String name) {
        this.name = name;
    }
}
```

监听器不需要在 web.xml 中部署,但为了测试该监听器,需要将 test.jsp 的代码修改如下:

```jsp
<%@ page language="java" contentType="text/html; charset=UTF-8"
pageEncoding="UTF-8"%>
<%@ page import="listener.MySessionListener"%>
<!DOCTYPE html>
<html>
<head>
<meta charset="UTF-8">
<title>Insert title here</title>
</head>
<body>
    test.jsp
    <%
        MySessionListener msl = new MySessionListener();
        msl.setName("测试 session 内监听器");
        //绑定到 session
        session.setAttribute("msl", msl);
        //从 session 中移除
        session.removeAttribute("msl");
    %>
```

```
</body>
</html>
```

启动服务器，通过 http://localhost:8080/ch8/test.jsp 测试例 8-7 监听器，测试结果如图 8.7 所示。

图 8.7　例 8-7 监听器的测试结果

8.5　监听器的实际应用——统计在线用户

做一个网站 ch8SS，实现在线用户的统计。可以通过 ServletContextListener 监听，当 Web 应用上下文启动时，在 ServletContext 中添加一个 List 用来存放在线的用户名。然后通过 HttpSessionAttributeListener 监听，当用户登录成功，将用户名设置到 Session 中，同时将用户名放到 ServletContext 的 List 中。当用户注销会话时，将用户名从应用上下文范围中的 List 列表中删除。具体步骤如下：

视频讲解

❶ **JSP 页面**

创建 Web 工程 ch8SS，并创建主页 index.jsp，具体代码如下：

```
<%@ page language="java" contentType="text/html; charset=UTF-8"
pageEncoding="UTF-8"%>
<!DOCTYPE html>
<html>
<head>
<meta charset="UTF-8">
<title>Insert title here</title>
</head>
<body>
    <form action="loginServlet" method="post">
        用户名：<input type="text" name="username">
        <input type="submit" value="登录">
    </form>
```

```
</body>
</html>
```

❷ 创建登录与注销 Servlet

在 ch8SS 的 src 目录下，创建一个名为 servlet 的包，并在该包中创建 LoginServlet 和 LogoutServlet。

LoginServlet 的代码如下：

```java
package servlet;
import java.io.IOException;
import java.io.PrintWriter;
import java.util.List;
import javax.servlet.ServletException;
import javax.servlet.annotation.WebServlet;
import javax.servlet.http.HttpServlet;
import javax.servlet.http.HttpServletRequest;
import javax.servlet.http.HttpServletResponse;
@WebServlet("/loginServlet")
public class LoginServlet extends HttpServlet {
    private static final long serialVersionUID = 1L;
    public void doGet(HttpServletRequest request, HttpServletResponse
    response) throws ServletException, IOException {
        this.doPost(request, response);
    }
    public void doPost(HttpServletRequest request, HttpServletResponse
    response) throws ServletException, IOException {
        request.setCharacterEncoding("utf-8");
                                    //获取请求参数中的用户名
        String username = request.getParameter("username");
//向session中添加属性,会触发HttpSessionAttributeListener 中的
//attributeAdded方法
        if (username != null && !username.equals("")) {
            request.getSession().setAttribute(username, username);
        }
        //从应用上下文中获取在线用户名列表
        List<String> online = (List<String>) getServletContext().
        getAttribute("online");
        response.setContentType("text/html;charset=utf-8");
        PrintWriter out = response.getWriter();
        out.println("");
        out.println("  <title>用户列表</title>");
        out.println("  ");
        out.println("当前用户是: " + username);
        out.print("    <hr><h3>在线用户列表</h3>");
        int size = online == null ? 0 : online.size();
        for (int i = 0; i < size; i++) {
            if (i > 0) {
```

```java
                out.println("<br>");
            }
            out.println(i + 1 + "." + online.get(i));
        }
        //注意:要对链接URL进行自动重写处理
        out.println("<hr/><a href=\"" + response.encodeURL("logoutServlet?id=" + username) + "\">注销</a>");
        out.println("  ");
        out.println("");
        out.flush();
        out.close();
    }
}
```

LogoutServlet 的代码如下:

```java
package servlet;
import java.io.IOException;
import java.io.PrintWriter;
import java.util.List;
import javax.servlet.ServletException;
import javax.servlet.annotation.WebServlet;
import javax.servlet.http.HttpServlet;
import javax.servlet.http.HttpServletRequest;
import javax.servlet.http.HttpServletResponse;
@WebServlet("/logoutServlet")
public class LogoutServlet extends HttpServlet {
    private static final long serialVersionUID = 1L;
    public void doGet(HttpServletRequest request, HttpServletResponse response)
            throws ServletException, IOException {
        this.doPost(request, response);
    }
    public void doPost(HttpServletRequest request, HttpServletResponse response)
            throws ServletException, IOException {
        request.setCharacterEncoding("utf-8");
        String username = request.getParameter("id");//即将注销的当前用户
        //从应用上下文中获取在线用户名列表
        List<String> online = (List<String>)getServletContext().
        getAttribute("online");
        online.remove(username);
        //销毁会话
        request.getSession().invalidate();
        response.setContentType("text/html;charset=utf-8");
        PrintWriter out = response.getWriter();
        out.println("");
        out.println("  <title>用户列表</title>");
        out.println("  ");
        out.print("    <h3>在线用户列表</h3>");
        int size = online == null ? 0 : online.size();
        for (int i = 0; i < size; i++) {
            if(i > 0){
```

```
                out.println("<br>");
            }
            out.println(i + 1 + "." + online.get(i));
        }
        out.println("<hr><a href=\"index.jsp\">主页</a>");
        out.println("  ");
        out.println("");
        out.flush();
        out.close();
    }
}
```

❸ 创建监听器

在 ch8SS 的 src 目录下,创建一个名为 listener 的包,并在该包中创建 OnlineListener 监听器,代码如下:

```
package listener;
import java.util.ArrayList;
import java.util.List;
import javax.servlet.ServletContext;
import javax.servlet.ServletContextEvent;
import javax.servlet.ServletContextListener;
import javax.servlet.http.HttpSessionAttributeListener;
import javax.servlet.http.HttpSessionBindingEvent;
public class OnlineListener implements ServletContextListener,
HttpSessionAttributeListener{
    private ServletContext application = null;
    /**
     * 应用上下文初始时会回调的方法
     */
    @Override
    public void contextInitialized(ServletContextEvent e) {
        //初始化一个application对象
        application = e.getServletContext();
        //设置一个列表属性,用于保存在线用户名
        this.application.setAttribute("online", new ArrayList<String>());
    }
    /**
     * 向会话中添加属性时的回调方法
     */
    @Override
    public void attributeAdded(HttpSessionBindingEvent e) {
        //取得用户名列表
        List<String> onlines = (List<String>) this.application.getAttribute
        ("online");
        onlines.add((String) e.getValue());
        //将添加后的列表重新设置到application属性中
        this.application.setAttribute("online", onlines);
    }
```

❹ 部署监听器

配置文件 web.xml 的代码如下:

```xml
<?xml version="1.0" encoding="UTF-8"?>
<web-app xmlns:xsi="http://www.w3.org/2001/XMLSchema-instance"
    xmlns="http://xmlns.jcp.org/xml/ns/javaee"
    xsi:schemaLocation="http://xmlns.jcp.org/xml/ns/javaee http://xmlns.jcp.org/xml/ns/javaee/web-app_3_1.xsd"
    id="WebApp_ID" version="3.1">
    <!--监听器 -->
    <listener>
        <listener-class>listener.OnlineListener</listener-class>
    </listener>
</web-app>
```

❺ 发布并测试应用

将 ch8SS 发布到 Tomcat 服务器，并启动服务器。在浏览器地址栏中可通过"http://localhost:8080/ch8SS/"测试该应用。

注意：关闭浏览器并不能减少在线人数，需要通过"注销"链接减少在线人数。

8.6 本章小结

本章讲解了过滤器与监听器的概念、原理以及实际应用。

过滤器是 Servlet 的一种特殊用法，完成通用的操作，如编码的过滤、事务的提交、用户的登录状态等。过滤器能够在客户端请求到达 Servlet 之前被截获，处理之后再发给被请求的资源；还可以响应截获，修改之后再发给用户。拦截分为处理前拦截和处理后拦截。

监听器监听客户端的请求、服务端的操作，监听器能够自动激发一些操作。监听器是 Web 开发的重要组成部分，对 Web 应用进行监督和控制，大大增强了 Web 应用的时间处理能力。

习题 8

1. 简述过滤器的运行原理。
2. Filter 接口中有哪些方法？它们分别具有什么功能？
3. 过滤器可以实现哪些常用功能？
4. 简述监听器的分类。
5. 哪种监听器不需要部署？

第9章 EL 与 JSTL

学习目的与要求

本章主要介绍表达式语言（Expression Language，EL）和 JSP 标准标签库（Java Server Pages Standard Tag Library，JSTL）的基本用法。通过本章的学习，读者能够掌握 EL 表达式语法，掌握 EL 隐含对象，了解什么是 JSTL，掌握 JSTL 的核心标签库。

本章主要内容

- EL
- JSTL

在 JSP 页面中可以使用 Java 代码来实现页面显示逻辑，但网页中夹杂着 HTML 与 Java 代码，给网页的设计与维护带来困难。可以使用 EL 来访问和处理应用程序的数据，也可以使用 JSTL 来替换网页中实现页面显示逻辑的 Java 代码。这样，JSP 页面就尽量减少了 Java 代码的使用，为以后的维护提供了方便。

本章涉及的 Java 源文件保存在工程 ch9 的 src 中，涉及的 JSP 页面保存在工程 ch9 的 WebContent 中。

9.1 EL

视频讲解

9.1.1 EL 简介

EL 是 JSP 2.0 规范中增加的语言，它的基本语法为：

```
${表达式}
```

EL 表达式类似于 JSP 表达式<%=表达式%>，EL 语句中的表达式值会被直接送到浏览器显示。通过 page 指令的 isELIgnored 属性来说明是否支持 EL 表达式。isELIgnored 属性值为 false 时，JSP 页面可以使用 EL 表达式；isELIgnored 属性值为 true 时，JSP 页面不能使用 EL 表达式（被当成字符串直接输出）。isELIgnored 属性值默认为 false。

EL 表达式语言最大的优势是可以方便地访问 JSP 的内置对象和 JavaBean 组件，完成使用<%%>或<%=%>完成的功能，使 JSP 页面从 HTML 代码中嵌入 Java 代码的混乱结构得以改善，提高了程序的可读性和易维护性。总之，EL 表达式具有如下特点：

① 可以访问 JSP 的内置对象（pageContext、request、session 以及 application 等）。
② 简化了对 JavaBean 以及集合的访问方式。
③ 可以对数据类型自动转换。
④ 可以通过各种运算符进行运算。
⑤ 可以使用自定义函数实现更加复杂的业务功能。

9.1.2 EL 语法

EL 的语法简单，使用方便。它以"${"开始，以"}"结束。

❶ EL 中的常量

EL 表达式中的常量包括布尔、整型、浮点数、字符串以及 NULL 等，例如${true}、${200}、${16.7}、${"hello"}、${null}。

❷ EL 中的"[ ]"与"."操作符

EL 使用"[ ]"和"."操作符来访问数据，主要使用 EL 获取对象的属性，包括获取 JavaBean 的属性值、获取数组中的元素以及获取集合对象中的元素。

1）获取 JavaBean 的属性值

假设在 JSP 页面中有这样一句话：

```
<jsp:getProperty property="age" name="user"/>
```

那么，可以使用 EL 获取 user 的属性 age，代码如下：

```
${user.age}
```

或

```
${user["age"]}
```

其中，点运算符前面为 JavaBean 对象 user，后面为该对象的属性 age，表示利用 user 对象的 getAge()方法取值并显示在网页上。

2）获取数组中的元素

假设在 JSP 页面或 Servlet 中有这样一段话：

```
String dogs[] = {"lili","huahua","guoguo"};
request.setAttribute("array", dogs);
```

那么，在 JSP 中可以使用 EL 取出数组中的元素，代码如下：

```
${array[0]}
```

```
${array[1]}
${array[2]}
```

3）获取集合对象中的元素

假设在 JSP 页面或 Servlet 中有这样一段话：

```
ArrayList<UserBean> users = new ArrayList<UserBean>();
UserBean ub1 = new UserBean("zhang",20);
UserBean ub2 = new UserBean("zhao",50);
users.add(ub1);
users.add(ub2);
request.setAttribute("array", users);
```

其中，UserBean 有两个属性：name 和 age，那么在 JSP 页面中可以使用 EL 取出 UserBean 中的属性，代码如下：

```
${array[0].name}  ${array[0].age}
${array[1].name}  ${array[1].age}
```

❸ EL 的错误处理机制

EL 为 JSP 页面的错误处理提供了比较友好的处理方式：不提供警告，只提供默认值和错误，默认值是空字符串，错误就是抛出一个异常。EL 的错误处理机制具体如下。

① 对于 null 值直接以空字符串显示，而不是 null，运算时也不会发生错误或空指针异常。所以在使用 EL 访问对象的属性时，无须判断对象是否为 null 对象。

② 在 EL 中访问一个存在对象的不存在属性，EL 表达式会抛出 PropertyNotFoundException 异常。

9.1.3 EL 运算符

可以在 EL 表达式中使用各种运算符。

❶ 算术运算符

在 EL 表达式中有 5 个算术运算符，如表 9.1 所示。

表 9.1 EL 的算术运算符

算术运算符	说明	示例	结果
+	加	${13+2}	15
-	减	${13-2}	11
*	乘	${13*2}	26
/（或 div）	除	${13/2} 或 ${13 div 2}	6.5
%（或 mod）	取模（求余）	${13%2} 或 ${13 mod 2}	1

❷ 关系运算符

在 EL 表达式中有 6 个关系运算符，如表 9.2 所示。

表 9.2　EL 的关系运算符

关系运算符	说　　明	示　　例	结　　果
==（或 eq）	等于	${13 == 2}或${13 eq 2}	false
!=（或 ne）	不等于	${13 != 2}或${13 ne 2}	true
<（或 lt）	小于	${13 < 2}或${13 lt 2}	false
>（或 gt）	大于	${13 > 2}或${13 gt 2}	true
<=（或 le）	小于或等于	${13 <= 2}或${13 le 2}	false
>=（或 ge）	大于或等于	${13 >= 2}或${13 ge 2}	true

❸ 逻辑运算符

在 EL 表达式中有 3 个逻辑运算符，如表 9.3 所示。

表 9.3　EL 的逻辑运算符

逻辑运算符	说　　明	示　　例	结　　果
&&（或 and）	逻辑与	如果 A 为 true，B 为 false，则 A && B（或 A and B）	false
‖（或 or）	逻辑或	如果 A 为 true，B 为 false，则 A ‖ B（或 A or B）	true
!（或 not）	逻辑非	如果 A 为 true，则!A（或 not A）	false

❹ empty 运算符

empty 运算符用于检测一个值是否为 null。例如，变量 A 不存在，则${empty A}返回的结果为 true。

❺ 条件运算符

EL 中的条件运算符是"? :"。例如，${A ? B:C}，如果 A 为 true，计算 B 并返回其结果；如果 A 为 false，计算 C 并返回其结果。

9.1.4　EL 隐含对象

EL 隐含对象共有 11 个，本书只介绍几个常用的 EL 隐含对象：pageScope、requestScope、sessionScope、applicationScope、param 和 paramValues。

❶ 与作用范围相关的隐含对象

与作用范围有关的 EL 隐含对象有 pageScope、requestScope、sessionScope 和 applicationScope，分别可以获取 JSP 内置对象 pageContext、request、session 和 application 中的数据。如果在 EL 中没有使用隐含对象指定作用范围，则会依次从 page、request、session、application 范围查找，找到就直接返回，不再继续找下去；如果查找所有范围都没有找到，就返回空字符串。获取数据的格式如下：

```
${EL 隐含对象.关键字对象.属性}
```

或

```
${EL 隐含对象.关键字对象}
```

例如：

```
<jsp:useBean id="user" class="bean.UserBean" scope="page"/>
```

第 9 章　EL 与 JSTL

```
<jsp:setProperty name="user" property="name" value="EL 隐含对象" />
name: ${pageScope.user.name}
```

再如，在 JSP 页面或 Servlet 中有这样一段话：

```
ArrayList<UserBean> users = new ArrayList<UserBean>();
UserBean ub1 = new UserBean("zhang",20);
UserBean ub2 = new UserBean("zhao",50);
users.add(ub1);
users.add(ub2);
request.setAttribute("array", users);
```

其中，UserBean 有两个属性：name 和 age，那么在 request 有效的范围内可以使用 EL 取出 UserBean 的属性，代码如下：

```
${requestScope.array[0].name}    ${requestScope.array[0].age}
${requestScope.array[1].name}    ${requestScope.array[1].age}
```

❷ 与请求参数相关的隐含对象

与请求参数相关的 EL 隐含对象有 param 和 paramValues。获取数据的格式如下：

```
${EL 隐含对象.参数名}
```

例如，input.jsp 的代码如下：

```
<form method = "post" action = "param.jsp">
    <p>姓名：<input type="text" name="username" size="15" /></p>
    <p>兴趣：
    <input type="checkbox" name="habit" value="看书"/>看书
    <input type="checkbox" name="habit" value="玩游戏"/>玩游戏
    <input type="checkbox" name="habit" value="旅游"/>旅游
    <p>
    <input type="submit" value="提交"/>
</form>
```

那么，在 param.jsp 页面中可以使用 EL 获取参数值，代码如下：

```
<%request.setCharacterEncoding("UTF-8");%>
<body>
<h2>EL 隐含对象 param、paramValues</h2>
姓名：${param.username}</br>
兴趣：
${paramValues.habit[0]}
${paramValues.habit[1]}
${paramValues.habit[2]}
```

例9-1　编写一个 Servlet，在该 Servlet 中使用 request 对象存储数据，然后从该 Servlet 转发到 show.jsp 页面，在 show.jsp 页面中显示 request 对象的数据。首先，运行 Servlet，在 IE 地址栏中输入"http://localhost:8080/ch9/saveServlet"，程序运行结果如图 9.1 所示。

图 9.1 使用 EL 内置对象获取 JSP 内置对象的数据

SaveServlet.java 的代码如下：

```java
package servlet;
import java.io.IOException;
import javax.servlet.RequestDispatcher;
import javax.servlet.ServletException;
import javax.servlet.annotation.WebServlet;
import javax.servlet.http.HttpServlet;
import javax.servlet.http.HttpServletRequest;
import javax.servlet.http.HttpServletResponse;
@WebServlet("/saveServlet")
public class SaveServlet extends HttpServlet {
    private static final long serialVersionUID = 1L;
    protected void doGet(HttpServletRequest request,
            HttpServletResponse response) throws ServletException,
            IOException {
        doPost(request, response);
    }
    protected void doPost(HttpServletRequest request,
            HttpServletResponse response) throws ServletException,
            IOException {
        String names[] = { "zhao", "qian", "sun", "li" };
        request.setAttribute("name", names);
        RequestDispatcher dis = request.getRequestDispatcher("show.jsp");
        dis.forward(request, response);
    }
}
```

页面文件 show.jsp 的代码如下：

```jsp
<%@ page language="java" contentType="text/html; charset=UTF-8"
pageEncoding="UTF-8"%>
<!DOCTYPE html>
<html>
<head>
<meta charset="UTF-8">
<title>EL 隐含对象</title>
</head>
<body>
```

```
从 Servlet 转发过来的 request 内置对象的数据如下：<br>
    ${requestScope.name[0]}<br>
    ${requestScope.name[1]}<br>
    ${requestScope.name[2]}<br>
    <!-- requestScope 可以省略 -->
    ${name[3]}<br>
</body>
</html>
```

9.2 JSTL

视频讲解

9.2.1 JSP 标准标签库 JSTL

JSTL 规范由 Sun 公司制定，Apache 的 Jakarta 小组负责实现。JSTL 标准标签库由 5 个不同功能的标签库组成，包括 Core、I18N、XML、SQL 和 Functions，本节简要介绍 JSTL 的 Core 和 Functions 标签库中几个常用的标签。

❶ 配置 JSTL

JSTL 现在已经是 Java EE5 的一个组成部分了，如果采用支持 Java EE5 或以上版本的集成开发环境开发 Web 应用程序时，就不需要再配置 JSTL 了。但本书采用的是 Eclipse 平台，因此需要配置 JSTL。配置 JSTL 的步骤如下。

1）复制 JSTL 的标准实现

在 Tomcat 的\webapps\examples\WEB-INF\lib 目录下，找到 taglibs-standard-impl-1.2.5.jar 和 taglibs-standard-spec-1.2.5.jar 文件，然后复制到 Web 工程的 WEB-INF\lib 目录下。

2）使用 taglib 标记定义前缀与 uri 引用

如果使用 Core 核心标签库，首先需要在 JSP 页面中使用 taglib 标记定义前缀与 uri 引用，代码如下：

```
<%@ taglib prefix="c" uri="http://java.sun.com/jsp/jstl/core"%>
```

如果使用 Functions 函数标签库，首先需要在 JSP 页面中使用 taglib 标记定义前缀与 uri 引用，代码如下：

```
<%@ taglib prefix="fn" uri="http://java.sun.com/jsp/jstl/functions"%>
```

❷ 核心标签库之通用标签

1）<c:out>标签

<c:out>用来显示数据的内容，与 <%= 表达式 %> 或${表达式}类似。格式如下：

```
<c:out value="输出的内容" [default="defaultValue"]/>
```

或

```
<c:out value="输出的内容">
    defaultValue
</c:out>
```

其中，value 值可以是一个 EL 表达式，也可以是一个字符串；default 可有可无，当 value 值不存在时，就输出 defaultValue。例如：

```
<c:out value="${param.data}" default="没有数据" />
<br>
<c:out value="${param.nothing}" />
<br>
<c:out value="这是一个字符串" />
```

输出的结果如图 9.2 所示。

图 9.2 <c:out>标签

2）<c:set>标签

（1）设置作用域变量。

可以使用<c:set>在 page、request、session、application 等范围内设置一个变量。格式如下：

```
<c:set value="value" var="varName" [scope="page|request|session|application"]/>
```

将 value 值赋值给变量 varName。例如：

```
<c:set value="zhao" var="userName" scope="session"/>
```

相当于

```
<% session.setAttribute("userName","zhao"); %>
```

（2）设置 JavaBean 的属性。

使用<c:set>设置 JavaBean 的属性时，必须使用 target 属性进行设置。格式如下：

```
<c:set value="value" target="target"  property="propertyName"/>
```

将 value 赋值给 target 对象（JavaBean 对象）的 propertyName 属性。如果 target 为 null 或没有 set 方法，则抛出异常。

3）<c:remove>标签

如果要删除某个变量，可以使用<c:remove>标签。例如：

```
<c:remove var="userName" scope="session"/>
```

相当于

```
<%session.removeAttribute("userName") %>
```

第 9 章　EL 与 JSTL

❸ **核心标签库之流程控制标签**

1）<c:if>标签

<c:if>标签实现 if 语句的作用，具体语法格式如下：

```
<c:if test="条件表达式">
    主体内容
</c:if>
```

其中，条件表达式可以是 EL 表达式，也可以是 JSP 表达式。如果表达式的值为 true，则会执行<c:if>的主体内容，但是没有相对应的<c:else>标签。如果想在条件成立时执行一块内容，不成立时执行另一块内容，则可以使用<c:choose>、<c:when>及<c:otherwise>标签。

2）<c:choose>、<c:when>及<c:otherwise>标签

<c:choose>、<c:when>及<c:otherwise>标签实现 if/elseif/else 语句的作用。具体语法格式如下：

```
<c:choose>
    <c:when test="条件表达式 1">
        主体内容 1
    </c:when>
    <c:when test="条件表达式 2">
        主体内容 2
    </c:when>
    <c:otherwise>
        表达式都不正确时，执行的主体内容
    </c:otherwise>
</c:choose>
```

例 9-2　编写一个 JSP 页面 ifelse.jsp，在该页面中使用<c:set>标签将两个字符串设置为 request 范围内的变量。使用<c:if>标签求出这两个字符串的最大值（按字典顺序比较大小），使用<c:choose>、<c:when>及<c:otherwise>标签求出这两个字符串的最小值。

页面文件 ifelse.jsp 的代码如下：

```
<%@ page language="java" contentType="text/html; charset=UTF-8"
pageEncoding="UTF-8"%>
<%@ taglib prefix="c" uri="http://java.sun.com/jsp/jstl/core"%>
<!DOCTYPE html>
<html>
<head>
<meta charset="UTF-8">
<title>ifelse.jsp</title>
</head>
<body>
    <c:set value="if" var="firstNumber" scope="request" />
```

```
        <c:set value="else" var="secondNumber" scope="request" />
        <c:if test="${firstNumber>secondNumber}">
            最大值为${firstNumber}
        </c:if>
        <c:if test="${firstNumber<secondNumber}">
            最大值为${secondNumber}
        </c:if>
        <c:choose>
            <c:when test="${firstNumber<secondNumber}">
                最小值为${firstNumber}
            </c:when>
            <c:otherwise>
                最小值为${secondNumber}
            </c:otherwise>
        </c:choose>
    </body>
</html>
```

<c:when>及<c:otherwise>必须放在<c:choose>中。当<c:when>的test结果为true时，会输出<c:when>的主体内容，而不理会<c:otherwise>的内容。<c:choose>中可有多个<c:when>，程序会从上到下进行条件判断，如果有个<c:when>的test结果为true，就输出其主体内容，之后的<c:when>就不再执行。如果所有的<c:when>的test结果都为false时，则会输出<c:otherwise>的内容。<c:if>与<c:choose>也可以嵌套使用，例如：

```
<c:set value="fda" var="firstNumber" scope="request"/>
<c:set value="else" var="secondNumber" scope="request"/>
<c:set value="ddd" var="threeNumber" scope="request"/>
<c:if test="${firstNumber>secondNumber}">
    <c:choose>
        <c:when test="${firstNumber<threeNumber}">
            最大值为${threeNumber}
        </c:when>
        <c:otherwise>
            最大值为${firstNumber}
        </c:otherwise>
    </c:choose>
</c:if>
<c:if test="${secondNumber>firstNumber}">
    <c:choose>
        <c:when test="${secondNumber<threeNumber}">
            最大值为${threeNumber}
        </c:when>
        <c:otherwise>
            最大值为${secondNumber}
        </c:otherwise>
```

```
    </c:choose>
</c:if>
```

❹ **核心标签库之迭代标签**

1）<c:forEach>标签

<c:forEach>标签可以实现程序中的 for 循环。语法格式如下：

```
<c:forEach var="变量名" items="数组或 Collection 对象">
    循环体
</c:forEach>
```

其中，items 属性可以是数组或 Collection 对象，每次循环读取对象中的一个元素，并赋值给 var 属性指定的变量，之后就可以在循环体使用 var 指定的变量获取对象的元素。例如，在 JSP 页面或 Servlet 中有这样一段代码：

```
ArrayList<UserBean> users = new ArrayList<UserBean>();
UserBean ub1 = new UserBean("zhao",20);
UserBean ub2 = new UserBean("qian",40);
UserBean ub3 = new UserBean("sun",60);
UserBean ub4 = new UserBean("li",80);
users.add(ub1);
users.add(ub2);
users.add(ub3);
users.add(ub4);
request.setAttribute("usersKey", users);
```

那么，在 JSP 页面中可以使用<c:forEach>循环遍历出数组中的元素。代码如下：

```
<table>
  <tr>
    <th>姓名</th>
    <th>年龄</th>
  </tr>
<c:forEach var="user" items="${requestScope.usersKey}">
  <tr>
    <td>${user.name}</td>
    <td>${user.age}</td>
  </tr>
</c:forEach>
</table>
```

有些情况下，需要为<c:forEach>标签指定 begin、end、step 和 varStatus 属性。begin 为迭代时的开始位置，默认值为 0；end 为迭代时的结束位置，默认值是最后一个元素；step 为迭代步长，默认值为 1；varStatus 代表迭代变量的状态，包括 count（迭代的次数）、index（当前迭代的索引，第一个索引为 0）、first（是否是第一个迭代对象）和 last（是否是最后一个迭代对象）。例如：

```
<table border=1>
    <tr>
        <th>Value</th>
        <th>Square</th>
        <th>Index</th>
    </tr>
    <c:forEach var="x" varStatus="status" begin="0" end="10" step="2">
        <tr>
            <td>${x}</td>
            <td>${x * x}</td>
            <td>${status.index}</td>
        </tr>
    </c:forEach>
</table>
```

上述程序运行结果如图9.3所示。

Value	Square	Index
0	0	0
2	4	2
4	16	4
6	36	6
8	64	8
10	100	10

图9.3 <c:forEach>标签

2）<c:forTokens>标签

<c:forTokens>用于迭代字符串中由分隔符分隔的各成员，它是通过 java.util.StringTokenizer 实例来完成字符串的分隔，属性 items 和 delims 作为构造 StringTokenizer 实例的参数。语法格式如下：

```
<c:forTokens var="变量名" items="要迭代的String对象" delims="指定分隔字符串的分隔符">
        循环体
</c:forTokens>
```

例如：

```
<c:forTokens items="chenheng:lulinlin:liuhaiyan" delims=":" var="name">
    ${name}<br>
</c:forTokens>
```

上述程序运行结果如图9.4所示。

第 9 章 EL 与 JSTL

chenheng
lulinlin
liuhaiyan

图 9.4 <c:forTokens>标签

<c:forTokens>标签与<c:forEach>标签一样，也有 begin、end、step 和 varStatus 属性，并且用法一样，这里就不再赘述。

❺ 核心标签库之 URL 标签

JSTL 中提供了<c:url>、<c:redirect>、<c:import>等与 URL 操作相关的标签。

1）<c:url>

<c:url>标签用于在 JSP 页面中构造一个 URL 地址，其语法格式如下：

```
<c:url value="指定要构造的URL"
    [var="varName"][scope="{page|request|session|application}"]
    [context="context"]>
    [<c:param name="paramName" value="paramValue"/>]
</c:url>
```

其中，value 指定要构造的 URL；var 指定构造的 URL 保存到范围域中的属性名称；scope 指定构造的 URL 结果保存到哪个范围域中；context 指定 URL 地址所属的同一 Web 容器下的 Web 应用上下文；<c:param>标签向指定的 URL 地址传递参数；[]表示可选。示例代码如下：

```
<c:url value="goto.jsp?username=chenheng&pwd=123" var="myURL"/>
<a href="${myURL}">去看看</a>
```

或

```
<c:url value="/goto.jsp" var="myURL" context="/ch9" scope="page">
      <c:param name="username" value="chenheng"/>
      <c:param name="pwd" value="123"/>
</c:url>
<a href="${myURL}">去看看</a>
```

上述示例代码中，如果使用 context 属性指定应用上下文，则 value 属性值必须是以"/"开头的相对地址。

2）<c:redirect>

<c:redirect>标签用于执行 response.sendRedirect()方法的功能，将当前访问请求重定向到其他资源，其语法格式如下：

```
<c:redirect url="重定向的URL地址" [context="context"]>
      [<c:param name="paramName" value="paramValue"/>]
</c:redirect>
```

其中，url 指定重定向的目标资源的 URL 地址；context 指定重定向地址所属的同一 Web 容器下的 Web 应用上下文；<c:param>标签向指定的 URL 地址传递参数；[]表示可选。示例代码如下：

```
<c:url value="goto.jsp?username=chenheng&pwd=123" var="myURL"/>
<c:redirect url="${myURL}"/>
```

3）<c:import>

<c:import>标签用于在 JSP 页面中导入一个 URL 地址指向的资源内容，可以是一个静态或动态文件，可以是当前应用或任何其他资源。语法格式如下：

```
<c:import url="url"
    [var="varName"][scope="{page|request|session|application}"]
    [context="context"][charEncoding="charEncoding"]>
    [<c:param name="paramName" value="paramValue"/>]
</c:import>
```

其中，url 指定要导入资源的 URL 地址；var 指定导入资源保存到范围域中的属性名称；scope 指定导入资源保存到哪个范围域中（默认为 page）；context 指定导入资源所属的同一 Web 容器下的 Web 应用上下文（默认为当前应用）；charEncoding 指定将导入资源内容转换成字符串时所使用的字符集编码；<c:param>标签向指定的导入资源文件传递参数；[]表示可选。要想在某页面中导入 head.jsp 并向 head.jsp 传递参数 id，示例代码如下：

```
<c:import url="head.jsp?id=1"/>
```

或

```
<c:import url="/head.jsp" context="/ch9" var="importURL" scope="page"
charEncoding="UTF-8">
    <c:param name="id" value="1"/>
</c:import>
```

<c:import>标签与<jsp:include>动作指令功能类似，但后者只能包含当前应用下的文件资源，而<c:import>标签可以包含任何其他应用或网站下的资源，例如<c:import url="http://www.sohu.com"/>。

❻ 函数标签库

在 JSP 页面中，调用 JSTL 中的函数时，需要使用 EL 表达式，调用语法格式如下：

```
${fn:函数名(参数1，参数2，…)}
```

下面介绍几个常用的函数。

1）contains 函数

contains 函数功能是判断一个字符串中是否包含指定的子字符串。如果包含，则返回 true，否则返回 false。其定义如下：

```
contains(string, substring)
```

该函数调用示例代码如下:

```
${fn:contains("I am studying", "am") }
```

上述 EL 表达式将返回 true。

2) containsIgnoreCase 函数

containsIgnoreCase 函数与 contains 函数功能相似,但判断是不区分大小写的。其定义如下:

```
containsIgnoreCase(string, substring)
```

该函数调用示例代码如下:

```
${fn:containsIgnoreCase("I AM studying", "am") }
```

上述 EL 表达式将返回 true。

3) endsWith 函数

endsWith 函数功能是判断一个字符串是否以指定的后缀结尾。其定义如下:

```
endsWith(string, suffix)
```

该函数调用示例代码如下:

```
${fn:endsWith("I AM studying", "am") }
```

上述 EL 表达式将返回 false。

4) indexOf 函数

indexOf 函数功能是返回指定子字符串在某个字符串中第一次出现时的索引,找不到时,将返回-1。其定义如下:

```
indexOf(string, substring)
```

该函数调用示例代码如下:

```
${fn:indexOf("I am studying", "am") }
```

上述 EL 表达式将返回 2。

5) join 函数

join 函数功能是将一个 String 数组中的所有元素合并成一个字符串,并用指定的分隔符分开。其定义如下:

```
join(array, separator)
```

例如,假设一个 String 数组 my,它有 3 个元素:"I""am"和"studying",那么,下列 EL 表达式:

```
${fn:join(my, ",") }
```

将返回"I,am,studying"。

6) length 函数

length 函数功能是返回集合中元素的个数,或者字符串中的字符个数。其定义如下:

```
length(input)
```

该函数调用示例代码如下：

```
${fn:length("aaa")}
```

上述 EL 表达式将返回 3。

7）replace 函数

replace 函数功能是将字符串中出现的所有 beforestring 用 afterstring 替换，并返回替换后的结果。其定义如下：

```
replace(string, beforestring, afterstring)
```

该函数调用示例代码如下：

```
${fn:replace("I am am studying", "am", "do") }
```

上述 EL 表达式将返回"I do do studying"。

8）split 函数

split 函数功能是将一个字符串，使用指定的分隔符 separator 分离成一个子字符串数组。其定义如下：

```
split(string, separator)
```

该函数调用示例代码如下：

```
<c:set var="my" value="${fn:split('I am studying', ' ') }"/>
<c:forEach var="myArrayElement" items="${my }">
    ${myArrayElement}<br>
</c:forEach>
```

上述示例代码显示结果如图 9.5 所示。

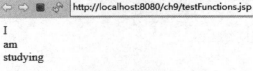

图 9.5 split 示例结果

9）startsWith 函数

startsWith 函数功能是判断一个字符串是否以指定的前缀开头。其定义如下：

```
startsWith(string, prefix)
```

该函数调用示例代码如下：

```
${fn:startsWith("I AM studying", "am") }
```

上述 EL 表达式将返回 false。

10）substring 函数

substring 函数功能是返回一个字符串的子字符串。其定义如下：

```
substring(string, begin, end)
```

该函数调用示例代码如下：

```
${fn:substring("abcdef", 1, 3)}
```

上述 EL 表达式将返回"bc"。

11）toLowerCase 函数

toLowerCase 函数功能是将一个字符串转换成它的小写版本。其定义如下：

```
toLowerCase(string)
```

该函数调用示例代码如下：

```
${fn:toLowerCase("I AM studying") }
```

上述 EL 表达式将返回"i am studying"。

12）toUpperCase 函数

toUpperCase 函数功能与 toLowerCase 函数功能相反，不再赘述。

13）trim 函数

trim 函数功能是将一个字符串开头和结尾的空格去掉。其定义如下：

```
trim(string)
```

该函数调用示例代码如下：

```
${fn:trim("   I AM studying    ") }
```

上述 EL 表达式将返回"I AM studying"。

9.2.2 实践环节——使用 JSTL 标签打印九九乘法表

编写一个 JSP 页面，在该页面中使用<c:forEach>标签打印九九乘法表。页面运行效果如图 9.6 所示。

图 9.6 使用<c:forEach>打印九九乘法表

9.3 本章小结

本章重点介绍了 EL 表达式和 JSTL 核心标签库的用法。 EL 与 JSTL 的应用大大提高了编程效率，并且降低了维护难度。

习题 9

1. 在 Web 应用程序中有以下程序代码段，执行后转发到某个 JSP 页面：

```
ArrayList<String> dogNames = new ArrayList<String>();
dogNames.add("goodDog");
request.setAttribute("dogs", dogNames);
```

下列选项中，（   ）可以正确地使用 EL 取得数组中的值。

  A．${ dogs .0} B．${ dogs [0]} C．${ dogs .[0]} D．${ dogs "0"}

2. JSTL 标签（   ）可以实现 Java 程序中的 if 语句功能。

  A．\<c:set\> B．\<c:out\> C．\<c:forEach\> D．\<c:if\>

3. （   ）不是 EL 的隐含对象。

  A．request   B．pageScope

  C．sessionScope   D．applicationScope

4. JSTL 标签（   ）可以实现 Java 程序中的 for 语句功能。

  A．\<c:set\> B．\<c:out\> C．\<c:forEach\> D．\<c:if\>

5. 与\<jsp:include\>功能类似的 JSTL 标签是（   ）。

  A．\<c:url\> B．\<c:redirect\> C．\<c:include\> D．\<c:import\>

第10章 文件的上传与下载

学习目的与要求

本章重点介绍 Servlet 的 HttpServletRequest 对文件上传的支持。通过本章的学习,读者能够理解文件上传的原理,掌握文件上传与下载的步骤。

本章主要内容

- 基于 Servlet 的文件上传
- 文件的下载

在 JSP 应用程序中实现文件上传有两种常见的组件:jspsmart 组件和 commons-fileupload 组件,但从 Servlet 3.0 版本开始可以直接支持文件上传,而且使用也非常简单。因此,本章仅介绍了基于 Servlet 的文件上传。

本章涉及的 Java 源文件保存在工程 ch10 的 src 中,涉及的 JSP 页面保存在工程 ch10 的 WebContent 中。

10.1 基于 Servlet 的文件上传

视频讲解

文件的上传与下载在 Web 应用程序中是很常见的功能。例如,在 OA 办公系统中,用户可以使用文件上传来提交公文。

Servlet 3.0 之前的版本不能对文件上传直接支持,需要使用第三方框架来实现,而且使用起来也不够简单。从 Servlet 3.0 已经提供了这个功能,而且使用也非常简单。

让 Servlet 支持文件上传,需要做两件事情:

① 给 Servlet 添加@MultipartConfig 注解。

② 从 HttpServletRequest 对象中获取 Part 文件对象。

10.1.1 文件上传的表单设置

表单元素:<input type="file">,在浏览器中会显示一个输入框和一个按钮,输入框可供用户填写本地文件的文件名和路径名，按钮可以让浏览器打开一个文件选择框供用户选择文件。

文件上传的表单例子如下：

```
<form action="upload" method="post" enctype="multipart/form-data">
    <input type="file" name="headImage"><br>
    <input type="submit" value="上传">
</form>
```

不要忘记设置文件上传的表单属性 enctype，并将它的值设置为 multipart/form-data。同时，表单的提交方式设置为 post。

10.1.2　@MultipartConfig 注解

@MultipartConfig 注解主要是为了辅助 Servlet 中 HttpServletRequest 提供对上传文件的支持。该注解标注在 Servlet 上面，以表示该 Servlet 希望处理请求的 MIME 类型是 multipart/form-data。另外，它还提供了若干属性用于简化对上传文件的处理。该注解的常用属性如表 10.1 所示。

表 10.1　@MultipartConfig 常用属性列表

属 性 名	类 型	是否可选	描　　述
location	string	是	指定上传文件存放的目录。当指定了 location 后，在调用 Part 的 write(String fileName)方法将文件写入到磁盘的时候，文件名称可以不用带路径，但是如果 fileName 带了绝对路径，那将以 fileName 所带路径为准将文件写入磁盘
maxFileSize	long	是	指定上传文件的最大值，单位是字节。默认值为-1，表示没有限制
maxRequestSize	long	是	指定上传文件的个数，应用在多文件上传。默认值为-1，表示没有限制

注：即使没有使用@MultipartConfig 注解设置属性，也要将该注解加到 Servlet 的上面。

10.1.3　Part 接口

每一个文件用一个 javax.servlet.http.Part 对象来表示。单个文件上传时，在 Servlet 中可以通过 HttpServletRequest 对象 request 调用方法 getPart(String name)获得 Part 文件对象。其中，参数 name 为文件域的名称。例如：

```
Part photo = request.getPart("resPath");
```

多文件上传时，在 Servlet 中可以通过 HttpServletRequest 对象 request 调用方法 getParts()获得 Part 文件对象集合。例如：

```
Collection<Part> photos = request.getParts();
```

Part 接口的常用方法如表 10.2 所示。

表 10.2 Part 接口的常用方法

方　　法	功　能　说　明
void delete()	删除任何相关的临时磁盘文件
String getContentType()	获得客户端浏览器设置的文件数据项的 MIME 类型
String getHeader(String name)	获得指定的 part 头的一个字符串。例如：getHeader ("content-disposition") 返回 form-data; name="xxx"; filename="xxx"
InputStream getInputStream( )	获得一个输入流，通过这个输入流来读取文件的内容
String getName( )	获得表单文件域的名称
long getSize( )	获得文件的大小
void write(String fileName)	将文件上传到 fileName 指定的目录中

10.1.4　单文件上传

Servlet 规范中不提供获取上传文件名的方法，但可以通过 Part 对象调用 getHeader("content-disposition")方法间接获取。根据 Part 对象获取文件名的方法如下：

```
public static String getFileName(Part part){
    if(part == null)
        return null;
    //fileName 形式为：form-data; name="resPath"; filename="20140920_
    //110531.jpg"
    String fileName = part.getHeader("content-disposition");
    //没有选择文件
    if(fileName.lastIndexOf("=") + 2 == fileName.length() - 1)
        return null;
    return fileName.substring(fileName.lastIndexOf("=") + 2, fileName.
    length() - 1);
}
```

将 getFileName(Part part)方法添加到本章的工具类 MyUtil 中，方便调用。下面通过例 10-1 讲述单文件上传的步骤。

例 10-1　单文件上传。

❶ 编写 **uploadHttpOne.jsp 页面**

页面 uploadHttpOne.jsp 的运行效果如图 10.1 所示。

图 10.1　uploadHttpOne.jsp 的运行效果

页面文件 uploadHttpOne.jsp 的代码如下：

```jsp
<%@ page language="java" contentType="text/html; charset=UTF-8"
    pageEncoding="UTF-8"%>
<!DOCTYPE html>
<html>
<head>
<meta charset="UTF-8">
<title>HttpServletRequest 对文件上传的支持</title>
</head>
<body>
<form action="uploadHttpOneServlet" method="post" enctype="multipart/form-data">
    <table>
        <tr>
            <td>文件描述: </td>
            <td><input type="text" name="filediscription"></td>
        </tr>
        <tr>
            <td>请选择文件: </td>
            <td><input type="file" name="resPath"></td>
        </tr>
        <tr>
            <td align="right"><input type="reset" value="重填"></td>
            <td><input type="submit" value="上传"></td>
        </tr>
    </table>
</form>
</body>
</html>
```

❷ 编写上传的 Servlet——UploadHttpOneServlet

UploadHttpOneServlet.java 的代码如下:

```java
package servlet;
import java.io.File;
import java.io.IOException;
import java.io.PrintWriter;
import javax.servlet.ServletException;
import javax.servlet.annotation.MultipartConfig;
import javax.servlet.annotation.WebServlet;
import javax.servlet.http.HttpServlet;
import javax.servlet.http.HttpServletRequest;
import javax.servlet.http.HttpServletResponse;
import javax.servlet.http.Part;
import util.MyUtil;
@WebServlet("/uploadHttpOneServlet")
@MultipartConfig(maxFileSize = 10*1024*1024)//设置上传文件的最大值为 10MB
```

```java
public class UploadHttpOneServlet extends HttpServlet {
    private static final long serialVersionUID = 1L;
    protected void doGet(HttpServletRequest request, HttpServletResponse response) throws ServletException, IOException {
        doPost(request, response);
    }
    protected void doPost(HttpServletRequest request, HttpServletResponse response) throws ServletException, IOException {
        //设置响应的内容类型
        response.setContentType("text/html;charset=utf-8");
        //取得输出对象
        PrintWriter out=response.getWriter();
        request.setCharacterEncoding("utf-8");
        //获得 Part 对象
        Part part = request.getPart("resPath");
        String filediscription = request.getParameter("filediscription");
        out.println("输入的文件描述: " + filediscription + "<br>");
        //指定上传的文件保存到服务器的 uploadFiles 目录中
        File uploadFileDir = new File(getServletContext().getRealPath("/uploadFiles"));
        if(!uploadFileDir.exists()){
            uploadFileDir.mkdir();
        }
        //获得原始文件名
        String oldName = MyUtil.getFileName(part);
        out.println("上传文件的原始名: " + oldName + "<br>");
        out.println("上传文件的大小: " + part.getSize() + "<br>");
        if(oldName != null){
            //上传到服务器的 uploadFiles 目录中
            part.write(uploadFileDir + File.separator + oldName);
        }
        out.println("文件上传到: "+uploadFileDir+File.separator+oldName+"<br>");
    }
}
```

UploadHttpOneServlet.java 的响应结果如图 10.2 所示。

输入的文件描述：重要文件
上传文件的原始名：陈恒_参与人模板.doc
上传文件的大小：32768
文件上传到：D:\eclipse-workspace\.metadata\.plugins\org.eclipse.wst.server.core\tmp0\

图 10.2　UploadHttpOneServlet 的响应结果

需要特别注意的是，使用 Eclipse IDE 内嵌的浏览器上传文件时可能会出现 FileNotFoundException 异常，这时使用常规浏览器（如谷歌）上传文件即可。

10.1.5　多文件上传

本节通过例 10-2 讲述多文件上传的步骤。

例 10-2　多文件上传。

❶ 编写 uploadHttpMulti.jsp 页面

页面 uploadHttpMulti.jsp 的运行效果如图 10.3 所示。

图 10.3　uploadHttpMulti.jsp 页面的运行效果

页面文件 uploadHttpMulti.jsp 的代码如下：

```jsp
<%@ page language="java" contentType="text/html; charset=UTF-8"
pageEncoding="UTF-8"%>
<!DOCTYPE html>
<html>
<head>
<meta charset="UTF-8">
<title>HttpServletRequest 对文件上传的支持</title>
</head>
<body>
    <form action="uploadHttpMultiServet" method="post" enctype="multipart/form-data">
        <table>
            <tr>
                <td>文件 1 描述：</td>
                <td><input type="text" name="filediscription1"></td>
            </tr>
            <tr>
                <td>请选择文件 1：</td>
                <td><input type="file" name="resPath1"></td>
            </tr>
            <tr>
                <td>文件 2 描述：</td>
```

```html
                <td><input type="text" name="filediscription2"></td>
            </tr>
            <tr>
                <td>请选择文件 2：</td>
                <td><input type="file" name="resPath2"></td>
            </tr>
            <tr>
                <td align="right"><input type="reset" value="重填"></td>
                <td><input type="submit" value="上传"></td>
            </tr>
        </table>
    </form>
</body>
</html>
```

❷ 编写上传的 Servlet——UploadHttpMultiServlet

UploadHttpMultiServlet.java 的代码如下：

```java
package servlet;
import java.io.File;
import java.io.IOException;
import java.io.PrintWriter;
import java.util.Collection;
import javax.servlet.ServletException;
import javax.servlet.annotation.MultipartConfig;
import javax.servlet.annotation.WebServlet;
import javax.servlet.http.HttpServlet;
import javax.servlet.http.HttpServletRequest;
import javax.servlet.http.HttpServletResponse;
import javax.servlet.http.Part;
import util.MyUtil;
@WebServlet("/uploadHttpMultiServlet")
@MultipartConfig
public class UploadHttpMultiServet extends HttpServlet {
    private static final long serialVersionUID = 1L;
    protected void doGet(HttpServletRequest request, HttpServletResponse response) throws ServletException, IOException {
        doPost(request, response);
    }
    protected void doPost(HttpServletRequest request, HttpServletResponse response) throws ServletException, IOException {
        //设置响应的内容类型
        response.setContentType("text/html;charset=utf-8");
        //取得输出对象
        PrintWriter out=response.getWriter();
        request.setCharacterEncoding("utf-8");
        String filediscription1 = request.getParameter("filediscription1");
```

```java
out.println("输入的文件1描述: " + filediscription1 + "<br>");
String filediscription2 = request.getParameter("filediscription2");
out.println("输入的文件2描述: " + filediscription2 + "<br>");
//指定上传的文件保存到服务器的uploadFiles目录中
File uploadFileDir = new File(getServletContext().getRealPath
("/uploadFiles"));
if(!uploadFileDir.exists()){
    uploadFileDir.mkdir();
}
//获得Part集合
Collection<Part> parts = request.getParts();
for (Part part : parts) {
    //没有选择文件或不是文件域
    if (part == null || !part.getName().contains("resPat")) {
        continue;
    }
    //获得原始文件名
    String oldName = MyUtil.getFileName(part);
    out.println("上传文件的原始名: " + oldName + "<br>");
    out.println("上传文件的大小: " + part.getSize() + "<br>");
    if(oldName != null){
        //上传到服务器的uploadFiles目录中
        part.write(uploadFileDir + File.separator + oldName);
    }
    out.println("文件上传到: " + uploadFileDir + File.separator +
    oldName + "<br>");
}
}
}
```

UploadHttpMultiServlet.java 的响应结果如图 10.4 所示。

```
输入的文件1描述: 重要文件1
输入的文件2描述: 重要文件2
上传文件的原始名: 新建文本文档.txt
上传文件的大小: 1614
文件上传到: D:\eclipse-workspace\.metadata\.plugins\org.eclipse.wst.server.core\tmp0\
上传文件的原始名: 新建文本文档 (2).txt
上传文件的大小: 45
文件上传到: D:\eclipse-workspace\.metadata\.plugins\org.eclipse.wst.server.core\tmp0\
```

图 10.4 UploadHttpMultiServlet 的响应结果

10.1.6 实践环节——上传文件的限定

尝试对 10.1.5 节中多文件上传实例文件大小进行限定。

10.2 文件的下载

视频讲解

10.2.1 文件下载的实现方法

实现文件下载经常有两种方法：一是通过超链接实现下载；一是利用程序编码实现下载。超链接实现下载固然简单，但暴露了下载文件的真实位置，并且只能下载存放在 Web 应用程序所在的目录下的文件。利用程序编码实现下载，可以增加安全访问控制，还可以从任意位置提供下载的数据，可以将文件存放到 Web 应用程序以外的目录中，也可以将文件保存到数据库中。

利用程序实现下载需要设置两个报头：

① Web 服务器需要告诉浏览器其所输出内容的类型不是普通文本文件或 HTML 文件，而是一个要保存到本地的下载文件。设置 Content-Type 的值为 application/x-msdownload。

② Web 服务器希望浏览器不直接处理相应的实体内容，而是由用户选择将相应的实体内容保存到一个文件中，这需要设置 Content-Disposition 报头。该报头指定了接收程序处理数据内容的方式，在 HTTP 应用中只有 attachment 是标准方式，attachment 表示要求用户干预。在 attachment 后面还可以指定 filename 参数，该参数是服务器建议浏览器将实体内容保存到文件中的文件名。

设置报头的示例如下：

```
response.setHeader("Content-Type", "application/x-msdownload" );
response.setHeader("Content-Disposition", "attachment; filename=" + filename);
```

10.2.2 文件下载示例

下面通过一个实例讲述利用程序实现下载的过程。

该实例要求：首先将 10.1 节上传到服务器 uploadFiles 目录中的文件列举到页面，然后单击页面中的下载链接下载文件。具体步骤如下。

❶ 编写列举被下载文件的 Servlet——ShowDownServlet.java

ShowDownServlet.java 的代码如下：

```
package servlet;
import java.io.File;
import java.io.IOException;
import javax.servlet.RequestDispatcher;
import javax.servlet.ServletException;
import javax.servlet.annotation.WebServlet;
import javax.servlet.http.HttpServlet;
import javax.servlet.http.HttpServletRequest;
import javax.servlet.http.HttpServletResponse;
```

```java
@WebServlet("/showDownServlet")
public class ShowDownServlet extends HttpServlet {
    private static final long serialVersionUID = 1L;
    protected void doGet(HttpServletRequest request, HttpServletResponse
    response) throws ServletException, IOException {
        //指定从服务器的uploadFiles目录中下载文件
        File downLoadFileDir = new File(getServletContext().getRealPath
        ("/uploadFiles"));
        //获得目录中的文件
        File[] list = downLoadFileDir.listFiles();
        request.setAttribute("fileList", list);
        RequestDispatcher dis = request.getRequestDispatcher("showInfo.jsp");
        dis.forward(request, response);
    }
    protected void doPost(HttpServletRequest request, HttpServletResponse
    response) throws ServletException, IOException {
        doGet(request, response);
    }
}
```

❷ **编写文件显示页面——showInfo.jsp**

通过地址"http://localhost:8080/ch10/showDownServlet"运行 ShowDownServlet，跳转到 showInfo.jsp 页面，运行效果如图 10.5 所示。

图 10.5 页面 showInfo.jsp 运行效果

页面文件 showInfo.jsp 的代码如下：

```jsp
<%@ page language="java" contentType="text/html; charset=UTF-8"
pageEncoding="UTF-8"%>
<%@ taglib prefix="c" uri="http://java.sun.com/jsp/jstl/core"%>
<!DOCTYPE html>
<html>
<head>
<meta charset="UTF-8">
<title>显示下载文件</title>
</head>
<body>
```

```
    <table border="1">
        <tr bgcolor="LightGreen">
            <th>文件名称</th>
            <th>下载课件</th>
        </tr>
    <c:forEach var="afile" items="${fileList}">
      <tr>
        <!-- afile.name 相当于 file.getName() -->
        <td>${afile.name}</td>
        <td align="center"><a href="downloadServlet?resPath=${afile.name}"
        style="text-decoration:none">下载</a></td>
      </tr>
    </c:forEach>
    </table>
</body>
</html>
```

页面文件 showInfo.jsp 中使用了 JSTL 标签，别忘记导入相关的 JAR 包（参考第 9 章内容）。

❸ **编写下载 Servlet——DownloadServlet.java**

DownloadServlet.java 的代码如下：

```
package servlet;
import java.io.File;
import java.io.FileInputStream;
import java.io.IOException;
import javax.servlet.ServletException;
import javax.servlet.ServletOutputStream;
import javax.servlet.annotation.WebServlet;
import javax.servlet.http.HttpServlet;
import javax.servlet.http.HttpServletRequest;
import javax.servlet.http.HttpServletResponse;
import util.MyUtil;
@WebServlet("/downloadServlet")
public class DownloadServlet extends HttpServlet {
    private static final long serialVersionUID = 1L;
    protected void doGet(HttpServletRequest request, HttpServletResponse response)
            throws ServletException, IOException {
        doPost(request, response);
    }
    protected void doPost(HttpServletRequest request, HttpServletResponse
    response)
            throws ServletException, IOException {
        //要下载的文件路径
        File downLoadFileDir = new File(getServletContext().getRealPath
        ("/uploadFiles"));
```

```java
            String aFileName = null;              //要下载的文件名
            FileInputStream in = null;            //输入流
            ServletOutputStream out = null;       //输出流
            try {
                aFileName = request.getParameter("resPath");
                //设置下载文件使用的报头
                response.setHeader("Content-Type", "application/x-msdownload");
                //MyUtil.toUTF8String(aFileName)解决下载的文件名中中文字符乱码问题
                response.setHeader("Content-Disposition","attachment; filename="
                    + MyUtil.toUTF8String(aFileName));
                //读入文件
                in = new FileInputStream(downLoadFileDir + File.separator +
                    aFileName);
                //得到响应对象的输出流,用于向客户端输出二进制数据
                out = response.getOutputStream();
                out.flush();
                int aRead = 0;
                byte b[] = new byte[1024];
                while ((aRead = in.read(b)) != -1 & in != null) {
                    out.write(b, 0, aRead);
                }
                out.flush();
                in.close();
                out.close();
            } catch (Throwable e) {
                e.printStackTrace();
            }
        }
    }
```

10.2.3 文件名乱码问题

如果下载的文件名中有中文字符,浏览器提示保存的文件名将显示为乱码。要解决这个乱码问题,需要对下载的文件名按照 UTF-8 进行编码。

首先,在本章工具类 MyUtil 中添加一个静态字符编码转换方法,代码如下:

```java
public static String toUTF8String(String str){
    StringBuffer sb = new StringBuffer();
    int len = str.length();
    for(int i = 0; i < len; i++){
        //取出字符中的每个字符
        char c = str.charAt(i);
        //Unicode 码值为 0~255,不作处理
        if(c >= 0 && c <= 255){
            sb.append(c);
```

```java
        }else{//转换为UTF-8编码
            byte b[];
            try {
                b = Character.toString(c).getBytes("UTF-8");
            } catch (UnsupportedEncodingException e) {
                e.printStackTrace();
                b = null;
            }
            //转换为%HH的字符串形式
            for(int j = 0; j < b.length; j ++){
                int k = b[j];
                if(k < 0){
                    k &= 255;
                }
                sb.append("%" + Integer.toHexString(k).toUpperCase());
            }
        }
    }
    return sb.toString();
}
```

然后，修改 10.2.2 节中的 DownloadServlet.java，对要下载的文件名调用 toUTF8String() 方法，代码片段如下：

```
…
aFileName = request.getParameter("resPath");
//设置下载文件使用的报头
response.setHeader("Content-Type", "application/x-msdownload" );
response.setHeader("Content-Disposition", "attachment; filename=" + MyUtil.toUTF8String(aFileName));
//读入文件
…
```

10.3 本章小结

本章主要介绍了基于 Servlet 3.0 上传文件的使用方法，列举了文件上传与下载的实例。

习题 10

1. 基于表单的文件上传需要注意什么？
2. 文件下载的实现方法有哪些？它们的优缺点是什么？

第 11 章 JDBC 数据库访问

学习目的与要求

本章主要介绍在 JSP 程序中如何访问数据库。通过本章的学习，用户能够熟练掌握 JSP 对数据库的操作，包括查询、添加、更新和删除。

本章主要内容

- 使用纯 Java 数据库驱动程序连接 MySQL
- 数据库操作
- 游动查询
- 使用连接池
- 其他典型数据库的连接
- 使用 PreparedStatement 语句访问数据库
- 分页查询
- 使用 CallableStatement 语句访问存储子程序
- Excel 与数据库互访

数据库在 Web 应用中扮演着越来越重要的作用。如果没有数据库，很多重要的应用，像电子商务、搜索引擎等都不可能实现。本章主要介绍在 JSP 程序中如何访问关系数据库，如 Oracle、SQL Server 和 MySQL 等。

本章涉及的 Java 源文件保存在工程 ch11 的 src 中，涉及的 JSP 页面保存在工程 ch11 的 WebContent 中。

11.1 使用纯 Java 数据库驱动程序连接 MySQL

视频讲解

在 JSP 页面中访问数据库，首先要与数据库建立连接，通过连接向数据库发送指令，并获得返回的结果。JDBC 连接数据库有两种常用方式：建立 JDBC-ODBC 桥接器和加载纯 Java 驱动程序。

JDBC（Java DataBase Connectivity）是 Java 平台核心类库的一部分，它由一组标准接口与类组成。经常使用 JDBC 完成三件事：

① 与指定的数据库建立连接；

② 向已连接的数据库发送 SQL 命令；

③ 处理 SQL 命令返回的结果。

ODBC（Open DataBase Connectivity）是由 Microsoft 主导的数据库连接标准，提供了通用的数据库访问平台。但是，使用 ODBC 连接数据库的应用程序移植性较差，因为应用程序所在的计算机必须提供 ODBC。所以，本章仅介绍了使用纯 Java 驱动程序连接数据库。

使用纯 Java 数据库驱动程序连接数据库需要两个步骤：

① 注册纯 Java 数据库驱动程序；

② 和指定的数据库建立连接。

下面以 MySQL 5.x 为例，讲解如何使用纯 Java 数据库驱动程序连接数据库。

❶ 注册纯 Java 数据库驱动程序

每种数据库都配有自己的纯 Java 数据库驱动程序。可从官网"https://dev.mysql.com/downloads/connector/j/"下载 MySQL 的 Java 数据库驱动程序，本书采用的是 mysql-connector-java-5.1.45-bin.jar。

为了连接 MySQL 数据库，需要将 mysql-connector-java-5.1.45-bin.jar 文件复制到 Web 应用程序的/WEB-INF/lib 目录下。另外，本章程序使用了 JSTL 标签，因此需要将 JSTL 标签的相关 JAR 包复制到 Web 应用程序的/WEB-INF/lib 目录下（参考第 9 章）。

然后，通过 java.lang.Class 类的 forName()，动态注册 MySQL 的纯 Java 驱动程序，代码如下：

```
try {
        Class.forName("com.mysql.jdbc.Driver");
} catch (ClassNotFoundException e) {
        e.printStackTrace();
}
```

❷ 和指定的数据库建立连接

和 MySQL 数据库建立连接的代码如下：

```
try {
    Connection con=DriverManager.getConnection("jdbc:mysql://localhost:3306/数据库名?characterEncoding=utf-8","用户名","密码");
} catch (SQLException e) {
    e.printStackTrace();
}
```

MySQL 默认端口号为 3306；"?characterEncoding=utf-8"解决中文字符乱码问题；用户名和密码是访问 MySQL 服务器的用户权限。

为方便测试程序，本章创建了 MySQL 数据库 test，并在 test 中新建数据表 goods。创建表的 SQL 语句如下：

```
DROP TABLE IF EXISTS 'goods';
CREATE TABLE 'goods' (
  'id' int(11) NOT NULL AUTO_INCREMENT,
  'gname' varchar(50) CHARACTER SET utf8 COLLATE utf8_unicode_ci DEFAULT NULL,
  'gprice' double DEFAULT NULL,
  PRIMARY KEY ('id')
) ENGINE=InnoDB AUTO_INCREMENT=1 DEFAULT CHARSET=latin1;
```

例11-1 创建 example11_1.jsp 页面，并在该页面中使用纯 Java 数据库驱动程序连接数据库 test，并查询数据表 goods 中的数据。查询结果如图 11.1 所示。

图 11.1　example11_1.jsp 页面运行效果

页面 example11_1.jsp 的代码如下：

```jsp
<%@ page language="java" contentType="text/html; charset=UTF-8"
    pageEncoding="UTF-8"%>
<%@ page import="java.sql.*"%>
<!DOCTYPE html>
<html>
<head>
<meta charset="UTF-8">
<title>访问MySQL数据库</title>
</head>
<body>
    <%
        Connection con = null;
        Statement st = null;
        ResultSet rs = null;
        //加载驱动
        try {
            Class.forName("com.mysql.jdbc.Driver");
        } catch (ClassNotFoundException e) {
            e.printStackTrace();
        }
        try {
            //建立连接
    con=DriverManager.getConnection("jdbc:mysql://localhost:3306/test?characterEncoding=utf-8","root","root");
            st = con.createStatement();
            //发送查询SQL语句，返回结果集
            rs = st.executeQuery("select * from goods ");
            while(rs.next()){
```

```
                    out.print(rs.getString("id") + "  ");
                    out.print(rs.getString("gname") + "  ");
                    out.print(rs.getString("gprice") + "  ");
                    out.print("<br>");
                }
            } catch (SQLException e) {
                e.printStackTrace();
            }finally{
                rs.close();
                st.close();
                con.close();
            }
        %>
    </body>
</html>
```

由例 11-1 的代码可以看出，访问数据库需要以下几个步骤：

1）导入 java.sql 包

java.sql 包含 Java 操作关系数据库的类和接口，因此在使用 JDBC 的程序中必须引入该包。

2）加载驱动程序

使用 Class 类（java.lang 包）中的方法 forName() 来加载该驱动程序的类定义 com.mysql.jdbc.Driver，从而创建了该驱动程序的一个实例。

3）连接数据库

完成上述操作后，就可以连接一个特定的数据库了。这需要创建 Connection 的一个实例，并使用 DriverManager 的方法 getConnection() 来尝试连接指定的数据库。示例代码如下：

```
con=DriverManager.getConnection("jdbc:mysql://localhost:3306/test?characterEncoding=utf-8","root","root");
```

4）访问数据库

首先用 Connection 的 createStatement() 方法从指定的数据库连接得到一个 Statement 实例，然后该 Statement 实例调用 executeQuery() 方法来执行一条 SQL 语句。示例代码如下：

```
st = con.createStatement();
rs = st.executeQuery("select * from goods");
```

5）处理返回的结果集

ResultSet 对象是 JDBC 中比较重要的一个对象，几乎所有的查询操作都将数据作为 ResultSet 对象返回。处理结果集 ResultSet 对象的代码如下：

```
while(rs.next()){
        out.print(rs.getString("id") + "  ");
        out.print(rs.getString("gname") + "  ");
        out.print(rs.getString("gprice") + "  ");
        out.print("<br>");
}
```

6）关闭数据库连接，释放资源

对数据库的操作完成之后，要及时关闭 ResultSet 对象、Statement 对象和 Connection 对象，从而释放占用的资源。示例代码如下：

```
rs.close();
st.close();
con.close();
```

关闭的顺序由先到后依次为 ResultSet 对象、Statement 对象和 Connection 对象。

11.2 数据库操作

视频讲解

与指定的数据库建立连接后，就可以使用 JDBC 提供的 API 对数据库进行操作，包括查询、新增、更新、删除等。

11.2.1 查询操作

和数据库建立连接后，对数据库表进行查询操作的步骤如下。

❶ 创建 Statement 对象

由已创建的 Connection 对象 con 调用 createStatement()方法创建 Statement 对象，代码如下：

```
Statement smt = con.createStatement();
```

❷ 执行 SQL 查询语句

创建 Statement 对象后，可以使用 Statement 对象调用 executeQuery(String sql)方法执行 SQL 查询语句。该方法返回 ResultSet 对象，代表查询结果。

❸ 处理查询结果

ResultSet 对象是 executeQuery(String sql)方法的返回值，被称为结果集，它代表符合 SQL 查询语句的所有行。

ResultSet 对象自动维护指向其当前数据行的游标。每调用一次 next()方法，游标向下移动一行。最初它位于结果集的第一行之前，因此第一次调用 next()，将把游标置于第一行上，使它成为当前行。随着每次调用 next()，游标向下移动一行，按照从上至下的次序获取 ResultSet 行，实现顺序查询。

ResultSet 对象包含 SQL 查询语句的执行结果。它通过一套 get()方法对这些行中的数据进行访问，即使用 getXxx()方法获得数据。getXxx()方法很多，究竟用哪一个，由列的数据类型来决定。使用 getXxx()方法时，需要注意以下两点：

① 无论列是何种数据类型，总可以使用 getString(int columnIndex)或 getString(String columnName)方法获得列值的字符串表示。

② 使用 getString(int columnIndex)方法查看一行记录时，不允许颠倒顺序，例如，不允许：

```
rs.getString(2);
rs.getString(1);
```

查询操作示例请参考例 11-1。

11.2.2 新增、更新与删除操作

和数据库建立连接后，对数据库表进行新增、更新与删除操作的具体步骤如下。

❶ **创建 Statement 对象**

Statement 对象的创建与查询操作一样，由已创建的 Connection 对象 con 调用 createStatement()方法创建。

❷ **执行 SQL 语句，完成新增、更新与删除操作**

创建 Statement 对象后，可以使用 Statement 对象调用 executeUpdate(String sql)方法执行 SQL 语句，完成新增、更新与删除操作。

executeUpdate(String sql)方法主要用于执行 INSERT、UPDATE 或 DELETE 语句以及 SQL DDL 语句，例如 CREATE TABLE 和 DROP TABLE。该方法返回一个整数（代表被更新的行数）对于 CREATE TABLE 和 DROP TABLE 等不操作行的指令，返回零。

【例】**11-2** 编写两个 JSP 页面：addGoods.jsp 和 showAllGoods.jsp。用户可以在 addGoods.jsp 页面中输入信息，单击"添加"按钮将信息添加到 goods 表中。在 showAllGoods.jsp 页面中显示所有商品信息。在该例中需要编写一个 Servlet（GoodsServlet.java）实现添加和查询记录，还需要实体类 Goods 封装查询的商品信息。页面运行效果如图 11.2 所示。

（a）添加记录

（b）查询记录

图 11.2 例 11-2 的运行效果

页面文件 addGoods.jsp 的代码如下：

```
<%@ page language="java" contentType="text/html; charset=UTF-8"
```

```
pageEncoding="UTF-8"%>
<!DOCTYPE html>
<html>
<head>
<meta charset="UTF-8">
<title>Insert title here</title>
</head>
<body>
    <form action="goodsServlet" method="post">
    <table border="1">
        <tr>
            <td>商品名称:</td>
            <td><input type="text" name="gname"/></td>
        </tr>
        <tr>
            <td>商品价格:</td>
            <td><input type="text" name="gprice"/></td>
        </tr>
        <tr>
            <td><input type="submit" value="添加"></td>
            <td><input type="reset" value="重置"></td>
        </tr>
    </table>
    </form>
</body>
</html>
```

实体类 Goods.java 的代码如下：

```
package entity;
public class Goods {
    private int id;
    private String gname;
    private double gprice;
    //此处省略了 setter 和 getter 方法
}
```

GoodsServlet.java 的代码如下：

```
package servlet;
import java.io.IOException;
import java.sql.Connection;
import java.sql.DriverManager;
import java.sql.ResultSet;
import java.sql.SQLException;
import java.sql.Statement;
import java.util.ArrayList;
```

```java
import javax.servlet.RequestDispatcher;
import javax.servlet.ServletException;
import javax.servlet.annotation.WebServlet;
import javax.servlet.http.HttpServlet;
import javax.servlet.http.HttpServletRequest;
import javax.servlet.http.HttpServletResponse;
import entity.Goods;
@WebServlet("/goodsServlet")
public class GoodsServlet extends HttpServlet {
private static final long serialVersionUID = 1L;
protected void doGet(HttpServletRequest request, HttpServletResponse response) throws ServletException, IOException {
    Connection con = null;
    Statement st = null;
    ResultSet rs = null;
    //将查询结果装到集合ArrayList<Goods>中，并返回页面显示
    ArrayList<Goods> allGoods = null;
    //加载驱动
    try {
        Class.forName("com.mysql.jdbc.Driver");
    } catch (ClassNotFoundException e) {
        e.printStackTrace();
    }
    try {
        //建立连接
con=DriverManager.getConnection("jdbc:mysql://localhost:3306/test?characterEncoding=utf-8","root","root");
        st = con.createStatement();
        //发送添加SQL语句，实现添加的功能
        request.setCharacterEncoding("utf-8");
        st.executeUpdate("insert into goods values(null,'" + request.getParameter("gname") + "','" + request.getParameter("gprice") + ")");
        //发送查询SQL语句，返回结果集
        rs = st.executeQuery("select * from goods ");
        //将查询结果装到集合ArrayList<Goods>中
        allGoods = new ArrayList<Goods>();
        while(rs.next()){
            Goods g = new Goods();
            g.setId(rs.getInt(1));
            g.setGname(rs.getString(2));
            g.setGprice(rs.getDouble(3));
            allGoods.add(g);
        }
        rs.close();
        st.close();
```

```
            con.close();
        } catch (SQLException e) {
            e.printStackTrace();
        }
        //存到 request 对象，以便在页面 showAllGoods.jsp 中显示
        request.setAttribute("allGoods", allGoods);
        //跳转到 showAllGoods.jsp 显示商品
        RequestDispatcher dis = request.getRequestDispatcher("showAllGoods.jsp");
        dis.forward(request, response);
    }
    protected void doPost(HttpServletRequest request, HttpServletResponse response) throws ServletException, IOException {
        doGet(request, response);
    }
}
```

页面文件 showAllGoods.jsp 的代码如下：

```
<%@ page language="java" contentType="text/html; charset=UTF-8"
    pageEncoding="UTF-8"%>
<%@ taglib prefix="c" uri="http://java.sun.com/jsp/jstl/core"%>
<!DOCTYPE html>
<html>
<head>
<meta charset="UTF-8">
<title>Insert title here</title>
</head>
<body>
    <table border="1">
        <tr>
            <th>商品编号</th>
            <th>商品名称</th>
            <th>商品价格</th>
        </tr>
    <c:forEach items="${allGoods }" var="goods">
        <tr>
            <td>${goods.id }</td>
            <td>${goods.gname }</td>
            <td>${goods.gprice }</td>
        </tr>
    </c:forEach>
    </table>
</body>
</html>
```

11.2.3 实践环节——使用 Statement 语句实现多条件查询

编写两个 JSP 页面：inputQuery.jsp 和 showGoods.jsp。用户可以在 inputQuery.jsp 页面输入查询条件，单击"查询"按钮后，在 showGoods.jsp 页面中显示符合查询条件的商品信息。需要编写一个 Servlet 类 SelectGoodsServlet.java，实现该题的条件查询功能，另外需要使用例 11-2 的实体类 Goods。页面运行效果如图 11.3 所示。

（a）输入条件

（b）符合条件的记录

图 11.3 页面运行效果

11.3 游动查询

视频讲解

有时候需要结果集的游标前后移动，这时可使用滚动结果集。为了获得滚动结果集，必须首先用下面的方法得到一个 Statement 对象：

```
Statement st = con.createStatement(int type, int concurrency);
```

根据 type 和 concurrency 的取值，当执行 ResultSet rs = st.executeQuery(String sql)时，会返回不同类型的结果集。

type 的取值决定滚动方式，它的取值如下。
- ResultSet.TYPE_FORWORD_ONLY：表示结果集只能向下滚动。
- ResultSet.TYPE_SCROLL_INSENSITIVE：表示结果集可以上下滚动，当数据库变化时，结果集不变。
- ResultSet.TYPE_SCROLL_SENSITIVE：表示结果集可以上下滚动，当数据库变化时，结果集同步改变。

concurrency 的取值表示是否可以用结果集更新数据库，它的取值如下。

- ResultSet.CONCUR_READ_ONLY：表示不能用结果集更新数据库表。
- ResultSet.CONCUR_UPDATETABLE：表示能用结果集更新数据库表。

游动查询经常用到 ResultSet 的方法如下。

- public boolean absolute(int row)：将游标移到参数 row 指定的行。如果 row 取负值，就是倒数的行，如-1 表示最后一行。当移到最后一行之后或第一行之前时，该方法返回 false。
- public void afterLast()：将游标移到结果集的最后一行之后。
- public void beforeFirst()：将游标移到结果集的第一行之前。
- public void first()：将游标移到结果集的第一行。
- public int getRow()：得到当前游标所指定的行号，如果没有行，则返回 0。
- public boolean isAfterLast()：判断游标是不是在结果集的最后一行之后。
- public boolean isBeforeFirst()：判断游标是不是在结果集的第一行之前。
- public void last()：将游标移到结果集的最后一行。
- public boolean previous()：将游标向上移动（和 next 方法相反），当移动到结果集的第一行之前时返回 false。

例 11-3 编写一个 Servlet（MoveSelectServlet.java），在 Servlet 中查询 goods 表中的记录，并在 showMoveGoods.jsp 页面中逆序显示偶数行的记录。在本例中需要使用例 11-2 的实体类 Goods，但需要为 Goods 添加一个属性 rowno 封装行号。首先，运行 Servlet，结果如图 11.4 所示。

图 11.4 逆序输出偶数行的记录

MoveSelectServlet.java 的代码如下：

```
package servlet;
import java.io.IOException;
import java.sql.Connection;
import java.sql.DriverManager;
import java.sql.ResultSet;
import java.sql.SQLException;
import java.sql.Statement;
import java.util.ArrayList;
import javax.servlet.RequestDispatcher;
import javax.servlet.ServletException;
import javax.servlet.annotation.WebServlet;
import javax.servlet.http.HttpServlet;
import javax.servlet.http.HttpServletRequest;
```

```java
import javax.servlet.http.HttpServletResponse;
import entity.Goods;
@WebServlet("/MoveSelectServlet")
public class MoveSelectServlet extends HttpServlet {
    private static final long serialVersionUID = 1L;
    protected void doGet(HttpServletRequest request, HttpServletResponse
    response) throws ServletException, IOException {
        Connection con = null;
        Statement st = null;
        ResultSet rs = null;
        //将查询结果装到集合 ArrayList<Goods>中,并返回页面显示
        ArrayList<Goods> allGoods = null;
        //加载驱动
        try {
            Class.forName("com.mysql.jdbc.Driver");
        } catch (ClassNotFoundException e) {
            e.printStackTrace();
        }
        try {
            //建立连接
con=DriverManager.getConnection("jdbc:mysql://localhost:3306/test?chara
cterEncoding=utf-8","root","root");
            st = con.createStatement(ResultSet.TYPE_SCROLL_SENSITIVE,
                    ResultSet.CONCUR_READ_ONLY);
            //发送查询 SQL 语句,返回结果集
            rs = st.executeQuery("select * from goods ");
            //将游标移动到最后一行
            rs.last();
            //获取最后一行的行号
            int lownumber = rs.getRow();
            //存到 request 对象,以便在页面 showMoveGoods.jsp 中显示
            request.setAttribute("lownumber", lownumber);
            //将查询结果装到集合 ArrayList<Goods>中
            allGoods = new ArrayList<Goods>();
            //为了逆序输出记录,需将游标移动到最后一行之后
            rs.afterLast();
            while (rs.previous()) {
                int i = rs.getRow();
                if (i % 2 == 0) {
                    Goods g = new Goods();
                    g.setRowno(i);
                    g.setId(rs.getInt(1));
                    g.setGname(rs.getString(2));
                    g.setGprice(rs.getDouble(3));
                    allGoods.add(g);
```

```java
            }
        }
        rs.close();
        st.close();
        con.close();
    } catch (SQLException e) {
        e.printStackTrace();
    }
    //存到request对象,以便在页面showMoveGoods.jsp中显示
    request.setAttribute("allGoods", allGoods);
    //跳转到showMoveGoods.jsp显示商品
    RequestDispatcher dis = request.getRequestDispatcher("showMoveGoods.jsp");
    dis.forward(request, response);
}
protected void doPost(HttpServletRequest request, HttpServletResponse response) throws ServletException, IOException {
    doGet(request, response);
}
}
```

showMoveGoods.jsp 页面的代码如下:

```jsp
<%@ page language="java" contentType="text/html; charset=UTF-8"
    pageEncoding="UTF-8"%>
<%@ taglib prefix="c" uri="http://java.sun.com/jsp/jstl/core"%>
<!DOCTYPE html>
<html>
<head>
<meta charset="UTF-8">
<title>Insert title here</title>
</head>
<body>
    该表共有${lownumber }条记录<br>
    现在逆序输出偶数行的记录:<br>
<table border="1">
    <tr>
        <th>行号</th>
        <th>商品编号</th>
        <th>商品名称</th>
        <th>商品价格</th>
    </tr>
<c:forEach items="${allGoods }" var="goods">
    <tr>
        <td>${goods.rowno }</td>
        <td>${goods.id }</td>
        <td>${goods.gname }</td>
```

```
            <td>${goods.gprice }</td>
         </tr>
      </c:forEach>
   </table>
</body>
</html>
```

11.4 使用连接池

视频讲解

与数据库建立连接是一个耗资源的活动，每次都要花费一定的时间。这个时间对于一次或几次数据库连接，系统的开销或许不明显。可是对于大型电子商务网站，同时有成千上万人频繁地进行数据库连接操作，势必占用很多系统资源，网站的响应速度必定下降，严重时甚至会造成服务器的崩溃。因此，合理地建立数据库连接是非常重要的。

数据库连接池的基本思想是：为数据库连接建立一个"缓冲池"。预先在"缓冲池"中放入一定数量的连接，当需要建立数据库连接时，只需从"缓冲池"中取出一个，使用完毕之后再放回去。可以通过设定连接池最大连接数来防止系统无限度地与数据库连接。更为重要的是，通过连接池的管理机制监视数据库连接的数量及使用情况，为系统开发、测试和性能调整提供依据。其工作原理如图 11.5 所示。

图 11.5　连接池的工作原理

第一次访问连接池时，需要耗费一定时间，这是因为在第一次访问时连接池中没有可用连接，但是在第二次访问时连接池中就有了一些可用的连接，可以直接从连接池中获得连接来访问数据库。

【例】11-4　编写一个 JSP 页面 conPool.jsp，在该页面中使用 scope 作为 application 的 Bean（由 ConnectionPool 负责创建）。该 Bean 创建时，将建立一定数量的连接对象。因此，所有的用户将共享这些连接对象。在 JSP 页面中使用 Bean 获得一个连接对象，然后使用该连接对象访问数据库中的 goods 表（查询出商品价格大于 50 的商品）。页面运行效果如图 11.6 所示。

商品编号	商品名称	商品价格
3	花生99	99
4	花生66	66

图 11.6　使用连接池连接数据库

ConnectionPool.java 的代码如下：

```java
package bean;
import java.sql.*;
import java.util.ArrayList;
public class ConnectionPool {
    //存放 Connection 对象的数组，数组被看成连接池
    ArrayList<Connection> list = new ArrayList<Connection>();
    //构造方法，创建 15 个连接对象，放到连接池中
    public ConnectionPool() {
        try {
            Class.forName("com.mysql.jdbc.Driver");
        } catch (ClassNotFoundException e) {
            e.printStackTrace();
        }
        for (int i = 0; i < 15; i++) {
            try {
Connection con = DriverManager.getConnection("jdbc:mysql://localhost:3306/test?characterEncoding=utf-8","root","root");
                list.add(con);
            } catch (SQLException e) {
                //TODO Auto-generated catch block
                e.printStackTrace();
            }
        }
    }
    //从连接池中取出一个连接对象
    public synchronized Connection getOneCon() {
        if (list.size() > 0) {
            //删除数组的第一个元素，并返回该元素中的连接对象
            return list.remove(0);
        } else {
            //连接对象被用完
            return null;
        }
    }
    //把连接对象放回连接池中
    public synchronized void releaseCon(Connection con) {
```

```
            list.add(con);
        }
    }
```

页面文件 conPool.jsp 的代码如下：

```jsp
<%@ page language="java" contentType="text/html; charset=UTF-8"
    pageEncoding="UTF-8"%>
<%@ page import="java.sql.*"%>
<!DOCTYPE html>
<html>
<head>
<meta charset="UTF-8">
<title>使用连接池连接数据库</title>
</head>
<body>
<jsp:useBean id="conpool" class="bean.ConnectionPool" scope="application" />
<body bgcolor="AliceBlue">
    <%
        Connection con = null;
        Statement st = null;
        ResultSet rs = null;
        try {
            //使用 conpool 对象调用 getOneCon 方法从连接池中获得一个连接对象
            con = conpool.getOneCon();
            if (con == null) {
                out.print("人数过多，稍后访问");
                return;
            }
            st = con.createStatement();
            rs = st.executeQuery("select * from goods where gprice>50");
            out.print("<table border=1>");
            out.print("<tr>");
            out.print("<th>商品编号</th>");
            out.print("<th>商品名称</th>");
            out.print("<th>商品价格</th>");
            out.print("</tr>");
            while (rs.next()) {
                out.print("<tr>");
                out.print("<td>" + rs.getString(1) + "</td>");
                out.print("<td>" + rs.getString(2) + "</td>");
                out.print("<td>" + rs.getString(3) + "</td>");
                out.print("</tr>");
            }
            out.print("</table>");
        } catch (SQLException e) {
```

```
                e.printStackTrace();
            } finally {
                try {
                    if (rs != null) {
                        rs.close();
                    }
                    if (st != null) {
                        st.close();
                    }
                    if (con != null) {
                        //使用 conpool 对象调用 releaseCon 方法将连接对象放回连接池中
                        conpool.releaseCon(con);
                    }
                } catch (SQLException e) {
                    e.printStackTrace();
                }
            }
        %>
    </body>
</html>
```

11.5 其他典型数据库的连接

视频讲解

使用纯 Java 数据库驱动程序连接不同类型数据库的程序流程和框架是基本相同的，需要重点关注的是在连接各数据库时驱动程序加载部分代码和连接部分代码。下面分别介绍加载纯 Java 数据库驱动程序连接 SQL Server 和 Oracle。

11.5.1 连接 SQL Server 数据库

连接 SQL Server 数据库的步骤如下。

❶ **获取纯 Java 数据库驱动程序**

可以登录微软的官方网站 http://www.microsoft.com 下载 Microsoft JDBC Driver 6.2（本书采用的版本是 6.2）for SQL Server（sqljdbc_6.2.2.1_enu.tar.gz），解压 sqljdbc_6.2.2.1_enu.tar.gz 后得到 mssql-jdbc-6.2.2.jre7.jar（对应 JDK1.7）和 mssql-jdbc-6.2.2.jre8.jar（对应 JDK1.8）文件。然后，根据 JDK 的版本将对应的 JAR 文件复制到 Web 应用程序的 /WEB-INF/lib 目录下。

❷ **加载驱动程序**

```
Class.forName("com.microsoft.sqlserver.jdbc.SQLServerDriver");
```

❸ **建立连接**

```
Connection con= DriverManager.getConnection(
```

```
"jdbc:sqlserver://localhost:1433; DatabaseName= mydatabase ","用户名","密码");
```

11.5.2 连接 Oracle 数据库

连接 Oracle 数据库的步骤如下。

❶ **获取纯 Java 数据库驱动程序**

以 Oracle 10g 为例，Oracle 10g 的纯 Java 驱动程序一般位于数据库安装目录 "\oracle\product\10.2.0\db_1\jdbc\lib" 下，名为 classes12.jar。为了连接 Oracle 10g 数据库，可以将 classes12.jar 文件复制到 Web 应用程序的/WEB-INF/lib 目录下。

❷ **加载驱动程序**

```
Class.forName("oracle.jdbc.driver.OracleDriver");
```

❸ **建立连接**

```
Connection con=DriverManager.getConnection("jdbc:oracle:thin:@主机:端口号:数据库名","用户名","密码");
```

其中，主机是安装 Oracle 服务器的 IP 地址，如果是本机，则为 localhost；Oracle 默认端口号为 1521；Oracle 默认数据库名为 orcl；用户名和密码是访问 Oracle 服务器的用户权限。

注意：应用程序连接 Oracle 数据库时，必须事先启动 Oracle 服务器的 OracleService ORCL 和 OracleOraDb10g_home1TNSListener 两个服务，否则会抛出连接异常。

11.6 使用 PreparedStatement 语句访问数据库

与 Statement 语句一样，PreparedStatement 语句同样可以完成向数据库发送 SQL 语句、获取数据库操作结果的功能。PreparedStatement 语句习惯地称为预处理语句。

视频讲解

11.6.1 预处理语句的优点

Statement 对象在每次执行 SQL 语句时都将该语句传送给数据库，然后数据库解释器负责将 SQL 语句转换成内部命令，并执行该命令，完成相应的数据库操作。这种机制每次向数据库发送一条 SQL 语句时，都要先转化成内部命令，如果不断地执行程序，就会加重解释器的负担，影响执行的速度。而 PreparedStatement 对象将 SQL 语句传送给数据库进行预编译，以后需要执行同一条语句时就不再需要重新编译，直接执行就可以了，这样就大大提高了数据库的执行速度。

11.6.2 使用通配符

可以使用 Connection 的对象 con 调用 preparedStatement(String sql)方法对参数 sql 指定

的 SQL 语句进行预先编译，生成数据库的底层命令，并将该命令封装在 PreparedStatement 对象中。对于 SQL 语句中变动的部分，可以使用通配符 "?" 代替。例如：

```
PreparedStatement ps = con.preparedStatement("insert into goodsInfo values (?,?,?,?)";
```

然后使用对应的 setXxx(int parameterIndex, xxx value) 方法设置 "?" 代表的值，其中，参数 parameterIndex 用来表示 SQL 语句中从左到右的第 parameterIndex 个通配符号，value 表示该通配符所代表的具体值。例如：

```
ps.setInt(1,9);
ps.setString(2, "手机");
ps.setDouble(3, 1900.8);
ps.setString(4, "通信");
```

若让 SQL 语句执行生效，需使用 PreparedStatement 的对象 ps 调用 executeUpdate() 方法。如果是查询的话，ps 就调用 executeQuery() 方法，并返回 ResultSet 对象。

Statement 对象在执行 executeQuery(String sql)、executeUpdate(String sql) 等方法时，如果 SQL 语句有些部分是动态的数据，必须使用 "+" 连字符组成完整的 SQL 语句，十分不便。例如，例 11-2 在添加商品时，必须按照如下方式组成 SQL 语句：

```
st.executeUpdate("insert into goods values(null,'" + request.getParameter("gname") + "'," + request.getParameter("gprice") + ")");
```

使用预处理语句不仅提高了数据库的访问效率，而且方便了程序的编写。预处理语句对象调用 executeUpdate() 和 executeQuery() 方法时不需要传递参数。例如：

```
int i = ps.executeUpdate();
```

或

```
ResultSet rs = ps.executeQuery();
```

例 11-5 将例 11-2 修改为使用 PreparedStatement 语句实现。本例中只需修改 Servlet 即可，Servlet 修改后的代码如下：

```
package servlet;
import java.io.IOException;
import java.sql.Connection;
import java.sql.DriverManager;
import java.sql.ResultSet;
import java.sql.SQLException;
import java.sql.PreparedStatement;
import java.util.ArrayList;
import javax.servlet.RequestDispatcher;
import javax.servlet.ServletException;
import javax.servlet.annotation.WebServlet;
import javax.servlet.http.HttpServlet;
import javax.servlet.http.HttpServletRequest;
```

```java
import javax.servlet.http.HttpServletResponse;
import entity.Goods;
@WebServlet("/goodsServlet")
public class GoodsServlet extends HttpServlet {
    private static final long serialVersionUID = 1L;
    protected void doGet(HttpServletRequest request, HttpServletResponse response) throws ServletException, IOException {
        Connection con = null;
        PreparedStatement ps = null;
        ResultSet rs = null;
        //将查询结果装到集合 ArrayList<Goods>中，并返回页面显示
        ArrayList<Goods> allGoods = null;
        //加载驱动
        try {
            Class.forName("com.mysql.jdbc.Driver");
        } catch (ClassNotFoundException e) {
            e.printStackTrace();
        }
        try {
            //建立连接
            con=DriverManager.getConnection("jdbc:mysql://localhost:3306/test?characterEncoding=utf-8","root","root");
            //预处理
            ps = con.prepareStatement("insert into goods values(null,?,?)");
            //发送添加SQL语句，实现添加的功能
            request.setCharacterEncoding("utf-8");
            ps.setString(1, request.getParameter("gname"));
            ps.setString(2, request.getParameter("gprice"));
            ps.executeUpdate();
            //预处理
            ps = con.prepareStatement("select * from goods ");
            //发送查询SQL语句，返回结果集
            rs = ps.executeQuery();
            //将查询结果装到集合 ArrayList<Goods>中
            allGoods = new ArrayList<Goods>();
            while(rs.next()){
                Goods g = new Goods();
                g.setId(rs.getInt(1));
                g.setGname(rs.getString(2));
                g.setGprice(rs.getDouble(3));
                allGoods.add(g);
            }
            rs.close();
            ps.close();
            con.close();
```

```
        } catch (SQLException e) {
            e.printStackTrace();
        }
        //存到 request 对象,以便在页面 showAllGoods.jsp 中显示
        request.setAttribute("allGoods", allGoods);
        //跳转到 showAllGoods.jsp 显示商品
        RequestDispatcher dis = request.getRequestDispatcher("showAllGoods.jsp");
        dis.forward(request, response);
    }
    protected void doPost(HttpServletRequest request, HttpServletResponse 
response) throws ServletException, IOException {
        doGet(request, response);
    }
}
```

11.6.3 实践环节——使用 PreparedStatement 语句实现多条件查询

将 11.2.3 节的实践环节修改为使用 PreparedStatement 语句实现。

11.7 分页查询

视频讲解

分页查询一般有两种实现方式:一次查询出所有数据,在页面中进行分页显示;每页查询一次,每次只查询本页显示的数据。本节采用第二种方式实现分页查询。下面通过一个实例介绍分页查询。

例11-6 编写一个 Servlet 类 SelectGoodsByPageServlet,在该 Servlet 类中实现分页查询 goods 表的数据,假设每页显示 3 条数据。另外,需要编写一个显示数据页面 showGoodsByPage.jsp。实现步骤如下:

❶ **编写 SelectGoodsByPageServlet 类**

在 Servlet 包中创建 SelectGoodsByPageServlet 类,在该类实现分页查询。首先获得查询数据总条数 totalCount,然后根据 totalCount 计算总页数 totalPage,最后计算起始位置 startIndex。具体代码如下:

```
package servlet;
import java.io.IOException;
import java.sql.Connection;
import java.sql.DriverManager;
import java.sql.PreparedStatement;
import java.sql.ResultSet;
import java.sql.SQLException;
import java.sql.Statement;
```

```java
import java.util.ArrayList;
import javax.servlet.RequestDispatcher;
import javax.servlet.ServletException;
import javax.servlet.annotation.WebServlet;
import javax.servlet.http.HttpServlet;
import javax.servlet.http.HttpServletRequest;
import javax.servlet.http.HttpServletResponse;
import entity.Goods;
@WebServlet("/selectGoodsByPageServlet")
public class SelectGoodsByPageServlet extends HttpServlet {
    private static final long serialVersionUID = 1L;
    protected void doGet(HttpServletRequest request, HttpServletResponse response) throws ServletException, IOException {
        Connection con = null;
        PreparedStatement ps = null;
        ResultSet rs = null;
        //将查询结果装到集合ArrayList<Goods>中，并返回页面显示
        ArrayList<Goods> allGoods = null;
        //加载驱动
        try {
            Class.forName("com.mysql.jdbc.Driver");
        } catch (ClassNotFoundException e) {
            e.printStackTrace();
        }
        try {
            //建立连接
con=DriverManager.getConnection("jdbc:mysql://localhost:3306/test?characterEncoding=utf-8","root","root");
            Statement stmt = con.createStatement(ResultSet.TYPE_SCROLL_INSENSITIVE, ResultSet.CONCUR_READ_ONLY);
            rs = stmt.executeQuery("select * from goods ");
            rs.last();                      //移到最后一行
            int totalCount = rs.getRow();   //分页查询，数据总数
            int totalPage = 0;
            if (totalCount == 0) {
                totalPage = 0;              //总页数
            } else {
                //返回大于或者等于指定表达式的最小整数
                totalPage = (int) Math.ceil((double) totalCount/3);
            }
            String pageCur11 = request.getParameter("pageCur");//获得当前页码
            if (pageCur11 == null) {        //首页面
                pageCur11 = "1";
            }
            int  pageCur = Integer.parseInt(pageCur11);
```

```java
            if ((pageCur - 1) * 3 > totalCount) {
                pageCur = pageCur - 1;
            }
            int startIndex = (pageCur - 1) * 3; //起始位置
            int perPageSize = 3;                //每页3个
            //预处理
            ps = con.prepareStatement("select * from goods limit ?, ?");
            ps.setInt(1, startIndex);
            ps.setInt(2, perPageSize);
            //发送查询SQL语句,返回结果集
            rs = ps.executeQuery();
            //将查询结果装到集合ArrayList<Goods>中
            allGoods = new ArrayList<Goods>();
            while(rs.next()){
                Goods g = new Goods();
                g.setId(rs.getInt(1));
                g.setGname(rs.getString(2));
                g.setGprice(rs.getDouble(3));
                allGoods.add(g);
            }
            //存到request对象,以便在页面showAllGoods.jsp中显示
            request.setAttribute("allGoods", allGoods);
            request.setAttribute("totalCount", totalCount);
            request.setAttribute("totalPage", totalPage);
            request.setAttribute("pageCur", pageCur);
            rs.close();
            ps.close();
            con.close();
        } catch (SQLException e) {
            e.printStackTrace();
        }
        //跳转到showGoodsByPage.jsp显示商品
        RequestDispatcher dis = request.getRequestDispatcher
        ("showGoodsByPage.jsp");
        dis.forward(request, response);
    }
    protected void doPost(HttpServletRequest request, HttpServletResponse
    response) throws ServletException, IOException {
        doGet(request, response);
    }
}
```

❷ 编写 **showGoodsByPage.jsp** 页面

编写 showGoodsByPage.jsp 页面,分页显示 goods 表中数据,具体代码如下:

```
<%@ page language="java" contentType="text/html; charset=UTF-8"
```

```jsp
pageEncoding="UTF-8"%>
<%@ taglib prefix="c" uri="http://java.sun.com/jsp/jstl/core"%>
<!DOCTYPE html>
<html>
<head>
<meta charset="UTF-8">
<title>Insert title here</title>
</head>
<body>
    <table border="1">
        <tr>
            <th>商品编号</th>
            <th>商品名称</th>
            <th>商品价格</th>
        </tr>
    <c:forEach items="${allGoods }" var="goods">
        <tr>
            <td>${goods.id }</td>
            <td>${goods.gname }</td>
            <td>${goods.gprice }</td>
        </tr>
    </c:forEach>
        <tr>
            <td colspan="3" align="right">
                   
       共${totalCount}条记录  共${totalPage}
    页  
                第${pageCur}页  
                <c:url var="url_pre" value="selectGoodsByPageServlet">
                    <c:param name="pageCur" value="${pageCur - 1 }"/>
                </c:url>
                <c:url var="url_next" value="selectGoodsByPageServlet">
                    <c:param name="pageCur" value="${pageCur + 1 }"/>
                </c:url>
                <!-- 第一页,没有上一页 -->
                <c:if test="${pageCur != 1 }">
                    <a href="${url_pre}">上一页</a>    
                </c:if>
                <!-- 最后一页,没有下一页 -->
                <c:if test="${pageCur != totalPage && totalPage !=0 }">
                    <a href="${url_next}">下一页</a>
                </c:if>
            </td>
        </tr>
    </table>
```

```
</body>
</html>
```

❸ 测试分页

重启 Tomcat 服务器后，通过地址"http://localhost:8080/ch11/selectGoodsByPageServlet"测试分页，运行结果如图 11.7 所示。

图 11.7　分页查询 1

单击图 11.7 中"下一页"超链接，显示如图 11.8 所示的结果。

图 11.8　分页查询 2

11.8　使用 CallableStatement 语句访问存储子程序

视频讲解

CallableStatement 语句对象可被用来访问数据库中的存储过程，可以通过连接对象调用 prepareCall()方法得到 CallableStatement 对象。CallableStatement 对象的获得一般有如下几种情况。

- 不带参数的存储过程：

```
CallableStatement cs = con.preparedCall ("{call 存储过程名()}");
```

- 带参数的存储过程：

```
CallableStatement cs = con.preparedCall ("{call 存储过程名(?,?,?)}");
```

- 函数：

```
CallableStatement cs = conn.preparedCall ("{?= call 存储函数名(?,?,?)}");//?
=接收函数的返回值
```

在执行存储过程或函数之前，凡是类型为 IN 的参数，使用 CallableStatement 对象的 setXxx( )方法来设置参数的值；凡是类型为 OUT 的参数，使用 registerOutParameter( )方法来完成注册。示例代码片段如下：

```
...
CallableStatement cs = con.preparedCall ("{call proc_getInfo(?,?)}");
cs.registerOutParameter(2, java.sql.Types.INTEGER);//第二个参数为 OUT 类型
cs.setInt(1, 888); //第一个参数为 IN 类型
cs.execute( );
int count = cs.getInt(2);//获得 OUT 类型的参数值
...
```

为了测试 CallableStatement 对象的使用，下面创建了 Oracle 10g 的存储过程，这些存储过程放在一个名为 PKG_TEST 的包中，脚本如下：

```
CREATE OR REPLACE PACKAGE PKG_TEST
AS
TYPE MYRCTYPE IS REF CURSOR; /*游标*/
/*添加商品的存储过程*/
    PROCEDURE TEST_ADDGOODS(P_GOODSID    IN NUMBER,
                            P_GOODSNAME  IN VARCHAR2,
                            P_GOODSPRICE IN NUMBER,
                            P_GOODSTYPE  IN VARCHAR2);
    /*统计商品价格大于 400 的存储过程*/
    PROCEDURE TEST_GETCOUNT( P_GOODSPRICE IN NUMBER,GOODSCOUNT OUT NUMBER);
    /*查询商品信息的存储过程，返回结果集*/
PROCEDURE TEST_GETGOODS (P_GOODSPRICE NUMBER, P_RC OUT MYRCTYPE);
END PKG_TEST;
/*包体的创建*/
CREATE OR REPLACE PACKAGE BODY PKG_TEST
    AS
/*添加商品的存储过程*/
    PROCEDURE TEST_ADDGOODS(P_GOODSID    IN NUMBER,
                            P_GOODSNAME  IN VARCHAR2,
                            P_GOODSPRICE IN NUMBER,
                            P_GOODSTYPE  IN VARCHAR2) IS
      BEGIN
            INSERT INTO GOODSINFO(GOODSID, GOODSNAME, GOODSPRICE, GOODSTYPE)
                VALUES  (P_GOODSID, P_GOODSNAME, P_GOODSPRICE, P_GOODSTYPE);
      END TEST_ADDGOODS;
    /*统计商品价格大于 400 的存储过程*/
    PROCEDURE TEST_GETCOUNT( P_GOODSPRICE IN NUMBER,GOODSCOUNT OUT NUMBER) IS
        BEGIN
            SELECT COUNT(*) INTO GOODSCOUNT FROM GOODSINFO  WHERE GOODSPRICE
            > P_GOODSPRICE;
        END TEST_GETCOUNT;
```

```
        /*查询商品信息的存储过程,返回结果集*/
        PROCEDURE TEST_GETGOODS(P_GOODSPRICE NUMBER, P_RC OUT MYRCTYPE) IS
            BEGIN
                OPEN P_RC FOR
                SELECT GOODSID, GOODSNAME, GOODSPRICE, GOODSTYPE
                FROM GOODSINFO WHERE GOODSPRICE > P_GOODSPRICE ;
            END TEST_GETGOODS;
    END PKG_TEST;
```

例11-7 编写 JSP 页面 testCallableStatement.jsp,在该页面中使用 CallableStatement 对象调用 Oracle 包 PKG_TEST 的存储过程。

页面文件 testCallableStatement.jsp 的代码如下:

```jsp
<%@ page language="java" contentType="text/html; charset=UTF-8"
    pageEncoding="UTF-8"%>
<%@ page import="java.sql.*"%>
<!DOCTYPE html>
<html>
<head>
<meta charset="UTF-8">
<title>CallableStatement 对象的使用</title>
</head>
<body>
<%
        Connection con = null;
        CallableStatement cs1 = null;
        CallableStatement cs2 = null;
        CallableStatement cs3 = null;
        ResultSet rs = null;
        try {
            Class.forName("oracle.jdbc.driver.OracleDriver");
        } catch (ClassNotFoundException e) {
            e.printStackTrace();
        }
        try {
            con = DriverManager.getConnection("jdbc:oracle:thin:@localhost:
            1521:orcl", "system","system");
            //调用存储过程 pkg_test.TEST_ADDGOODS 添加一个商品
            cs1 = con.prepareCall("call pkg_test.TEST_ADDGOODS(?,?,?,?)");
            //设置 IN 类型的参数值
            cs1.setInt(1, 9);
            cs1.setString(2, "XX 牙刷");
            cs1.setInt(3, 500);
            cs1.setString(4, "日用品");
```

```jsp
            cs1.execute();
            out.print("调用存储过程pkg_test.TEST_ADDGOODS成功添加一个商品。" +
            "<br>");
            //调用存储过程pkg_test.TEST_GETCOUNT统计商品价格大于400
            cs2=con.prepareCall("call pkg_test.TEST_GETCOUNT(?,?)");
            //设置IN类型的参数值
            cs2.setInt(1, 400);
            //注册OUT类型的参数
            cs2.registerOutParameter(2, Types.INTEGER);//java.sql包中的Types
            cs2.execute();
            //getXxx(index)中的index需要和上面registerOutParameter的index对应
            int count = cs2.getInt(2);
    out.print("调用存储过程pkg_test.TEST_GETCOUNT统计商品价格大于400的商
    品共有： " + count + "个。<br>");
            //调用存储过程pkg_test.TEST_GETGOODS获得商品价格大于400的信息
            cs3=con.prepareCall("call pkg_test.TEST_GETGOODS(?,?)");
            //设置IN类型的参数值
            cs3.setInt(1, 400);
            cs3.registerOutParameter(2, oracle.jdbc.OracleTypes.CURSOR);
            //Oracle游标
            cs3.execute();
            //由游标转换为结果集
            rs=(ResultSet)cs3.getObject(2);
            //遍历价格大于400的商品
    out.print("调用存储过程pkg_test.TEST_GETGOODS获得价格大于400的商品
    有： " + "<br>");
            while(rs.next()){
                out.print("GOODSID: " + rs.getString(1) + " "
                        + "GOODSNAME: " + rs.getString(2) + " "
                        + "GOODSPRICE: " + rs.getString(3) + " "
                        + "GOODSTYPE: " + rs.getString(4) + "<br>");
            }
            cs1.close();
            cs2.close();
            cs3.close();
            rs.close();
            con.close();
        } catch (SQLException e) {
            e.printStackTrace();
        }
    }
%>
</body>
</html>
```

11.9　Excel 与数据库互访

视频讲解

Java 读写 Excel 文件有两种途径：一个是 POI（Apache POI）；一个是 jExcelAPI（即 jxl）。POI 是 Apache 公司的一个子项目，主要提供一组操作 Windows 文档的 Java API。jxl 是一个开放源码项目，通过它 Java 开发人员可以读写、更新已经存在的 Excel 文件。

目前，jxl 只能处理 Excel 2007 之前的版本，即 xls 文件；而 POI 方式可以处理 xls 和 xlsx 文件。本节使用 POI 方式读写 Excel 文件。

可以从官网 http://poi.apache.org/download.html 下载 POI 的 JAR 包。本书下载的是 poi-bin-3.17-20170915.tar.gz 版本，解压后目录如图 11.9 所示。

图 11.9　poi-3.17 目录

读写 xls 格式时需要导入 poi-3.17.jar；读写 xlsx 格式时需要导入 poi-ooxml-3.17.jar 和 poi-ooxml-schemas-3.17.jar。当操作 Word、PPT、Viso、Outlook 等文档时需要导入 poi-scratchpad-3.17.jar。除了导入这些 JAR 包外，还需要将它们的依赖包导入，即将 lib 和 ooxml-lib 目录下的 JAR 包导入。因此，ch11 需要导入的 JAR 包如图 11.10 所示。

图 11.10　导入的 JAR 包

11.9.1 向数据库导入 Excel

本节以一个实例详细讲解如何将 Excel 文件内容导入数据库的数据表中。该实例的功能需求是：将 Excel 文件中的商品信息导入数据表 goods 中，具体步骤如下。

❶ 创建 Excel 文件

分别创建 goods.xls 和 goods.xlsx 文件。文件内容如图 11.11 所示。

	A	B
1	姓名	价格
2	花生101	101
3	花生102	102
4	花生103	103
5	花生104	104

图 11.11 被导入的文件内容

❷ 创建选择文件页面

创建选择文件页面 inputFile.jsp，具体代码如下：

```jsp
<%@ page language="java" contentType="text/html; charset=UTF-8" pageEncoding="UTF-8"%>
<!DOCTYPE html>
<html>
<head>
<meta charset="UTF-8">
<title>从 Excel 导入到数据库</title>
</head>
<body>
    选择导入的Excel文件：
    <form action="readExcelFileServlet" method="post" enctype="multipart/form-data">
        <input type="file" name="filePath">
        <input type="submit" value="导入">
    </form>
</body>
</html>
```

❸ 创建 Servlet

在 Servlet 包中创建一个 Servlet 类 ReadExcelFileServlet，在 Servlet 类中首先使用工具类 ReadExcelFile 将 Excel 文件内容读取到 List<Goods>中，然后将 List<Goods>中数据批量添加到数据表 goods 中。

ReadExcelFileServlet 的代码如下：

```java
package servlet;
import java.io.File;
import java.io.IOException;
import java.io.PrintWriter;
```

```java
import java.sql.Connection;
import java.sql.DriverManager;
import java.sql.PreparedStatement;
import java.sql.SQLException;
import java.util.List;
import javax.servlet.ServletException;
import javax.servlet.annotation.WebServlet;
import javax.servlet.http.HttpServlet;
import javax.servlet.http.HttpServletRequest;
import javax.servlet.http.HttpServletResponse;
import entity.Goods;
import util.ReadExcelFile;
@WebServlet("/readExcelFileServlet")
@MultipartConfig
public class ReadExcelFileServlet extends HttpServlet {
    private static final long serialVersionUID = 1L;
    protected void doGet(HttpServletRequest request, HttpServletResponse 
    response) throws ServletException, IOException {
        request.setCharacterEncoding("utf-8");
        //生产环境
        String savePath = request.getServletContext().getRealPath("/uploadFile");
        //如果文件目录不存在,创建目录
        File f = new File(savePath);
        if(!f.exists()) {
            f.mkdirs();
        }
        Part part = request.getPart("filePath");
        //获得文件名
        String fileName = MyUtil.getFileName(part);
        String newFileName = null;
        //选择了上传文件
        if(fileName != null && fileName.length() > 0) {
            int lastIndex = fileName.lastIndexOf(".");
            String fileType = fileName.substring(lastIndex);
            Date now = new Date();
            SimpleDateFormat sdf = new SimpleDateFormat("YYYYMMDDHHmmssSSS");
            newFileName = sdf.format(now)+ fileType;
            //上传文件
            part.write(savePath + File.separator + newFileName);
        }
        ReadExcelFile ref = new ReadExcelFile();
        //读取 Excel 文件的内容并存入 List
        List<Goods> ls = ref.getExcelInfo(new File(savePath, newFileName));
        //将 List 内容批量保存到数据库
        Connection con = null;
        PreparedStatement ps = null;
        //加载驱动
        try {
```

```java
            Class.forName("com.mysql.jdbc.Driver");
        } catch (ClassNotFoundException e) {
            e.printStackTrace();
        }
        try {
            //建立连接
con=DriverManager.getConnection("jdbc:mysql://localhost:3306/test?characterEncoding=utf-8","root","root");
            ps = con.prepareStatement("insert into goods values(null,?,?)");
            if(ls != null){
                for(int i = 0; i < ls.size(); i++){//多条记录
                    ps.setString(1, ls.get(i).getGname());
                    ps.setDouble(2, ls.get(i).getGprice());
                    ps.addBatch();
                }
                //批量保存
                ps.executeBatch();
            }
            ps.close();
            con.close();
        } catch (SQLException e) {
            e.printStackTrace();
        }
        //设置响应的内容类型
        response.setContentType("text/html;charset=utf-8");
        //取得输出对象
        PrintWriter out=response.getWriter();
        out.print("导入成功，共导入" + ls.size() + "条数据！");
    }
    protected void doPost(HttpServletRequest request, HttpServletResponse response) throws ServletException, IOException {
        doGet(request, response);
    }
}
```

工具类 ReadExcelFile 的代码如下：

```java
package util;
import java.io.File;
import java.io.FileInputStream;
import java.io.IOException;
import java.util.ArrayList;
import java.util.List;
import org.apache.poi.hssf.usermodel.HSSFWorkbook;
import org.apache.poi.ss.usermodel.Cell;
import org.apache.poi.ss.usermodel.Row;
import org.apache.poi.ss.usermodel.Sheet;
import org.apache.poi.ss.usermodel.Workbook;
import org.apache.poi.xssf.usermodel.XSSFWorkbook;
```

```java
import entity.Goods;
public class ReadExcelFile {
    //总行数
    private int totalRows = 0;
    //总条数
    private int totalCells = 0;
    //错误信息接收器
    private String errorMsg;
    //构造方法
    public ReadExcelFile() {
    }
    //获取总行数
    public int getTotalRows() {
        return totalRows;
    }
    //获取总列数
    public int getTotalCells() {
        return totalCells;
    }
    //获取错误信息
    public String getErrorInfo() {
        return errorMsg;
    }
    /**
     * 读Excel文件，获取信息集合
     * @param fielName
     * @return
     */
    public List<Goods> getExcelInfo(File mFile) {
        String fileName = mFile.getName();//获取文件名
        List<Goods> goodsList = null;
        try {
            if (!validateExcel(fileName)) {//验证文件名是否合格
                return null;
            }
            boolean isExcel2003 = true;//根据文件名判断文件是2003版本还是2007版本
            if (isExcel2007(fileName)) {
                isExcel2003 = false;
            }
            goodsList = createExcel(mFile, isExcel2003);
        } catch (Exception e) {
            e.printStackTrace();
        }
        return goodsList;
    }
    /**
     * 根据Excel里面的内容读取信息
     * @param mFile 文件
```

```java
 * @param is Excel 2003  Excel 是 2003 还是 2007 版本
 * @return
 * @throws IOException
 */
public List<Goods> createExcel(File mFile, boolean isExcel2003) {
    List<Goods> goodsList = null;
    try {
        Workbook wb = null;
        if (isExcel2003) {//当 Excel 是 2003 时,创建 excel2003
            wb = new HSSFWorkbook(new FileInputStream(mFile));
        } else {//当 Excel 是 2007 时,创建 excel2007
            wb = new XSSFWorkbook(mFile);
        }
        goodsList = readExcelValue(wb);//读取 Excel 中客户的信息
    } catch (Exception e) {
        e.printStackTrace();
    }
    return goodsList;
}
/**
 * 读取 Excel 中的信息
 * @param wb
 * @return
 */
private List<Goods> readExcelValue(Workbook wb) {
    //得到第一个 shell
    Sheet sheet = wb.getSheetAt(0);
    //得到 Excel 的行数
    this.totalRows = sheet.getPhysicalNumberOfRows();
    //得到 Excel 的列数(前提是有行数)
    if (totalRows > 1 && sheet.getRow(0) != null) {
        this.totalCells = sheet.getRow(0).getPhysicalNumberOfCells();
    }
    List<Goods> goodsList = new ArrayList<Goods>();
    //循环 Excel 行数,第一行是表头,不需要读
    for (int r = 1; r < totalRows; r++) {
        Row row = sheet.getRow(r);
        if (row == null) {
            continue;
        }
        Goods goods = new Goods();
        //循环 Excel 的列
        for (int c = 0; c < this.totalCells; c++) {
            Cell cell = row.getCell(c);
            if (null != cell) {
                if (c == 0) {
                    goods.setGname(cell.getStringCellValue());//商品名称
                } else if (c == 1) {
```

```
                    goods.setGprice(cell.getNumericCellValue());//商品价格
                }
            }
        }
        //添加到list
        goodsList.add(goods);
    }
    return goodsList;
}
/**
 * 验证Excel文件
 * @param filePath
 * @return
 */
public boolean validateExcel(String filePath) {
    if (filePath == null || !(isExcel2003(filePath) || isExcel2007
        (filePath))) {
        errorMsg = "文件名不是Excel格式";
        return false;
    }
    return true;
}
/**
 * 是否是2003的Excel,返回true是2003
 * @param filePath
 * @return
 */
public static boolean isExcel2003(String filePath) {
    return filePath.matches("^.+\\.(?i)(xls)$");
}
/**
 * 是否是2007的Excel,返回true是2007
 * @param filePath
 * @return
 */
public static boolean isExcel2007(String filePath) {
    return filePath.matches("^.+\\.(?i)(xlsx)$");
}
```

需要注意的是，向数据库导入Excel文件时，Excel中的数据类型有严格要求，例如，上例中商品名不能是数字类型，否则使用cell.getStringCellValue()语句获取单元格内容时，将会报"Cannot get a STRING value from a NUMERIC cell"错误。

11.9.2 从数据库导出Excel

从数据库导出Excel文件与向数据库导入Excel文件流程类似，一个是从Excel文件读数据，一个是向Excel文件写数据。本节继续通过一个实例讲解从数据库导出Excel文件。

本节实例的功能需求是：将数据表 goods 中的商品信息导入 Excel 文件中，具体步骤如下。

❶ 创建 Servlet

在 Servlet 包中创建 SelectAllGoodServlet 类，在该类中使用工具类 GetAllGoods 获得所有商品信息，并跳转到 showExportGoods.jsp 页面，显示被导出的商品信息。

SelectAllGoodServlet 的代码如下：

```java
package servlet;
import java.io.IOException;
import java.util.ArrayList;
import javax.servlet.RequestDispatcher;
import javax.servlet.ServletException;
import javax.servlet.annotation.WebServlet;
import javax.servlet.http.HttpServlet;
import javax.servlet.http.HttpServletRequest;
import javax.servlet.http.HttpServletResponse;
import entity.Goods;
import util.GetAllGoods;
@WebServlet("/selectAllGoodServlet")
public class SelectAllGoodServlet extends HttpServlet {
    private static final long serialVersionUID = 1L;
    protected void doGet(HttpServletRequest request, HttpServletResponse response) throws ServletException, IOException {
        //将查询结果装到集合 ArrayList<Goods>中，并返回页面显示
        ArrayList<Goods> allGoods = GetAllGoods.getAllGoods();
        //存到 request 对象，以便在页面 showAllGoods.jsp 中显示
        request.setAttribute("allGoods", allGoods);
        //跳转到 showExportGoods.jsp 显示商品
        RequestDispatcher dis = request.getRequestDispatcher("showExportGoods.jsp");
        dis.forward(request, response);
    }
    protected void doPost(HttpServletRequest request, HttpServletResponse response) throws ServletException, IOException {
        doGet(request, response);
    }
}
```

工具类 GetAllGoods 的代码如下：

```java
package util;
import java.sql.Connection;
import java.sql.DriverManager;
import java.sql.PreparedStatement;
import java.sql.ResultSet;
import java.sql.SQLException;
import java.util.ArrayList;
import entity.Goods;
```

```java
public class GetAllGoods {
    public static ArrayList<Goods> getAllGoods(){
        Connection con = null;
        PreparedStatement ps = null;
        ResultSet rs = null;
        //将查询结果装到集合 ArrayList<Goods>中,并返回页面显示
        ArrayList<Goods> allGoods = null;
        //加载驱动
        try {
            Class.forName("com.mysql.jdbc.Driver");
        } catch (ClassNotFoundException e) {
            e.printStackTrace();
        }
        try {
            //建立连接
con=DriverManager.getConnection("jdbc:mysql://localhost:3306/test?chara cterEncoding=utf-8","root","root");
            //预处理
            ps = con.prepareStatement("select * from goods ");
            //发送查询 SQL 语句,返回结果集
            rs = ps.executeQuery();
            //将查询结果装到集合 ArrayList<Goods>中
            allGoods = new ArrayList<Goods>();
            while(rs.next()){
                Goods g = new Goods();
                g.setId(rs.getInt(1));
                g.setGname(rs.getString(2));
                g.setGprice(rs.getDouble(3));
                allGoods.add(g);
            }
            rs.close();
            ps.close();
            con.close();
        } catch (SQLException e) {
            e.printStackTrace();
        }
        return allGoods;
    }
}
```

❷ 创建商品信息显示页面

创建 showExportGoods.jsp 页面显示被导出的商品信息,具体代码如下:

```
<%@ page language="java" contentType="text/html; charset=UTF-8"
    pageEncoding="UTF-8"%>
<%@ taglib prefix="c" uri="http://java.sun.com/jsp/jstl/core"%>
<!DOCTYPE html>
<html>
```

```html
<head>
<meta charset="UTF-8">
<title>导出数据</title>
</head>
<body>
    <table border="1">
        <tr>
            <th>商品编号</th>
            <th>商品名称</th>
            <th>商品价格</th>
        </tr>
        <c:forEach items="${allGoods }" var="goods">
        <tr>
            <td>${goods.id }</td>
            <td>${goods.gname }</td>
            <td>${goods.gprice }</td>
        </tr>
        </c:forEach>
    </table>
    <a href="exportExcelServlet">导出到 Excel</a>
</body>
</html>
```

❸ 创建实现导出功能的 **Servlet**

创建实现导出功能的 Servlet 类 ExportExcelServlet，在该 Servlet 类中首先使用工具类 GetAllGoods 从数据表 goods 中获得所有商品信息保存到 ArrayList<Goods>集合中，然后使用工具类 WriteExcelFile 将 ArrayList<Goods>集合中的数据写到 Excel 文件中，最后实现 Excel 文件下载。

ExportExcelServlet 的代码如下：

```java
package servlet;
import java.io.File;
import java.io.FileInputStream;
import java.io.IOException;
import java.text.SimpleDateFormat;
import java.util.ArrayList;
import java.util.Date;
import javax.servlet.ServletException;
import javax.servlet.ServletOutputStream;
import javax.servlet.annotation.WebServlet;
import javax.servlet.http.HttpServlet;
import javax.servlet.http.HttpServletRequest;
import javax.servlet.http.HttpServletResponse;
import entity.Goods;
import util.GetAllGoods;
```

```java
import util.WriteExcelFile;
@WebServlet("/exportExcelServlet")
public class ExportExcelServlet extends HttpServlet {
    private static final long serialVersionUID = 1L;
    protected void doGet(HttpServletRequest request, HttpServletResponse
    response) throws ServletException, IOException {
        //获得要导出的商品信息
        ArrayList<Goods> allGoods = GetAllGoods.getAllGoods();
        //生成的Excel在服务器的uploadFiles目录中,以便下载文件
        File downLoadFileDir = new File(getServletContext().getRealPath
        ("/uploadFiles"));
        if(!downLoadFileDir.exists()){
            downLoadFileDir.mkdir();
        }
        //生成日期字符串,定义Excel文件名
        Date date=new Date();
        SimpleDateFormat sdf=new SimpleDateFormat("yyyyMMddHHmmssSSS");
        String fileName = sdf.format(date)+".xls";
        WriteExcelFile.writeExcel(allGoods, downLoadFileDir + File.
        separator + fileName);
        //将生成的Excel文件从服务器下载
        FileInputStream in = null;           //输入流
        ServletOutputStream out = null;      //输出流
        try {
            //设置下载文件使用的报头
            response.setHeader("Content-Type", "application/x-msdownload");
            response.setHeader("Content-Disposition", "attachment;
            filename=" + fileName);
            //读入文件
            in = new FileInputStream(downLoadFileDir + File.separator + fileName);
            //得到响应对象的输出流,用于向客户端输出二进制数据
            out = response.getOutputStream();
            out.flush();
            int aRead = 0;
            byte b[] = new byte[1024];
            while ((aRead = in.read(b)) != -1 & in != null) {
                out.write(b, 0, aRead);
            }
            out.flush();
            in.close();
            out.close();
        } catch (Throwable e) {
            e.printStackTrace();
        }
    }
}
```

```
    protected void doPost(HttpServletRequest request, HttpServletResponse
    response) throws ServletException, IOException {
        doGet(request, response);
    }
}
```

工具类 WriteExcelFile 的代码如下：

```
package util;
import java.io.ByteArrayOutputStream;
import java.io.File;
import java.io.FileOutputStream;
import java.io.IOException;
import java.io.OutputStream;
import java.util.ArrayList;
import org.apache.poi.hssf.usermodel.HSSFCell;
import org.apache.poi.hssf.usermodel.HSSFRow;
import org.apache.poi.hssf.usermodel.HSSFSheet;
import org.apache.poi.hssf.usermodel.HSSFWorkbook;
import entity.Goods;
public class WriteExcelFile {
    private static HSSFWorkbook wb;
    public static void writeExcel(ArrayList<Goods> result, String path) {
        if (result == null) {
            return;
        }
        wb = new HSSFWorkbook();
        //Excel 文件 sheet1 表
        HSSFSheet sheet = wb.createSheet("sheet1");
        //第一行表头
        HSSFRow row0 = sheet.createRow(0);
        row0.createCell(0).setCellValue("id");
        row0.createCell(1).setCellValue("gname");
        row0.createCell(2).setCellValue("gprice");
        //将 ArrayList 中数据写到每一行
        for (int i = 0; i < result.size(); i++) {
            HSSFRow row = sheet.createRow(i + 1);
            if (result.get(i) != null) {
                    HSSFCell cell0 = row.createCell(0);
                    cell0.setCellValue(result.get(i).getId());
                    HSSFCell cell1 = row.createCell(1);
                    cell1.setCellValue(result.get(i).getGname());
                    HSSFCell cell2 = row.createCell(2);
                    cell2.setCellValue(result.get(i).getGprice());
            }
        }
```

```
            ByteArrayOutputStream os = new ByteArrayOutputStream();
            try {
                wb.write(os);
            } catch (IOException e) {
                e.printStackTrace();
            }
            byte[] content = os.toByteArray();
            //Excel 文件生成后存储的位置
            File file = new File(path);
            OutputStream fos = null;
            try {
                fos = new FileOutputStream(file);
                fos.write(content);
                os.close();
                fos.close();
            } catch (Exception e) {
                e.printStackTrace();
            }
        }
    }
```

❹ 测试导出功能

发布 ch11 应用到 Tomcat 服务器，并启动服务器。通过地址 http://localhost:8080/ch11/selectAllGoodServlet 显示被导出的商品信息，如图 11.12 所示。

单击图 11.12 中的"导出到 Excel"链接，显示如图 11.13 所示的效果。

图 11.12　被导出的商品信息　　　　　图 11.13　导出到 Excel

可以单击图 11.13 中的"打开"按钮，直接打开 Excel 文件；也可以单击"保存"按钮，将 Excel 文件另存到本地。

11.10　本章小结

本章重点介绍了使用加载纯 Java 驱动程序连接数据库。使用加载纯 Java 驱动程序的方式连接 MySQL、SQL Server 以及 Oracle 等主流关系数据库的步骤基本一样，这些主流

第 11 章 JDBC 数据库访问

关系数据库的新增、更新、删除与查询等操作也基本一样。读者连接不同数据库时，多注意驱动程序加载部分代码和连接部分代码。

习题 11

1. JDBC 能完成哪些工作？
2. 使用纯 Java 数据库驱动程序访问数据库时，有哪些步骤？
3. JDBC 连接数据库的常用方式有哪些？
4. 下列选项中，（　　）不是 JDBC 的工作任务。

  A．与数据库建立连接　 B．操作数据库，处理数据库返回的结果
  C．在网页中生成表格　 D．向数据库管理系统发送 SQL 语句

5. 在 JDBC 中，（　　）接口不能被 Connection 创建。

  A．Statement　 B．PreparedStatement
  C．CallableStatement　 D．RowsetStatement

6. 查询结果集 ResultSet 对象是以统一的行列形式组织数据的，执行"ResultSet rs = stmt.executeQuery ("select bid,name,author,publish,price from book");"语句，得到的结果集 rs 的列数为（　　）。

  A．4　 B．5　 C．6　 D．不确定

7. 给出了如下的查询条件字符串"String condition="insert book values(?,?,?,?,?)";"，（　　）接口适合执行该 SQL 查询。

  A．Statement　 B．PreparedStatement
  C．CallableStatement　 D．不确定

第12章 Ajax、JSON 以及 jQuery 技术

学习目的与要求

本章重点介绍 Ajax、JSON 以及 jQuery。通过本章的学习，用户能够理解 Ajax 技术原理，掌握 Ajax、JSON 以及 jQuery 等技术应用。

本章主要内容

- Ajax 技术
- JSON 技术
- jQuery 技术

随着电子商务、电子政务等 Web 应用的广泛普及和人们对用户体验度的追求，传统的"发送请求—等待—响应新页面或刷新整个当前页面"交互模式的弊端逐渐暴露出来。例如，我们在注册邮箱时，如果该邮箱已经被注册，而注册者并不知情，等到信息填写完毕后单击"注册"按钮，才提示该邮箱已经被注册，于是又回到注册界面，但原来填写的注册信息没有了，需要重新填写一遍。这样的用户体验实在不可取。本章讲解 Ajax 技术的目的就是解决类似不合理的用户体验。

本章涉及的 Java 源文件保存在工程 ch12 的 src 中，涉及的 JSP 页面保存在工程 ch12 的 WebContent 中。

12.1 Ajax 技术

在 2005 年，Google 通过其 Google Suggest 使 Ajax 技术变得流行起来。Google Suggest 使用 Ajax 技术创造出动态性极强的 Web 界面：在 Google 的搜索框输入关键字时，JavaScript 会将这些字符发送到服务器，然后服务器会返回一个搜索建议的列表。后来 Google 又推出了 Google Maps 这种典型的富客户端应用。Google Maps 的地图支持鼠标的拖动、放大和缩小，地图随着鼠标的拖动进行新数据的加载，但页面本身无须重新加

载。这种整页无刷新下的动态交互性效果，使 Web 应用达到了近似桌面应用的效果，Ajax 技术随之迅速发展。

12.1.1 Ajax 简介

Ajax（Asynchronous Javascript And XML，异步 JavaScript 和 XML）是一种创建交互式网页应用的网页开发技术，使得"不刷新页面向服务器发起请求"成为可能。Ajax 应用与传统 Web 应用的请求过程有什么不同呢？下面从图 12.1 进行比较。

（a）传统 Web 应用　　　　（b）Ajax 应用

图 12.1　传统 Web 应用与 Ajax 应用请求过程比较

由图 12.1 可以看出，Ajax 应用与传统 Web 应用相比较，客户端浏览器通过 JavaScript 事件触发对 Ajax 引擎的调用，Ajax 引擎在 Ajax 应用中担负着一个中间层的任务，负责收集数据并通过 Ajax 的 XMLHttpRequest 对象向服务器发送 HTTP 请求，服务器处理完成后返回 XML、JSON 以及文本等类型的响应结果，Ajax 引擎根据响应结果类型进行解析后再配合 HTML 和 CSS 渲染，将结果显示到客户端页面。

12.1.2　XMLHttpRequest 对象

XMLHttpRequest 对象是 Ajax 的核心，Ajax 引擎使用 XMLHttpRequest 和服务器进行

异步通信，是 Ajax 技术得以实现的一个重要的 JavaScript 对象。

所有现代浏览器（IE7+、Firefox、Chrome、Safari 以及 Opera）均支持 XMLHttpRequest 对象（new XMLHttpRequest()），老版本的 Internet Explorer（IE5 和 IE6）使用 ActiveXObject 对象（new ActiveXObject("Microsoft.XMLHTTP")）。XMLHttpRequest 对象创建如例 12-1 所示。

例 12-1 创建 JSP 页面 example12_1.jsp，在该页面的 JavaScript 代码中创建 XMLHttpRequest 对象。

example12_1.jsp 页面的代码如下：

```jsp
<%@ page language="java" contentType="text/html; charset=UTF-8" pageEncoding="UTF-8"%>
<!DOCTYPE html>
<html>
<head>
<meta charset="UTF-8">
<title>example12_1.jsp</title>
<script type="text/javascript">
    function createXHR(){
        var xmlhttp;
        if (window.XMLHttpRequest){//支持 XMLHttpRequest
            //IE7+、Firefox、Chrome、Opera、Safari 浏览器执行代码
            xmlhttp = new XMLHttpRequest();
            alert("新浏览器");
        }else{
            //IE6、IE5 浏览器执行代码
            xmlhttp = new ActiveXObject("Microsoft.XMLHTTP");
            alert("老旧浏览器");
        }
    }
</script>
</head>
<body>
    <a href="javascript:createXHR()">创建 XMLHttpRequest 对象</a>
</body>
</html>
```

12.1.3　Ajax 向服务器发送请求

创建 XMLHttpRequest 对象后，Ajax 就可以使用 XMLHttpRequest 对象的 open() 和 send() 方法将请求发送到服务器，请求可分为 GET 方式请求和 POST 方式请求。

❶ **GET 请求**

GET 方式请求示例代码如下：

```
xmlhttp.open("GET","AjaxTestServlet",true);
xmlhttp.send();
```

其中，open(method,url,async)方法的参数 method 规定请求的类型，url 指定请求的 URL 地址，async 指定是否异步处理请求，true 表示请求本质上是异步的，默认值是 true；send(content)方法将请求发送到服务器，如果请求声明为异步，该方法就立即返回，否则将会等待直到接收到响应为止，可选参数 content 可以是 DOM 对象的实例、输入流或者字符串，传入参数的内容作为请求体的一部分发送，content 参数仅用于 POST 请求。

为了避免浏览器的缓存情况，可以向请求的 URL 添加一个唯一的 ID，上述发送请求示例代码可以修改如下：

```
xmlhttp.open("GET","AjaxTestServlet?t=" + Math.random(),true);
xmlhttp.send();
```

如果希望通过 GET 方式发送信息，请向 URL 添加信息，示例如下：

```
xmlhttp.open("GET","AjaxTestServlet?uname=chenheng&upwd=123456",true);
xmlhttp.send();
```

❷ POST 请求

POST 方式请求示例代码如下：

```
xmlhttp.open("POST","AjaxTestServlet",true);
xmlhttp.send();
```

如果需要像 HTML 表单那样提交 POST 数据，首先使用 setRequestHeader()来添加 HTTP 头，然后在 send()方法中规定希望发送的数据，示例代码如下：

```
xmlhttp.open("POST","AjaxTestServlet",true);
xmlhttp.setRequestHeader("Content-type","application/x-www-form-urlencoded");
xmlhttp.send("uname=chenheng&upwd=123456");
```

其中，setRequestHeader(header,value)方法的 header 参数规定头的名称，value 规定头的值。

是使用 POST 请求还是 GET 请求呢？我们看看它们的区别。GET 请求简单且更快，在大部分情况下都能用；然而，在以下情况中，请使用 POST 请求：
① 无法使用缓存文件（如更新服务器上的文件或数据库）；
② 向服务器发送大量数据（因为 POST 没有数据量限制）；
③ 发送包含未知字符的用户输入时（因为 POST 比 GET 更稳定、更可靠）。

12.1.4 Ajax 获得服务器的响应

Ajax 向服务器发送请求后，如需获得服务器的响应，可以使用 XMLHttpRequest 对象的 responseText 或 responseXML 属性。示例代码如下：

```
document.getElementById("myResponse").innerHTML=xmlhttp.responseText;
```

responseText 属性包含客户端接收到服务器的 HTTP 响应的文本内容。当 readyState 属性值为 0（请求未初始化）、1（服务器连接已建立）或 2（请求已接收）时，responseText 包含一个空字符串；当 readyState 属性值为 3（请求处理中）时，responseText 包含还未完成的响应信息；当 readyState 属性值为 4（请求已完成，且响应已就绪）时，responseText 包含完整的响应信息。

responseXML 属性用于接收 XML 形式的响应信息。此时，Content-Type 头部指定 MIME 类型为 text/xml、application/xml 或以+xml 结尾。如果 Content-Type 头部不包含这些 MIME 类型之一，那么 responseXML 属性值为 null。但无论何时，readyState 属性值不为 4 时，responseXML 属性值也为 null。

Ajax 获得服务器的响应时，除了用到 responseText、responseXML 和 readyState 属性外，还经常用到 onreadystatechange 和 status 属性。下面分别介绍它们。

onreadystatechange 属性指定每次状态改变所触发事件的事件处理函数。也就是说，每当 readyState 属性改变时，就会调用 onreadystatechange 属性指定的函数。

status 属性描述了 HTTP 状态码。需要注意的是只有 readyState 的值为 3 或 4 时，才能使用 status 属性。常用的状态码有 200（请求成功）、202（请求被接收但处理未完成）、400（错误请求）、404（请求资源未找到）和 500（内部服务器错误）。

在 onreadystatechange 事件中，我们规定当服务器响应已做好被处理的准备时所执行的任务。当 readyState 等于 4 且状态为 200 时，表示响应已就绪，代码如下：

```
xmlhttp.onreadystatechange=function(){
    if (xmlhttp.readyState==4 && xmlhttp.status==200){
        document.getElementById("myResponse").innerHTML=xmlhttp.responseText;
    }
}
```

12.1.5　Ajax 示例

一个 Ajax 应用的实现一般需要以下几个步骤：
① 在页面中定义 Ajax 请求的触发事件。
② 创建 Ajax 对象"var xmlhttp =new XMLHttpRequest();"。
③ 确定请求地址和请求参数。
④ 调用 XMLHttpRequest 对象的 open()方法建立对服务器的调用："xmlhttp.open("GET","AjaxTestServlet",true);"。
⑤ 通过 XMLHttpRequest 对象的 onreadystatechange 属性指定响应事件的处理函数。
⑥ 在函数中根据响应状态进行数据获取和数据处理工作。
⑦ 调用 XMLHttpRequest 对象的 send()方法发送请求"xmlhttp.send(null)"。

例 12-2　创建 JSP 页面 example12_2.jsp，在该页面中演示在用户输入完学号时触发 Ajax 异步请求，从服务器获取学号所对应的学生信息，并对页面中相应的学生信息进行更新填充。example12_2.jsp 的运行效果如图 12.2 所示。

第 12 章 Ajax、JSON 以及 jQuery 技术

图 12.2　Ajax 示例运行效果

页面 example12_2.jsp 的代码如下：

```jsp
<%@ page language="java" contentType="text/html; charset=UTF-8" pageEncoding=
"UTF-8"%>
<!DOCTYPE html>
<html>
<head>
<meta charset="UTF-8">
<title>example12_2.jsp</title>
<script type="text/javascript">
function createXHR(){
    var xmlhttp;
    if (window.XMLHttpRequest){//支持 XMLHttpRequest
        xmlhttp = new XMLHttpRequest();
    }else{
        xmlhttp = new ActiveXObject("Microsoft.XMLHTTP");
    }
    return xmlhttp;
}
function ajaxFunction(sno){
    //创建 XMLHttpRequest 对象
    xhr = createXHR();
    //设定请求地址
    var url = "AjaxTestServlet?sno=" + sno;
    //建立对服务器的调用
    xhr.open("GET", url, true);
    //指定响应事件处理函数
    xhr.onreadystatechange = function(){
        //当 readyState = 4 且状态为 200 时，表示响应已就绪
        if(xhr.readyState ==4 && xhr.status == 200){
            //对响应结果进行处理
            var resData = xhr.responseText.split(",");
            //将响应数据更新到页面控件中显示
```

```
                    document.getElementById("sname").value = resData[0];
                    document.getElementById("sage").value = resData[1];
                    document.getElementById("sdept").value = resData[2];
                }
            };
            //向服务器发出请求
            xhr.send();
        }
    </script>
</head>
<body>
    <h1>获取学号对应的学生信息</h1>
    <p>学号：<input type="text" name="sno" onblur="ajaxFunction(this.value)"/></p>
    <p>姓名：<input type="text" id="sname" name="sname" readonly/></p>
    <p>年龄：<input type="text" id="sage" name="sage" readonly/></p>
    <p>院系：<input type="text" id="sdept" name="sdept" readonly/></p>
</body>
</html>
```

AjaxTestServlet.java 的代码如下：

```java
package servlet;
import java.io.IOException;
import java.util.HashMap;
import java.util.Map;
import javax.servlet.ServletException;
import javax.servlet.annotation.WebServlet;
import javax.servlet.http.HttpServlet;
import javax.servlet.http.HttpServletRequest;
import javax.servlet.http.HttpServletResponse;
@WebServlet("/AjaxTestServlet")
public class AjaxTestServlet extends HttpServlet {
    private static final long serialVersionUID = 1L;
    protected void doGet(HttpServletRequest request, HttpServletResponse response) throws ServletException, IOException {
        //设置请求和响应字符编码
        request.setCharacterEncoding("UTF-8");
        response.setContentType("text/html;charset=UTF-8");
        //使用 Map 模拟学生信息的数据库
        Map<String, String> stus = new HashMap<String, String>();
        stus.put("20190101", "陈恒01,77,计算机学院");
        stus.put("20190102", "陈恒02,88,软件学院");
        stus.put("20190103", "陈恒03,99,信息学院");
        //获取 Ajax 请求的数据
        String sno = request.getParameter("sno");
        //根据学号从模拟数据库中查询学生信息
```

```
        String stu = stus.get(sno);
        if(stu == null) {
            //3个error对应学生的3个信息
            stu = "error,error,error";
        }
        //将请求结果数据响应输出
        response.getWriter().print(stu);
    }
    protected void doPost(HttpServletRequest request, HttpServletResponse response) throws ServletException, IOException {
        doGet(request, response);
    }
}
```

12.2 JSON 技术

视频讲解

JSON（JavaScript Object Notation，JS 对象标记）是一种轻量级的数据交换格式。

12.2.1 JSON 的数据结构

与 XML 一样，JSON 也是基于纯文本的数据格式。它有两种数据结构。

❶ 对象结构

对象结构以"{"开始，以"}"结束。中间部分由 0 个或多个以英文","分隔的 key/value 对构成，key 和 value 之间以英文":"分隔。对象结构的语法结构如下：

```
{
    key1:value1,
    key2:value2,
    …
}
```

其中，key 必须为 String 类型，value 可以是 String、Number、Object、Array 等数据类型。例如，一个 person 对象包含姓名、密码、年龄等信息，使用 JSON 的表示形式如下：

```
{
    "pname":"陈恒",
    "password":"123456",
    "page":40
}
```

❷ 数组结构

数组结构以"["开始，以"]"结束。中间部分由 0 个或多个以英文","分隔的值的列表组成。数组结构的语法结构如下：

```
[
    value1,
    value2,
    …
]
```

上述两种（对象、数组）数据结构也可以分别组合构成更为复杂的数据结构。例如，一个 student 对象包含 sno、sname、hobby 和 college 对象，其 JSON 的表示形式如下：

```
{
    "sno":"2019022299999",
    "sname":"陈恒",
    "hobby":["篮球","足球"],
    "college":{
        "cname":"清华大学",
        "city":"北京"
    }
}
```

12.2.2　JSON 在 JavaScript 中的使用

JSON 是 JavaScript 的原生格式，在 JavaScript 中处理 JSON 数据不需要任何 API 或工具包。

例 12-3　创建 JSP 页面 example12_3.jsp，在该页面的 JavaScript 中处理 JSON 数据。

页面 example12_3.jsp 的代码如下：

```
<%@ page language="java" contentType="text/html; charset=UTF-8" pageEncoding="UTF-8"%>
<!DOCTYPE html>
<html>
<head>
<meta charset="UTF-8">
<title>example12_3.jsp</title>
<script type="text/javascript">
function opJSON(){
    //将 JSON 数据赋值给变量 stu
    var stu = {
            "sno":"2019022299999",
            "sname":"陈恒",
            "hobby":["篮球","足球"],
            "college":{
                "cname":"清华大学",
                "city":"北京"
            }
    };
```

```
            //获取JSON数据中的对象信息
            alert(stu.sno);
            alert(stu.hobby[0]);
            alert(stu.college.cname);
        }
    </script>
</head>
<body>
    <a href="javascript:opJSON()">在JavaScript中处理JSON数据</a>
</body>
</html>
```

在JavaScript中可以使用eval()函数将JSON文本转换为JavaScript对象。例如，对Ajax请求响应结果（假设响应数据格式为JSON）的转换。示例代码如下：

```
//定义JSON文本变量
    var jsonText = "{\"sname\":\"陈恒\", \"sage\":\"88\"}";
    //将JSON文本转换为JavaScript对象
    //两边加圆括号将参数值无条件地视为表达式进行解析
    var jsObj = eval("("+ jsonText + ")");
    //获取对象属性
    alert(jsObj.sname);
```

在JavaScript中可以使用JSON.stringify()方法将JavaScript对象转换为字符串。示例代码如下：

```
var obj = { "sname":"陈恒", "sno":20190101, "sage":88};
var myJSON = JSON.stringify(obj);
alert(myJSON);
```

12.2.3　JSON在Ajax中的使用

从12.2.2节可以看出拼写一个JSON文本变量非常烦琐且易出错。若服务器需要响应一个结构复杂、数据量大的数据对象，那么转换过程将更加费时费力。因此，在实际开发中，Java对象和JSON数据之间的互相转换通常使用第三方插件来协助完成，例如JSON-Lib、Jackson、Gson和FastJson等。本节以Jackson为例，讲解JSON在Ajax中的使用。

Jackson开源包及其描述如下所示。

- jackson-annotations.jar：JSON转换注解包。
- jackson-core.jar：JSON转换核心包。
- jackson-databind.jar：JSON转换数据绑定包。

以上3个Jackson开源包，本书采用的版本是2.9.4，读者可通过地址"http://mvnrepository.com/artifact/com.fasterxml.jackson.core"下载得到。

在使用Jackson插件进行Java对象和JSON数据之间的互相转换时，需要将以上3个

Jackson 开源包复制到 Web 应用的/WEB-INF/lib 目录下，本章将以上 3 个 Jackson 开源包复制到 Web 应用 ch12 的/WEB-INF/lib 目录下。

下面使用 Jackson 插件将例 12-2 的代码进行改进。

修改后的 example12_2.jsp 代码如下：

```jsp
<%@ page language="java" contentType="text/html; charset=UTF-8" pageEncoding="UTF-8"%>
<!DOCTYPE html>
<html>
<head>
<meta charset="UTF-8">
<title>example12_2.jsp</title>
<script type="text/javascript">
function createXHR(){
    var xmlhttp;
    if (window.XMLHttpRequest){//支持 XMLHttpRequest
        xmlhttp = new XMLHttpRequest();
    }else{
        xmlhttp = new ActiveXObject("Microsoft.XMLHTTP");
    }
    return xmlhttp;
}
//在 Ajax 中使用 JSON
function ajaxFunction(sno){
    //创建 XMLHttpRequest 对象
    xhr = createXHR();
    //设定请求地址
    var url = "AjaxTestServlet?sno=" + sno;
    //建立对服务器的调用
    xhr.open("GET", url, true);
    //指定响应事件处理函数
    xhr.onreadystatechange = function(){
        //当 readyState = 4 且状态为 200 时，表示响应已就绪
        if(xhr.readyState ==4 && xhr.status == 200){
            //将响应的 JSON 格式转换为 JavaScript 对象
            var resData = eval("(" + xhr.responseText + ")");
            //将响应数据更新到页面控件中显示
            document.getElementById("sname").value = resData.sname;
            document.getElementById("sage").value = resData.sage;
            document.getElementById("sdept").value = resData.sdept;
        }
    };
    //向服务器发出请求
    xhr.send();
}
```

```
</script>
</head>
<body>
    <h1>获取学号对应的学生信息</h1>
    <p>学号：<input type="text" name="sno" onblur="ajaxFunction(this.value)"/></p>
    <p>姓名：<input type="text" id="sname" name="sname" readonly/></p>
    <p>年龄：<input type="text" id="sage" name="sage" readonly/></p>
    <p>院系：<input type="text" id="sdept" name="sdept" readonly/></p>
</body>
</html>
```

修改后的 AjaxTestServlet.java 代码如下：

```
package servlet;
import java.io.IOException;
import java.util.HashMap;
import java.util.Map;
import javax.servlet.ServletException;
import javax.servlet.annotation.WebServlet;
import javax.servlet.http.HttpServlet;
import javax.servlet.http.HttpServletRequest;
import javax.servlet.http.HttpServletResponse;
import com.fasterxml.jackson.databind.ObjectMapper;
import entity.Student;
@WebServlet("/AjaxTestServlet")
public class AjaxTestServlet extends HttpServlet {
    private static final long serialVersionUID = 1L;
    protected void doGet(HttpServletRequest request, HttpServletResponse
    response) throws ServletException, IOException {
        //设置请求和响应字符编码
        request.setCharacterEncoding("UTF-8");
        response.setContentType("text/html;charset=UTF-8");
        //定义3个学生信息对象
        Student s1 = new Student("20190101","陈恒01",77,"计算机学院");
        Student s2 = new Student("20190102","陈恒02",88,"软件学院");
        Student s3 = new Student("20190103","陈恒03",99,"信息学院");
        //使用Map模拟学生信息的数据库
        Map<String, Student> stus = new HashMap<String, Student>();
        stus.put(s1.getSno(), s1);
        stus.put(s2.getSno(), s2);
        stus.put(s3.getSno(), s3);
        //获取Ajax请求的数据
        String sno = request.getParameter("sno");
        //根据学号从模拟数据库中查询学生信息
        Student stu = stus.get(sno);
        if(stu == null) {
```

```
            stu = new Student("error","error", 0,"error");
        }
        //创建Jackson插件的ObjectMapper对象
        ObjectMapper mapper = new ObjectMapper();
        //将一个Java对象转换成JSON
        mapper.writeValue(response.getWriter(), stu);
    }
    protected void doPost(HttpServletRequest request, HttpServletResponse response) throws ServletException, IOException {
        doGet(request, response);
    }
}
```

新增 Student 实体类，代码如下：

```
package entity;
public class Student {
    String sno;
    String sname;
    int sage;
    String sdept;
    public Student(String sno, String sname, int sage, String sdept) {
        super();
        this.sno = sno;
        this.sname = sname;
        this.sage = sage;
        this.sdept = sdept;
    }
    //省略 get 和 set 方法
}
```

12.3　jQuery 技术

12.3.1　jQuery 简介

视频讲解

jQuery 是一个快速、简洁的 JavaScript 框架，是继 Prototype 之后又一个优秀的 JavaScript 代码库（或 JavaScript 框架）。jQuery 设计的宗旨是"Write Less，Do More"，即倡导写更少的代码，做更多的事情。它封装 JavaScript 常用的功能代码，提供一种简便的 JavaScript 设计模式，优化 HTML 文档操作、事件处理、动画设计和 Ajax 交互。

jQuery 的核心特性可以总结为：具有独特的链式语法和短小清晰的多功能接口；具有高效灵活的 CSS 选择器，并且可对 CSS 选择器进行扩展；拥有便捷的插件扩展机制和丰富的插件。jQuery 兼容各种主流浏览器，如 IE 6.0+、FF 1.5+、Safari 2.0+、Opera 9.0+等。

12.3.2　jQuery 的安装

首先从 jQuery 官方网站 http://jquery.com/download/ 下载 jQuery 插件 jquery-3.2.1.min.js，将其复制到 Web 项目开发目录的 WebContent/js 目录下。然后在 JSP 页面中，通过<script type="text/javascript" src="js/jquery-3.2.1.min.js"></script>代码将 jquery-3.2.1.min.js 引入当前页面中。

12.3.3　jQuery 语法

jQuery 语法是选取 HTML 元素，并对选取的元素执行某些操作，语法如下：

```
$(selector).action()
```

其中，美元符号$定义 jQuery；选择符（selector）"查询"和"查找"HTML 元素；jQuery 的 action()执行对元素的操作。示例如下：

$("#test").hide()：隐藏所有 id="test"的元素。
$("#test").text("Hello world!")：设置 id="test"元素的文本内容为"Hello world！"。
$("#test").html("Hello world!")：设置 id="test"元素的内容（包括 HTML 标记）。
$("#test").val("chenheng")：设置 id="test"的表单字段的值。
$("#test").text()：返回 id="test"元素的文本内容。
$("#test").html()：返回 id="test"元素的内容（包括 HTML 标记）。
$("#test").val()：获得 id="test"的表单字段的值。

12.3.4　jQuery 对 Ajax 的实现

jQuery 提供多个与 Ajax 有关的方法，下面结合实例介绍两个常用的方法：ajax( )方法和 load( )方法。

❶ **ajax()方法**

ajax()方法只有一个参数：key/value 对象，包含各配置及回调函数信息。示例如例 12-4 所示。

【例】**12-4**　编写 JSP 页面 example12_4.jsp，在该页面中使用 ajax()方法向服务器发送 Ajax 异步请求，请求地址为"/TestjQueryAjaxServlet"，请求参数为表单参数，在响应成功时，用响应数据更新 id 值为 myInput 元素的内容。

页面 example12_4.jsp 的代码如下：

```
<%@ page language="java" contentType="text/html; charset=UTF-8" pageEncoding=
"UTF-8"%>
<!DOCTYPE html>
```

```html
<html>
<head>
<meta charset="UTF-8">
<title>example12_4.jsp</title>
<script type="text/javascript" src="js/jquery-3.2.1.min.js"></script>
<script type="text/javascript">
function testjQueryAjax(){
    $.ajax({
        //发送请求的地址
        url:"TestjQueryAjaxServlet",
        //请求方式,默认为get
        type:"post",
        //预期服务器返回的数据类型。如果不指定,自动根据MIME类型返回信息
        dataType:"text",
        //发送到服务器的数据
        data:{
            sno:$("#sno").val(),
            sname:$("#sname").val(),
            sage:$("#sage").val(),
            sdept:$("#sdept").val()
        },
        //请求成功后回调函数,参数为服务器返回的数据
        success:function(data){
            //将JSON文本转换为JavaScript对象
            var stu = eval("("+ data + ")");
            $("#myInput").html("学号: " + stu.sno + "<br>姓名: " + stu.sname
                + "<br>年龄: " + stu.sage + "<br>院系: " + stu.sdept);
        },
        //请求失败时调用的函数
        error:function(){
            alert("处理异常");
        }
    });
}
</script>
</head>
<body>
    <form id="myForm">
        <p>学号: <input type="text"  id="sno" name="sno"/></p>
        <p>姓名: <input type="text" id="sname" name="sname"/></p>
        <p>年龄: <input type="text" id="sage" name="sage"/></p>
        <p>院系: <input type="text" id="sdept" name="sdept"/></p>
        <p><input type="button" value="测试Ajax" onclick="testjQueryAjax()"/></p>
```

```
        <p><span id="myInput"></span></p>
    </form>
</body>
</html>
```

TestjQueryAjaxServlet.java 的代码如下：

```
package servlet;
import java.io.IOException;
import javax.servlet.ServletException;
import javax.servlet.annotation.WebServlet;
import javax.servlet.http.HttpServlet;
import javax.servlet.http.HttpServletRequest;
import javax.servlet.http.HttpServletResponse;
import com.fasterxml.jackson.databind.ObjectMapper;
import entity.Student;
@WebServlet("/TestjQueryAjaxServlet")
public class TestjQueryAjaxServlet extends HttpServlet {
    private static final long serialVersionUID = 1L;
    protected void doGet(HttpServletRequest request, HttpServletResponse response) throws ServletException, IOException {
        //设置请求和响应字符编码
        request.setCharacterEncoding("UTF-8");
        response.setContentType("text/html;charset=UTF-8");
        String sno = request.getParameter("sno");
        String sname = request.getParameter("sname");
        String sage = request.getParameter("sage");
        String sdept = request.getParameter("sdept");
        int isage = Integer.parseInt(sage);
        //创建学生对象
        Student s = new Student(sno, sname,isage,sdept);
        //创建 Jackson 插件的 ObjectMapper 对象
        ObjectMapper mapper = new ObjectMapper();
        //将一个 Java 对象转换成 JSON
        mapper.writeValue(response.getWriter(), s);
    }
    protected void doPost(HttpServletRequest request, HttpServletResponse response) throws ServletException, IOException {
        doGet(request, response);
    }
}
```

在例 12-4 中，Ajax 可以根据表单的 id 提交整个表单参数，需要注意的是表单参数的 id 和 name 相同。修改后的 JS 函数 testjQueryAjax()的代码如下：

```javascript
function testjQueryAjax(){
    $.ajax({
        //发送请求的地址
        url:"TestjQueryAjaxServlet",
        //请求方式，默认为get
        type:"post",
        //预期服务器返回的数据类型。如果不指定，自动根据MIME类型返回信息
        dataType:"text",
        //发送到服务器的数据
        data: $("#myForm").serializeArray(),
        //请求成功后回调函数，参数为服务器返回的数据
        success:function(data){
            //将JSON文本转换为JavaScript对象
            var stu = eval("("+ data + ")");
            $("#myInput").html("学号: " + stu.sno + "<br>姓名: " + stu.sname
                +"<br>年龄: " + stu.sage + "<br>院系: " + stu.sdept);
        },
        //请求失败时调用的函数
        error:function(){
            alert("处理异常");
        }
    });
}
```

运行 example12_4.jsp 页面并输入数据，然后单击"测试 Ajax"按钮，效果如图 12.3 所示。

图 12.3　ajax( )方法运行效果图

❷ load()方法

load( )方法从服务器加载数据，并将返回的数据放入被选元素中。语法如下：

```
$(selector).load(URL,data,callback);
```

其中，必需的 URL 参数规定希望加载的 URL；可选的 data 参数规定与请求一同发送的查询字符串键/值对集合；可选的 callback 参数规定当 load( )方法完成后所要允许的回调函数。回调函数可以设置不同的参数：responseTxt 表示包含调用成功时的结果内容；statusTxt 表示包含调用的状态；xhr 表示包含 XMLHttpRequest 对象。

例 12-5 编写 JSP 页面 example12_5.jsp，在该页面中使用 load()方法加载服务器 LoadTestServlet 响应的结果，并显示在页面上。运行 example12_5.jsp 页面，效果如图 12.4 所示。

```
← → ■ ⌕   http://localhost:8080/ch12/example12_5.jsp
学号：001
姓名：陈恒
年龄：88
院系：信息学院
```

图 12.4　load()方法运行效果图

页面 example12_5.jsp 的代码如下：

```
<%@ page language="java" contentType="text/html; charset=UTF-8" pageEncoding="UTF-8"%>
<!DOCTYPE html>
<html>
<head>
<meta charset="UTF-8">
<title>example12_5.jsp</title>
<script type="text/javascript" src="js/jquery-3.2.1.min.js"></script>
</head>
<body>
<span id="myReceive"></span>
<script type="text/javascript">
//加载 LoadTestServlet 响应数据同时，传递数据 sname
$("#myReceive").load("LoadTestServlet",{sname:"陈恒"},function(responseTxt,
statusTxt, xhr){
    if(statusTxt=="success"){
        //将 JSON 文本转换为 JavaScript 对象
        var stu = eval("("+ responseTxt + ")");
        $("#myReceive").html("学号：" + stu.sno + "<br>姓名：" + stu.sname
                    + "<br>年龄：" + stu.sage + "<br>院系：" + stu.sdept);
    }
    if(statusTxt=="error")
        alert("Error: "+xhr.status+": "+xhr.statusText);
});
</script>
```

```
</body>
</html>
```

LoadTestServlet.java 的代码如下：

```java
package servlet;
import java.io.IOException;
import javax.servlet.ServletException;
import javax.servlet.annotation.WebServlet;
import javax.servlet.http.HttpServlet;
import javax.servlet.http.HttpServletRequest;
import javax.servlet.http.HttpServletResponse;
import com.fasterxml.jackson.databind.ObjectMapper;
import entity.Student;
@WebServlet("/LoadTestServlet")
public class LoadTestServlet extends HttpServlet {
    private static final long serialVersionUID = 1L;
    protected void doGet(HttpServletRequest request, HttpServletResponse response) throws ServletException, IOException {
        request.setCharacterEncoding("UTF-8");
        response.setContentType("text/html;charset=UTF-8");
        String sname = request.getParameter("sname");
        System.out.println("sname:" + sname);
        //创建学生对象
        Student s = new Student("001","陈恒",88,"信息学院");
        //创建 Jackson 插件的 ObjectMapper 对象
        ObjectMapper mapper = new ObjectMapper();
        //将一个 Java 对象转换成 JSON
        mapper.writeValue(response.getWriter(), s);
    }
    protected void doPost(HttpServletRequest request, HttpServletResponse response) throws ServletException, IOException {
        doGet(request, response);
    }
}
```

12.4　实践环节——注册邮箱的唯一性验证

使用 jQuery Ajax 技术实现注册邮箱的唯一性验证。

12.5　本章小结

本章学习了 Ajax、JSON 以及 jQuery 技术，重点掌握三者结合的应用，即 JSON 在 Ajax 中的应用以及 jQuery 对 Ajax 的实现。

习题 12

1. XMLHttpRequest 对象有（　　）个返回状态值。
   A. 3　　　　　B. 4　　　　　C. 5　　　　　D. 6
2. 对象 XMLHttpRequest 的属性 readyState 值为（　　）表示异步访问服务器通信已经完成。
   A. 1　　　　　B. 2　　　　　C. 3　　　　　D. 4

第13章 电子商务平台的设计与实现

学习目的与要求

本章通过一个小型的电子商务平台，讲述如何使用 MVC（JSP+JavaBean+Servlet）模式来开发一个 Web 应用。通过本章的学习，掌握 Java Web 应用开发的流程、方法以及技术。

本章主要内容
- 系统设计
- 数据库设计
- 系统管理
- 组件设计
- 系统实现

视频讲解

本章系统使用 Servlet MVC 模式实现各个模块，JSP 引擎为 Tomcat 9.0，数据库采用的是 MySQL 5.5，集成开发环境为 Eclipse IDE for Java EE Developers。

13.1 系统设计

电子商务平台分为两个子系统，一是后台管理子系统，一是电子商务子系统。下面分别说明这两个子系统的功能需求与模块划分。

13.1.1 系统功能需求

❶ 后台管理子系统

后台管理子系统要求管理员登录成功后，才能对商品进行管理，包括添加商品、查询商品、修改商品以及删除商品。除商品管理外，管理员还需要对商品类型、注册用户、用户的订单以及网站公告等进行管理。

❷ 电子商务子系统

1）非注册用户

非注册用户或未登录用户具有的功能如下：浏览首页、查看商品详情和查看公告。

2）用户

成功登录的用户除具有未登录用户具有的功能外，还具有购买商品、查看购物车、关注商品以及查看用户中心的功能。

13.1.2　系统模块划分

❶ 后台管理子系统

管理员登录成功后，进入后台管理主页面（main.jsp），可以对商品及商品类型、注册用户、用户的订单以及网站公告进行管理。后台管理子系统的模块划分如图 13.1 所示。

图 13.1　后台管理子系统

❷ 电子商务子系统

非注册用户只可以浏览商品、查看公告，不能购买商品、关注商品、查看购物车和查看用户中心。成功登录的用户可以完成电子商务子系统的所有功能，包括购买商品、支付等。电子商务子系统的模块划分如图 13.2 所示。

图 13.2　电子商务子系统

13.2 数据库设计

系统采用加载纯 Java 数据库驱动程序的方式连接 MySQL 5.5 数据库。在 MySQL 5.5 中创建数据库 shop，并在 shop 中创建 9 张与系统相关的数据表：ausertable、busertable、carttable、focustable、goodstable、goodstype、orderdetil、orderbasetable 和 noticetable。

13.2.1 数据库概念结构设计

根据系统设计与分析，可以设计出如下数据结构。

❶ 管理员

管理员包括用户名和密码。管理员的用户名和密码由数据库管理员预设，不需要注册。

❷ 用户

用户包括用户 ID、邮箱和密码。注册用户的邮箱不能相同，用户 ID 唯一。

❸ 商品类型

商品类型包括类型 ID 和类型名称。商品类型由数据库管理员管理，包括新增和删除管理。

❹ 商品

商品包括商品编号、名称、原价、现价、库存、图片以及类型。其中，商品编号唯一，类型与"3．商品类型"关联。

❺ 购物车

购物车包括购物车 ID、用户 ID、商品编号以及购买数量。其中，购物车 ID 唯一，用户 ID 与"2．用户"关联，商品编号与"4．商品"关联。

❻ 关注商品

关注商品包括 ID、用户 ID、商品编号以及关注时间。其中，ID 唯一，用户 ID 与"2．用户"关联，商品编号与"4．商品"关联。

❼ 订单基础信息

订单基础信息包括订单编号、用户 ID、订单金额、订单状态以及下单时间。其中，订单编号唯一，用户 ID 与"2．用户"关联。

❽ 订单详情

订单详情包括订单编号、商品编号以及购买数量。其中，订单编号与"7．订单基础信息"关联，商品编号与"4．商品"关联。

❾ 公告

公告包括 ID、标题、内容以及公告时间。其中，ID 唯一。

根据以上的数据结构，结合数据库设计的特点，可以画出如图 13.3 所示的数据库概念结构图。

图 13.3　数据库概念结构图

13.2.2　数据库逻辑结构设计

将数据库概念结构图转换为 MySQL 数据库所支持的实际数据模型，即数据库的逻辑结构。

管理员信息表（ausertable）的设计如表 13.1 所示。

表 13.1　管理员信息表

字　段	含　义	类　型	长　度	是否为空
aname	用户名（PK）	varchar	50	no
apwd	密码	varchar	50	no

用户信息表（busertable）的设计如表 13.2 所示。

表 13.2　用户信息表

字　段	含　义	类　型	长　度	是否为空
id	用户 ID（PK 自增）	int	11	no
bemail	E-mail	varchar	50	no
bpwd	密码	varchar	50	no

商品类型表（goodstype）的设计如表 13.3 所示。

表 13.3　商品类型表

字　　段	含　　义	类　　型	长　　度	是否为空
id	类型 ID（PK 自增）	int	11	no
typename	类型名称	varchar	50	no

商品信息表（goodstable）的设计如表 13.4 所示。

表 13.4　商品信息表

字　　段	含　　义	类　　型	长　　度	是否为空
id	商品编号（PK 自增）	int	11	no
gname	名称	varchar	50	no
goprice	原价	double		no
grprice	现价	double		no
gstore	库存	int	11	no
gpicture	图片	varchar	50	no
goodstype_id	类型（FK）	int	11	no

购物车表（carttable）的设计如表 13.5 所示。

表 13.5　购物车表

字　　段	含　　义	类　　型	长　　度	是否为空
id	购物车 ID（PK 自增）	int	11	no
busertable_id	用户 ID（FK）	int	11	no
goodstable_id	商品编号（FK）	int	11	no
shoppingnum	购买数量	int	11	no

商品关注表（focustable）的设计如表 13.6 所示。

表 13.6　关注表

字　　段	含　　义	类　　型	长　　度	是否为空
id	ID（PK 自增）	int	11	no
goodstable_id	商品编号（FK）	int	11	no
busertable_id	用户 ID（FK）	int	11	no
focustime	关注时间	datetime		no

订单基础表（orderbasetable）的设计如表 13.7 所示。

表 13.7　订单基础表

字　　段	含　　义	类　　型	长　　度	是否为空
id	订单编号（PK 自增）	int	11	no
busertable_id	用户 ID（FK）	int	11	no
amount	订单金额	double		no
status	订单状态	tinyint	4	no
orderdate	下单时间	datetime		no

订单详情表（orderdetail）的设计如表 13.8 所示。

表 13.8 订单详情表

字　段	含　义	类　型	长　度	是否为空
id	ID（PK 自增）	int	11	no
orderbasetable_id	订单编号（FK）	int	11	no
goodstable_id	商品编号（FK）	int	11	no
shoppingnum	购买数量	int	11	no

公告表（noticetable）的设计如表 13.9 所示。

表 13.9 公告表

字　段	含　义	类　型	长　度	是否为空
id	公告 ID（PK 自增）	int	11	no
ntitle	标题	varchar	100	no
ncontent	内容	varchar	500	no
ntime	公告时间	datetime		

13.2.3 创建数据表

根据 13.2.2 小节的逻辑结构，创建数据表。篇幅有限，创建数据表的代码请读者参考本书提供的源代码 shop.sql。

13.3 系统管理

13.3.1 导入相关的 jar 包

新建一个 Web 工程 ch13，在 ch13 工程中开发本系统。在本系统的所有 JSP 页面中尽量使用 EL 表达式和 JSTL 标签。又因为本系统采用纯 Java 数据库驱动程序连接 MySQL5.5，所以需要将 mysql-connector-java-5.1.45-bin.jar、taglibs-standard-impl-1.2.5.jar 和 taglibs-standard-spec-1.2.5.jar 复制到 ch13/ WebContent/WEB-INF/lib 目录中。

13.3.2 JSP 页面管理

本系统由后台管理和电子商务两个子系统组成，为了方便管理，这两个子系统的 JSP 页面分开存放。在 WebContent/admin 目录下存放与后台管理子系统相关的 JSP 页面、CSS、JavaScript 和图片；在 WebContent/beforeUser 目录下存放与电子商务子系统相关的 JSP 页面、CSS、JavaScript 和图片。由于篇幅受限，本章只是附上 JSP 和 Java 文件的核心代码。

❶ 后台管理子系统

管理员在浏览器的地址栏中输入"http://localhost:8080/ch13/admin/login.jsp"访问登录页面，登录成功后，进入后台管理主页面（main.jsp）。main.jsp 的运行效果如图 13.4 所示。

图 13.4　后台管理主页面

后台管理主页面 main.jsp 的核心代码如下：

```
...
<body>
    <div id="header">
        <br>
        <br>
        <h1>欢迎您进入后台管理页面！</h1>
    </div>
    <div id="navigator">
        <ul>
            <li><a>商品管理</a>
                <ul>
                    <li><a href="admin/addGoods.jsp" target="center">添加商品
                    </a></li>
                    <li><a href="admin_selectGoods?act=deleteSelect" target="center">
                    删除商品</a></li>
```

```html
            <li><a href="admin_selectGoods?act=updateSelect" target="center">
            修改商品</a></li>
            <li><a href="admin_selectGoods?act=select" target="center">
            查询商品</a></li>
                </ul>
        </li>
        <li><a>类型管理</a>
            <ul>
        <li><a href="admin/addType.jsp" target="center">添加类型</a></li>
        <li><a href="admin_Type?act=deleteSelect" target="center">
        删除类型</a></li>
                </ul>
        </li>
        <li><a>用户管理</a>
            <ul>
        <li><a href="admin_userManager?act=manager" target="center">
        删除用户</a></li>
                </ul>
        </li>
        <li><a>订单管理</a>
            <ul>
        <li><a href="admin_orderManager?act=manager" target="center">
        删除订单</a></li>
                </ul>
        </li>
        <li><a>公告管理</a>
            <ul>
       <li><a href="admin_noticeManager?act=manager" target="center">
       添加公告</a></li>
                </ul>
        </li>
            <li><a href="admin_loginOut?act=admin">安全退出</a></li>
        </ul>
    </div>
    <div id="content">
        <iframe src="admin_selectGoods?act=select" name="center" frameborder="0">
        </iframe>
    </div>
    <div id="footer">Copyright ©XXXXXXX</div>
</body>
...
```

❷ 电子商务子系统

注册用户或游客在浏览器地址栏中输入"http://localhost:8080/ch13"就可以访问电子

商务子系统的首页（index.jsp）。index.jsp 的运行效果如图 13.5 所示。

图 13.5　电子商务子系统的首页

电子商务子系统的首页 index.jsp 的核心代码如下：

```
...
<div class="top10List clearfix">
    <c:forEach var="sg" items="${salelist}" varStatus="status">
        <ul class="clearfix">
            <img class="iteration" src="beforeUser/images/top_${status.index+1}.gif" />
            <li class="topimg">
                <a href="before_detail?gno=${sg.id}">
                <img class="samllimg" alt="" src="uploadFile/${sg.gpicture}"/>
                </a></li>
            <li class="iteration1">
              <a title="" href="before_detail?gno=${sg.id}">${sg.gname}
              </a><br /> 售价：<font
                class="f1">¥${sg.grprice}元</font><br /></li>
        </ul>
    </c:forEach>
</div>
...
<div class="top10List clearfix">
    <c:forEach var="sg" items="${focuslist}"
```

```html
                    varStatus="status">
            <ul class="clearfix">
                <img class="iteration" src="beforeUser/images/top_${status.index+1}.gif" />
                <li class="topimg"><a href="before_detail?gno=${sg.id}"><img
                        class="samllimg" alt="" src="uploadFile/${sg.gpicture}"/>
                        </a></li>
                <li class="iteration1"><a title=""
                        href="before_detail?gno=${sg.id}">${sg.gname}</a><br />售价：
                        <font class="f1">￥${sg.grprice}元 </font><br /></li>
            </ul>
        </c:forEach>
</div>
...
<div class="post_list ared">
    <ul>
        <c:forEach var="nt" items="${noticelist}">
            <li>
            <a href="javascript:openNotice('before_noticeDetail?nid=${nt.id}');">
            ${nt.ntitle}</a>
            </li>
        </c:forEach>
    </ul>
</div>
...
<div class="itemgood_nr clearfix">
    <ul>
        <c:forEach var="sg" items="${lastedlist}">
            <li>
                <div>
                    <p class="pic">
                        <a href="before_detail?gno=${sg.id}"><img
                            src="uploadFile/${sg.gpicture}" /></a>
                    </p>
                    <p class="wz">
                        <strong>
            <a href="before_detail?gno=${sg.id}">${sg.gname}</a></strong>
                            <em>现价:<span>￥${sg.grprice}</span></em>
                    </p>
                </div>
            </li>
        </c:forEach>
    </ul>
</div>
...
```

13.3.3 组件与 Servlet 管理

本系统使用的组件与 Servlet 包层次结构如图 13.6 所示。

图 13.6 包层次结构图

❶ dao 包

dao 包中存放的 Java 程序是实现数据库的操作。其中，BaseDao 是一个父类，该类负责连接数据库，执行增、删、改、查 SQL 语句；AdminDao 是 BaseDao 的一个子类，有关后台管理子系统的数据访问在该类中；BeforeUserDao 是 BaseDao 的另一个子类，有关电子商务子系统的数据访问在该类中。

❷ dto 包

dto 包中的类是实现数据封装的实体 bean（实体模型）。

❸ filters 包

filters 包中有两个过滤器，一是解决中文乱码，一是实现登录验证。

❹ service 包

service 包中有两个类：AdminService 和 BeforeUserService。AdminService 是后台管理子系统的业务模型；BeforeUserService 是电子商务子系统的业务模型。

❺ servlet 包

servlet 包有两个子包：admin 和 before。有关后台管理子系统的 Servlet 在 admin 子包中；有关电子商务子系统的 Servlet 在 before 子包中。

❻ util 包

util 包中存放的是系统的工具类，包括获取时间字符串以及获取上传文件的文件名。

13.4 组件设计

本系统的组件包括过滤器、验证码、实体模型、数据库操作（dao）、业务模型（service）以及工具类。

13.4.1 过滤器

本系统使用了两个过滤器：设置字符编码过滤器和登录验证过滤器。

❶ 设置字符编码过滤器

当用户提交请求时，在请求处理之前，系统使用过滤器对用户提交的信息进行解码与编码，避免了乱码的出现。

SetCharacterEncodingFilter.java 的核心代码如下：

```
...
public void doFilter(ServletRequest request, ServletResponse response,
FilterChain chain) throws IOException, ServletException {
    request.setCharacterEncoding("UTF-8");
    chain.doFilter(request, response);
}
...
```

❷ 登录验证过滤器

从系统分析得知，用户成功登录后，才能购买商品、关注商品；管理员成功登录后，才能管理商品、用户、订单以及公告。系统使用过滤器判别用户或管理员是否登录。

LoginValidateFilter.java 的核心代码如下：

```
...
@WebFilter("/*")
public class LoginValidateFilter implements Filter {
    public void destroy() {
    }
    public void doFilter(ServletRequest request, ServletResponse response,
    FilterChain chain)
            throws IOException, ServletException {
        HttpServletRequest req = (HttpServletRequest) request;
        HttpServletResponse resp = (HttpServletResponse) response;
        HttpSession session = req.getSession(true);
        resp.setContentType("text/html;");
        resp.setCharacterEncoding("UTF-8");
        //得到用户请求的 URI
        String request_uri = req.getRequestURI();
        //得到 Web 应用程序的上下文路径
        String ctxPath = req.getContextPath();
        //去除上下文路径，得到剩余部分的路径
        String uri = request_uri.substring(ctxPath.length());
        boolean adminflag = false;
        boolean beforeflag = false;
        //默认主页或 CSS 文件、图片文件、JS
```

```java
        if ("/".equals(uri) || uri.contains("css") || uri.contains("images")
                || uri.contains("jsjs")
                || uri.contains("uploadFile")) {
            chain.doFilter(request, response);
        }
        //JSP 请求
        if (uri.contains(".jsp")) {
            //admin 文件夹
            if (uri.contains("admin/")) {
                if (uri.contains("login.jsp") || null != session.getAttribute("admin")
                        || uri.contains("loginerror.jsp")) {
                    chain.doFilter(request, response);
                } else {
                    adminflag = true;
                }
            }
            //before 文件夹
            else if (uri.contains("beforeUser/")) {
                if (null != session.getAttribute("user") || uri.contains("register.jsp") || //注册页面
                        uri.contains("userLogin.jsp") || //登录
                        uri.contains("searchResult.jsp")//搜索结果页面
                        ) {
                    chain.doFilter(request, response);
                } else {
                    beforeflag = true;
                }
            }
            //除 admin 和 beforeUser 以外的 JSP
            else {
                chain.doFilter(request, response);
            }
        } //JSP 请求结束
        //Servlet 请求
        else {
            //后台管理的 Servlet 请求
            if (uri.contains("admin_")) {
                if (uri.contains("_login") || null != session.getAttribute("admin")) {
                    chain.doFilter(request, response);
                } else {
                    adminflag = true;
                }
            }
```

```
            //前台的 Servlet 请求
            else if (uri.contains("before_")) {
                if (null != session.getAttribute("user") || uri.contains
                ("_first") || //首页
                        uri.contains("_detail") ||              //商品详情
                        uri.contains("_userLogin") ||           //登录
                        uri.contains("_search") ||              //头部搜索
                        uri.contains("_noticeDetail") ||        //公告详情
                        uri.contains("_register") ||            //注册
                        uri.contains("_validateCode")           //验证码
                ) {
                    chain.doFilter(request, response);
                } else {
                    beforeflag = true;
                }
            }
        } //Servlet 请求结束
        if (beforeflag) {//前台没有登录
            String pah = "userLogin.jsp";
            PrintWriter out = resp.getWriter();
            out.println("您没有登录,请先<a href=beforeUser/" + pah +
            "target=_top>登录</a>! ");
            return;
        }
        if (adminflag) {//后台没有登录
            String pah = "login.jsp";
            PrintWriter out = resp.getWriter();
            out.println("您没有登录,请先<a href=admin/" + pah + " target=_top>
            登录</a>! ");
            return;
        }
    }
    public void init(FilterConfig fConfig) throws ServletException {
    }
}
```

13.4.2 验证码

本系统验证码的使用步骤如下。
❶ 产生验证码
使用 Servlet 类 ValidateCode 产生验证码,具体代码如下:

```
...
@WebServlet("/before_validateCode")
public class ValidateCode extends HttpServlet {
```

```java
private static final long serialVersionUID = 1L;
private char code[] = { 'a', 'b', 'c', 'd', 'e', 'f', 'g', 'h', 'i', 'j',
        'k', 'm', 'n', 'p', 'q', 'r', 's', 't', 'u', 'v',
        'w', 'x', 'y', 'z', 'A', 'B', 'C', 'D', 'E', 'F',
        'G', 'H', 'J', 'K', 'L', 'M', 'N', 'P', 'Q', 'R',
        'S', 'T', 'U', 'V', 'W', 'X', 'Y', 'Z', '2', '3',
        '4', '5', '6', '7', '8', '9' };
private static final int WIDTH = 50;
private static final int HEIGHT = 20;
private static final int LENGTH = 4;
protected void doGet(HttpServletRequest request, HttpServletResponse
response) throws ServletException, IOException {
    //TODO Auto-generated method stub
    //设置响应报头信息
    response.setHeader("Pragma", "No-cache");
    response.setHeader("Cache-Control", "no-cache");
    response.setDateHeader("Expires", 0);
    //设置响应的 MIME 类型
    response.setContentType("image/jpeg");
    BufferedImage image = new BufferedImage(WIDTH, HEIGHT,
            BufferedImage.TYPE_INT_RGB);
    Font mFont = new Font("Arial", Font.TRUETYPE_FONT, 18);
    Graphics g = image.getGraphics();
    Random rd = new Random();
    //设置背景颜色
    g.setColor(new Color(rd.nextInt(55) + 200, rd.nextInt(55) + 200, rd
            .nextInt(55) + 200));
    g.fillRect(0, 0, WIDTH, HEIGHT);
    //设置字体
    g.setFont(mFont);
    //画边框
    g.setColor(Color.black);
    g.drawRect(0, 0, WIDTH - 1, HEIGHT - 1);
    //随机产生的验证码
    String result = "";
    for (int i = 0; i < LENGTH; ++i) {
        result += code[rd.nextInt(code.length)];
    }
    HttpSession se = request.getSession();
    se.setAttribute("rand", result);
    //画验证码
    for (int i = 0; i < result.length(); i++) {
        g.setColor(new Color(rd.nextInt(200), rd.nextInt(200), rd
                .nextInt(200)));
        g.drawString(result.charAt(i) + "", 12 * i + 1, 16);
```

```java
        }
        //随机产生两个干扰线
        for (int i = 0; i < 2; i++) {
            g.setColor(new Color(rd.nextInt(200), rd.nextInt(200), rd
                    .nextInt(200)));
            int x1 = rd.nextInt(WIDTH);
            int x2 = rd.nextInt(WIDTH);
            int y1 = rd.nextInt(HEIGHT);
            int y2 = rd.nextInt(HEIGHT);
            g.drawLine(x1, y1, x2, y2);
        }
        //释放图形资源
        g.dispose();
        try {
            OutputStream os = response.getOutputStream();
            //输出图像到页面
            ImageIO.write(image, "JPEG", os);
        } catch (IOException e) {
            e.printStackTrace();
        }
    }
    protected void doPost(HttpServletRequest request, HttpServletResponse
    response) throws ServletException, IOException {
        doGet(request, response);
    }
}
```

❷ 在 JSP 页面显示验证码

显示验证码示例代码如下：

```
...
function refreshCode(){
   document.getElementById("code").src = "before_validateCode?t="+Math.random();
}
 ...
<tr>
   <td>
      <img id="code" src="before_validateCode"/>
   </td>
   <td class="ared">
      <a href="javascript:refreshCode();"><font color="blue">看不清，换一
      个！</font></a>
   </td>
   <td></td>
</tr>
...
```

13.4.3 实体模型

在控制层（Servlet）使用实体模型封装 JSP 页面提交的信息，然后由控制层将实体模型传递给业务层（Service），最后由业务层将实体模型传递给数据访问层（DAO）。实现实体模型的类中只有 get 和 set 方法，代码非常简单，不再赘述。

13.4.4 数据库操作

本系统有关数据库操作的 Java 类位于包 dao 中，为了方便管理，减少代码的冗余，所有执行数据库的命令方法都由 BaseDao 实现。有关后台管理子系统的数据访问由 AdminDao 实现；有关电子商务子系统的数据访问由 BeforeUserDao 实现。

❶ **BaseDao.java 的代码**

```java
...
public class BaseDao {
    //存放 Connection 对象的数组，数组被看成连接池
    static ArrayList<Connection> list = new ArrayList<Connection>();
    /**
     * @discription 获得链接
     */
    public synchronized static Connection getConnection() {
        Connection con = null;
        //如果连接池中有连接对象
        if (list.size() > 0) {
            return list.remove(0);
        }
        //连接池没有连接对象，创建连接放到连接池中
        else {
            for (int i = 0; i < 5; i++) {
                try {
                    Class.forName("com.mysql.jdbc.Driver");
                } catch (ClassNotFoundException e) {
                    //TODO Auto-generated catch block
                    e.printStackTrace();
                }
                //创建连接
                try {
                    con = DriverManager.getConnection(
                        "jdbc:mysql://localhost:3306/shop?characterEncoding=utf-8","root","root");
                    list.add(con);
                } catch (SQLException e) {
```

```java
                //TODO Auto-generated catch block
                e.printStackTrace();
            }
        }
    }
    return list.remove(0);
}
/**
 * @discription 关闭结果集
 */
public static void close(ResultSet rs) {
    if (rs != null) {
        try {
            rs.close();
        } catch (SQLException e) {
            //TODO Auto-generated catch block
            e.printStackTrace();
        }
    }
}
/**
 * @discription 关闭预处理
 */
public static void close(PreparedStatement pst) {
    if (pst != null) {
        try {
            pst.close();
        } catch (SQLException e) {
            //TODO Auto-generated catch block
            e.printStackTrace();
        }
    }
}
/**
 * @discription 关闭连接
 */
public synchronized static void close(Connection con) {
    if (con != null)
        list.add(con);
}
/**
 * @discription 关闭所有连接有关的对象
 */
public static void close(ResultSet rs,PreparedStatement ps,Connection con) {
    close(rs);
```

```java
        close(ps);
        close(con);
}
/**
 * @discription 新增、修改、删除处理
 * @param sql SQL 语句
 * @param param SQL 语句中的通配符对应的值,如果 SQL 语句无通配符,该数组为 null
 */
public boolean upadateByParams(String sql, Object param[]) {
    boolean flag = false;
    Connection con = getConnection();
    PreparedStatement ps = null;
    try {
        ps = con.prepareStatement(sql);
        if(param != null){
            for(int i = 1; i <= param.length; i++){
                ps.setObject(i, param[i-1]);
            }
        }
        int n = ps.executeUpdate();
        if(n > 0)
            flag = true;
    } catch (SQLException e) {
        //TODO Auto-generated catch block
        e.printStackTrace();
    }finally{
        close(null, ps, con);
    }
    return flag;
}
/**
 * @discription 新增、修改、删除多条记录(批量处理)
 * @param sql SQL 语句
 * @param param SQL 语句中的通配符对应的值,
 * 数组的长度代表处理的记录数,
 * {{X,X,X},{X,X,X},{X,X,X}},
 * 其中,{X,X,X}是每个 SQL 语句中的参数值
 * 如果 SQL 语句无通配符,该数组为 null
 * @return boolean
 */
public boolean BatchUpadateByParams(String sql, Object param[][]){
    Connection con = getConnection();
    PreparedStatement ps = null;
    try {
        ps = con.prepareStatement(sql);
```

```java
            if(param != null){
                for(int i = 0; i < param.length; i++){//多条记录
                    for(int j = 1; j <= param[i].length; j++){//每条记录里
                                                              //的参数替换
                        ps.setObject(j, param[i][j-1]);
                    }
                    ps.addBatch();
                }
                ps.executeBatch();
            }
            return true;
        } catch (SQLException e) {
            //TODO Auto-generated catch block
            e.printStackTrace();
            return false;
        }finally{
            close(null, ps, con);
        }
    }
    /**
     * @discription 查询
     * @param sql SQL 语句
     * @param param SQL 语句中的通配符对应的值,如果 SQL 语句无通配符,该数组为 null
     * @return List<Map<String, Object>>
     */
    public static List<Map<String, Object>> select(String sql,Object[] param){
        Connection con = getConnection();
        PreparedStatement ps = null;
        ResultSet rs = null;
        List<Map<String, Object>> list = new ArrayList<Map<String,Object>>();
        try {
            ps = con.prepareStatement(sql);
            if(param != null){
                for(int i = 1; i <= param.length; i++){
                    ps.setObject(i, param[i-1]);
                }
            }
            rs = ps.executeQuery();
            ResultSetMetaData rm = rs.getMetaData();
            //列数
            int count = rm.getColumnCount();
            while(rs.next()){
                Map<String,Object> map = new HashMap<String, Object>();
```

```java
            for(int i = 1; i <= count; i++){
                //key 为列名，value 为列值
            map.put(rm.getColumnName(i).toLowerCase(), rs.getObject(rm.getColumnName(i)));
            }
            list.add(map);
        }
    } catch (SQLException e) {
        //TODO Auto-generated catch block
        e.printStackTrace();
    }finally{
        close(rs, ps, con);
    }
    return list;
}
/**
 * 获得最后一个id
 */
public int getLastId(String sql, String sql1, Object[] param) {
    Connection con = getConnection();
    PreparedStatement ps = null;
    PreparedStatement ps1 = null;
    ResultSet rs = null;
    int id = 0;
    try {
        ps = con.prepareStatement(sql);
        if(param != null){
            for(int i = 1; i <= param.length; i++){
                ps.setObject(i, param[i-1]);
            }
        }
        ps.executeUpdate();
        ps1 = con.prepareStatement(sql1);
        rs = ps1.executeQuery();
        if(rs.next())
            id = rs.getInt(1);
        close(ps1);
    } catch (SQLException e) {
        e.printStackTrace();
    }finally{
        close(rs, ps, con);
    }
    return id;
```

            }
        }

❷ AdminDao.java 的代码

```java
package dao;
import java.util.ArrayList;
import java.util.List;
import java.util.Map;
import dto.AdminDTO;
import dto.GoodsDTO;
import dto.NoticeDTO;
public class AdminDao extends BaseDao{
    /**
     * @discription 管理员登录判定
     */
    public boolean adminLogin(AdminDTO adt){
        String sql = "select * from AUSERTABLE where ANAME = ? and APWD = ?";
        Object param[] = {adt.getAname(), adt.getApwd()};
        List<Map<String, Object>> list = select(sql, param);
        if(list.size() > 0){
            return true;
        }else{
            return false;
        }
    }
    /**
     * 添加类型
     */
    public boolean addType(String typename){
        String sql = "insert into goodstype values(null,?)";
        Object param[] = {typename};
        return upadateByParams(sql, param);
    }
    /**
     * 获得关联类型
     */
    public List<Map<String, Object>> getReType(String id){
        String sql = "select * from GOODSTABLE where goodstype_id = ?";
        Object param[] = {id};
        return select(sql, param);
    }
    /**
     * @discription 查询所有商品
     */
    public List<Map<String, Object>> selectGoods(){
```

```java
    String sql = "select * from GOODSTABLE gt,GOODSTYPE gy where
gt.goodstype_id = gy.id ";
    Object param[] = null;
    return select(sql, param);
}
/**
 * @discription 分页查询商品
 */
public List<Map<String, Object>> selectAllGoodsByPage(int startIndex,
int perPageSize){
    String sql = "select * from GOODSTABLE gt,GOODSTYPE gy where
    gt.goodstype_id = gy.id order by gt.id limit ?, ?";
    Object param[] = {startIndex, perPageSize};
    return select(sql, param);
}
/**
 * @discription 获得商品类型列表
 */
public List<Map<String, Object>> getGoodsType(){
    String sql = "select * from GOODSTYPE";
    Object param[] = null;
    return select(sql, param);
}
/**
 * 删除类型
 */
public boolean deleteType(String id) {
    String sql = "delete  from GOODSTYPE where id = ? ";
    Object param[] = {id};
    return upadateByParams(sql, param);
}
/**
 * @discription 添加商品
 */
public boolean addGoods(GoodsDTO gd){
    String sql = "insert into GOODSTABLE values(null,?,?,?,?,?,?)";
    Object param[] = {
            gd.getGname(),
            gd.getGoprice(),
            gd.getGrprice(),
            gd.getGstore(),
            gd.getGpicture(),
            gd.getGoodstype_id()
        };
    return upadateByParams(sql, param);
```

```java
}
/**
 * @discription 查询一个商品
 */
public List<Map<String, Object>> selectAGoods(String gno){
    String sql = "select * from GOODSTABLE gt,GOODSTYPE gy where
    gt.goodstype_id = gy.id and gt.id = ?";
    Object param[] = {gno};
    return select(sql, param);
}
/**
 * @discription 修改商品
 */
public boolean updateGoods(GoodsDTO gd){
    String sql = "update GOODSTABLE set GNAME = ?, GOPRICE = ?, GRPRICE
    = ?, GSTORE = ?, GPICTURE = ?, goodstype_id = ? where id = ?";
    Object param[] = {
                    gd.getGname(),
                    gd.getGoprice(),
                    gd.getGrprice(),
                    gd.getGstore(),
                    gd.getGpicture(),
                    gd.getGoodstype_id(),
                    gd.getId()
                    };
    return upadateByParams(sql, param);
}
/**
 * @discription 删除商品
 */
public boolean deleteAgoods(String gno){
    String sql = "delete  from GOODSTABLE where id = ? ";
    Object param[] = {gno};
    return upadateByParams(sql, param);
}
/**
 * @discription 条件查询
 */
public List<Map<String, Object>> conditionSelectGoods(String gname,
String typeid){
    StringBuffer sql = new StringBuffer("select * from GOODSTABLE
    gt,GOODSTYPE gy where gt.goodstype_id = gy.id and 1 = 1 ");
    ArrayList<String> param = new ArrayList<String>();
    //输入商品名的情况
    if(null != gname && gname.trim().length() > 0){
```

```java
            sql.append(" and GNAME like ? ");
            param.add("%" + gname + "%");
        }
        //选中商品类型的情况
        if(!"all".equals(typeid)){
            sql.append(" and gt.goodstype_id = ? ");
            param.add(typeid);
        }
        return select(sql.toString(), param.toArray());
    }
    /**
     * @discription 分页条件查询
     */
    public List<Map<String, Object>> conditionSelectGoodsByPage(String gname, String typeid, int startIndex, int perPageSize){
        StringBuffer sql = new StringBuffer("select * from GOODSTABLE gt,GOODSTYPE gy where gt.goodstype_id = gy.id and 1 = 1 ");
        ArrayList<Object> param = new ArrayList<Object>();
        //输入商品名的情况
        if(null != gname && gname.trim().length() > 0){
            sql.append(" and GNAME like ? ");
            param.add("%" + gname + "%");
        }
        //选中商品类型的情况
        if(!"all".equals(typeid)){
            sql.append(" and gt.goodstype_id = ? ");
            param.add(typeid);
        }
        param.add(startIndex);
        param.add(perPageSize);
        sql.append(" limit ?, ? ");
        return select(sql.toString(), param.toArray());
    }
    /**
     * @discription 查询公告
     */
    public  List<Map<String, Object>> selectNotice(){
        String sql = "select id, NTITLE, NCONTENT,  NTIME from NOTICETABLE";
        return select(sql, null);
    }
    /**
     * @discription 添加公告
     */
    public boolean addNotice(NoticeDTO nd){
        String sql = "insert into NOTICETABLE values(null,?,?,SYSDATE)";
```

```java
        Object param[] = {
                nd.getNtitle(),
                nd.getNcontent()
        };
        return upadateByParams(sql, param);
}
/**
 * @discription 查询一个公告
 */
public List<Map<String, Object>> selectAnotice(String nid){
    String sql = "select ID, NTITLE, NCONTENT, NTIME from NOTICETABLE where ID = ?";
    Object param[] = {nid};
    return select(sql, param);
}
/**
 * @discription 删除一个公告
 */
public boolean deleteAnotice(String nid){
    String sql = "delete from NOTICETABLE where ID = ?";
    Object param[] = {nid};
    return upadateByParams(sql, param);
}
/**
 * @discription 删除商品时，判断有无关联
 */
public boolean isDelete(String gno){
    String sql1 = "select * from CARTTABLE where goodstable_id=?";
    String sql2 = "select * from FOCUSTABLE where goodstable_id=?";
    String sql3 = "select * from ORDERDETAIL where goodstable_id=?";
    Object param[] = {gno};
    //有子记录
    if(select(sql1, param).size() > 0 ||
            select(sql2, param).size() > 0 ||
            select(sql3, param).size() > 0){
        return false;
    }else{
        return true;
    }
}
/**
 * @discription 查询用户
 */
public List<Map<String, Object>> selectUsers(){
    String sql = "select * from BUSERTABLE ";
```

```java
        return select(sql, null);
    }
    /**
     * @discription 删除用户时,判断有无关联
     */
    public boolean isDeleteUser(String bid){
        String sql1 = "select * from CARTTABLE where busertable_id=?";
        String sql2 = "select * from FOCUSTABLE where busertable_id=?";
        String sql3 = "select * from ORDERBASETABLE where busertable_id=?";
        Object param[] = {bid};
        //有子记录
        if(select(sql1, param).size() > 0 ||
                select(sql2, param).size() > 0 ||
                select(sql3, param).size() > 0){
            return false;
        }else{
            return true;
        }
    }
    /**
     * @discription 删除用户
     */
    public boolean deleteUser(String bid){
        String sql = "delete from BUSERTABLE where BID=?";
        Object param[] = {bid};
        return upadateByParams(sql, param);
    }
    /**
     * @discription 查询用户
     */
    public List<Map<String, Object>> selectOrders(){
        String sql = "select ot.id, ot.AMOUNT, ot.STATUS, "
                + " ORDERDATE, bt.BEMAIL, ot.busertable_id "
                + " from ORDERBASETABLE ot, BUSERTABLE bt where "
                + " ot.busertable_id=bt.id ";
        return select(sql, null);
    }
    /**
     * @discription 删除订单明细
     */
    public boolean deleteOrderdetail(String ordersn){
        String sql = "delete from ORDERDETAIL where orderbasetable_id=?";
        Object param[] = {ordersn};
        return upadateByParams(sql, param);
    }
```

```java
/**
 * @discription 删除订单
 */
public boolean deleteOrderbasetable(String ordersn){
    String sql = "delete from ORDERBASETABLE where id=?";
    Object param[] = {ordersn};
    return upadateByParams(sql, param);
}
}
```

❸ **BeforeUserDao.java 的代码**

```java
package dao;
import java.sql.Connection;
import java.sql.PreparedStatement;
import java.sql.ResultSet;
import java.sql.SQLException;
import java.util.List;
import java.util.Map;
import dto.UserDTO;
public class BeforeUserDao  extends BaseDao{
    /**
     * @discription 获得最新商品列表
     */
    public List<Map<String, Object>> getLastedGoods(String typeid){
        //首页
        if("0".equals(typeid)){
            String sql = "select * from GOODSTABLE gt,GOODSTYPE gy where gt.goodstype_id = gy.id order by gt.id desc limit 15";
            return select(sql, null);
        }else{
            String sql = "select * from GOODSTABLE gt,GOODSTYPE gy where gt.goodstype_id = gy.id and gt.goodstype_id = ? order by gt.id desc limit 15";
            Object param[] = {typeid};
            return select(sql, param);
        }
    }
    /**
     * @discription 获得销售排行商品
     */
    public List<Map<String, Object>> getSaleOrder(){
        String sql = "select sum(od.SHOPPINGNUM) shopnumber," +
            " gd.id id, " +
            " gd.GNAME gname, " +
            " gd.GOPRICE goprice, " +
```

```java
                " gd.GRPRICE grprice, " +
                " gd.GPICTURE  gpicture " +
                " from GOODSTABLE gd  LEFT JOIN ORDERDETAIL  od  ON
                od.goodstable_id=gd.id" +
                " group by " +
                " gd.id, " +
                " gd.GNAME, " +
                " gd.GOPRICE, " +
                " gd.GRPRICE, " +
                " gd.GPICTURE " +
                " order by shopnumber desc limit 10 ";
        return select(sql, null);
    }
    /**
     * @discription 获得人气排行商品
     */
    public List<Map<String, Object>> getFocusOrder(){
        String sql ="select count(ft.goodstable_id)fn,gt.id id,gt.gname gname,"+
                " gt.grprice grprice, gt.gpicture gpicture " +
                " from GOODSTABLE gt " +
                " LEFT JOIN FOCUSTABLE ft  ON ft.goodstable_id=gt.id "+
                " group by gt.id, gt.gname, gt.grprice, gt.gpicture " +
                " order by fn desc limit 10 " ;
        return select(sql, null);
    }
    /**
     * @discription 获得公告
     */
    public List<Map<String, Object>> getNotice(){
        String sql = " select * from noticetable order by ntime desc limit 13 ";
        return select(sql, null);
    }
    /**
     * @discription 注册
     */
    public boolean register(UserDTO ud){
        String sql = "insert into BUSERTABLE values(null,?,?) ";
        Object param[] = {ud.getBemail(), ud.getBpwd()};
        return upadateByParams(sql, param);
    }
    /**
     * @discription 查询E-mail
     */
    public List<Map<String, Object>> isExit(UserDTO ud){
        String sql = "select * from BUSERTABLE where BEMAIL = ? ";
```

```java
        Object param[] = {ud.getBemail()};
        return select(sql, param);
}
/**
 * @discription 登录判定
 */
public List<Map<String, Object>> isLogin(UserDTO ud){
    String sql = "select * from BUSERTABLE where BEMAIL = ? and BPWD = ? ";
    Object param[] = {ud.getBemail(), ud.getBpwd()};
    return select(sql, param);
}
/**
 * @discription 关注商品
 */
public boolean focusGoods(String gno, int bid){
    String sql = "insert into FOCUSTABLE values(null,?,?,now())";
    Object param[] = {gno, bid};
    return upadateByParams(sql, param);
}
/**
 * @discription 查看是否已关注
 */
public List<Map<String, Object>> isFocus(String gno, int bid){
    String sql = "select * from FOCUSTABLE where goodstable_id = ? and busertable_id = ? ";
    Object param[] = {gno, bid};
    return select(sql, param);
}
/**
 * @discription 放入购物车
 */
public boolean putCart(String gno, int bid, String mknum){
    String sql1 = "select * from carttable where busertable_id=? and goodstable_id=? ";
    Object param1[] = {bid, gno};
    String sql2 = "update carttable set shoppingnum=shoppingnum+? where busertable_id=? and goodstable_id=?";
    Object param2[] = {mknum, bid, gno};
    String sql3 = "insert into carttable values(null,?,?,?)";
    Object param3[] = {bid, gno, mknum};
    //购物车中已存在
    if(select(sql1, param1).size() > 0){
        return updateByParams(sql2, param2);
    }else{//购物车中不存在
        return updateByParams(sql3, param3);
```

```java
        }
    }
    /**
     * 修改购物车
     */
    public boolean updateCart(String shoppingnum,String gno, int bid){
        String sql2 = "update carttable set shoppingnum=? where busertable_id=? and goodstable_id=?";
        Object param2[] = {shoppingnum, bid, gno};
        return updateByParams(sql2, param2);
    }
    /**
     * @discription 查询购物车
     */
    public List<Map<String, Object>> selectCart(int bid){
        String sql = " select gt.id, gt.gname, gt.gpicture, gt.grprice, ct.shoppingnum, ct.shoppingnum*gt.grprice smallsum " +
" from GOODSTABLE gt, CARTTABLE ct where gt.id=ct.goodstable_id and ct.busertable_id=? ";
        Object param[] = {bid};
        return select(sql, param);
    }
    /**
     * @discription 删除购物车
     */
    public boolean deleteCart(int bid, String gno){
        String sql1 = "delete from CARTTABLE where busertable_id=?";
        Object param1[] = {bid};
        String sql2 = "delete from CARTTABLE where busertable_id=? and goodstable_id=?";
        Object param2[] = {bid, gno};
        if(gno == null){
            //清空购物车
            return updateByParams(sql1, param1);
        }else{
            //删除一个商品
            return updateByParams(sql2, param2);
        }
    }
    /**
     * @discription 向订单表插入记录
     */
    public int addOrder(int bid, String amount){
        String sql = "insert into ORDERBASETABLE values(null,?,?,?,now())";
        Object param[] = {bid, amount, "0"};//0代表没有支付
```

```java
        String sql1 = " select LAST_INSERT_ID() from ORDERBASETABLE ";
        return getLastId(sql, sql1, param);
}
/**
 * @discription 向订单表详情插入记录
 */
public boolean addOrderDetail(int ordersn,int bid){
    String sql = "insert into ORDERDETAIL (orderbasetable_id, goodstable_id,
    SHOPPINGNUM) select " + ordersn + ", goodstable_id, SHOPPINGNUM from
    CARTTABLE where busertable_id = ? ";
    Object param[] = {bid};
    return updateByParams(sql, param);
}
/**
 * @discription 更新商品库存
 */
public boolean updateGoodsStore(int ordersn){
    boolean b=false;
    Connection con = getConnection();
    PreparedStatement ps=null;
    Connection con1=getConnection();
    PreparedStatement ps1=null;
    ResultSet rs=null;
    try {
        ps1=con1.prepareStatement("select shoppingnum, goodstable_id
        from orderdetail where orderbasetable_id=?");
        ps1.setInt(1, ordersn);
        rs=ps1.executeQuery();
    ps=con.prepareStatement("update GOODSTABLE set GSTORE=GSTORE-?
    where id=? ");
        //批量更新
        while(rs.next()){
            ps.setInt(1, rs.getInt(1));
            ps.setString(2, rs.getString(2));
            ps.addBatch();
        }
        int n[]=ps.executeBatch();
        if(n[0]>0)
            b=true;
    } catch (SQLException e) {
        e.printStackTrace();
    }finally{
        close(rs,ps1,con1);
        close(null,ps,con);
    }
}
```

```java
        return b;
    }
    /**
     * @discription 订单支付
     */
    public boolean pay(String ordersn){
        String sql = "update ORDERBASETABLE set STATUS=1 where id = ?";
        Object param[] = {ordersn};
        return upadateByParams(sql, param);
    }
    /**
     * @discription 头部搜索框
     */
    public List<Map<String, Object>> search(String keyword){
        String sql = "select * from GOODSTABLE where GNAME like ? ";
        Object param[] = {"%" + keyword + "%"};
        return select(sql, param);
    }
    /**
     * @discription 我的订单
     */
    public List<Map<String, Object>> myOrder(int bid){
        String sql = " select id, amount, busertable_id, status, orderdate from ORDERBASETABLE where busertable_id = ? ";
        Object param[] = {bid};
        return select(sql, param);
    }
    /**
     * @discription 我的关注
     */
    public List<Map<String, Object>> myFocus(int bid){
        String sql = " select gt.id, gt.gname, gt.goprice, gt.grprice, gt.gpicture from FOCUSTABLE ft, GOODSTABLE gt " + " where ft.goodstable_id=gt.id and  ft.busertable_id = ? ";
        Object param[] = {bid};
        return select(sql, param);
    }
    /**
     * @discription 我的订单详情
     */
    public List<Map<String, Object>> myOrderDetail(String ordersn){
        String sql =" select gt.id, gt.gname, gt.goprice, gt.grprice, gt.gpicture, odt.shoppingnum from GOODSTABLE gt, ORDERDETAIL odt " +
                " where  odt.orderbasetable_id = ? and gt.id=odt.goodstable_id ";
        Object param[] = {ordersn};
```

```
        return select(sql, param);
    }
}
```

13.4.5 业务模型

本系统有关业务的 Java 类位于包 service 中，业务层需要调用数据访问层才能完成功能的实现。有关后台管理子系统的业务由 AdminService 实现；有关电子商务子系统的业务由 BeforeUserService 实现。

❶ **AdminService.java 的代码**

```java
package service;
import java.util.List;
import java.util.Map;
import dao.AdminDao;
import dto.AdminDTO;
import dto.GoodsDTO;
import dto.NoticeDTO;
public class AdminService {
    AdminDao ad = new AdminDao();
    /**
     * @discription 管理员登录判定
     */
    public boolean adminLogin(AdminDTO adt){
        return ad.adminLogin(adt);
    }
    /**
     * @discription 查询所有商品
     */
    public List<Map<String, Object>> selectGoods(){
        return ad.selectGoods();
    }
    public List<Map<String, Object>> selectAllGoodsByPage(int startIndex, int perPageSize){
        return ad.selectAllGoodsByPage(startIndex, perPageSize);
    }
    /**
     * @discription 获得商品类型列表
     */
    public List<Map<String, Object>> getGoodsType(){
        return ad.getGoodsType();
    }
    /**
     * 添加类型
```

```java
     */
    public boolean addType(String typename) {
        return ad.addType(typename);
    }
    /**
     * @discription 获得关联类型列表
     */
    public List<Map<String, Object>> getReType(String id){
        return ad.getReType(id);
    }
    /**
     * 删除类型
     */
    public String deleteType(String id) {
        if(getReType(id).size() > 0) {
            return "有子记录，无法删除！3秒钟返回";
        }else {
            ad.deleteType(id);
            return "删除成功！3秒钟返回";
        }
    }
    /**
     * @discription 添加商品
     */
    public boolean addGoods(GoodsDTO gd){
        return ad.addGoods(gd);
    }
    /**
     * @discription 查询一个商品
     */
    public List<Map<String, Object>> selectAGoods(String gno){
        return ad.selectAGoods(gno);
    }
    /**
     * @discription 修改商品
     */
    public boolean updateGoods(GoodsDTO gd){
        return ad.updateGoods(gd);
    }
    /**
     * @discription 删除商品
     */
    public String deleteAgoods(String gno){
        if(!ad.isDelete(gno)){//有子记录，不能删除
            return "nodelete";
```

```java
        }else{
            ad.deleteAgoods(gno);
            return "ok";
        }
    }
    /**
     * @discription 删除多个商品
     */
    public String deleteManygoods(String gnos[]){
        int count = 0;
        for(int i = 0; i < gnos.length; i++){
            if(!ad.isDelete(gnos[i])){
                break;
            }else{
                count++;
            }
        }
        //没有子记录
        if(count == gnos.length){
            for(int i = 0; i < gnos.length; i++){
                ad.deleteAgoods(gnos[i]);
            }
            return "ok";
        }else{//有子记录
            return "nodelete";
        }
    }
    /**
     * @discription 条件查询
     */
    public List<Map<String, Object>> conditionSelectGoods(String gname, String typeid){
        return ad.conditionSelectGoods(gname, typeid);
    }
    /**
     * @discription 分页条件查询
     */
    public List<Map<String, Object>> conditionSelectGoodsByPage(String gname, String typeid, int startIndex, int perPageSize){
        return ad.conditionSelectGoodsByPage(gname, typeid, startIndex, perPageSize);
    }
    /**
     * @discription 查询公告
     */
```

```java
    public List<Map<String, Object>> selectNotice(){
        return ad.selectNotice();
    }
    /**
     * @discription 添加公告
     */
    public boolean addNotice(NoticeDTO nd){
        return ad.addNotice(nd);
    }
    /**
     * @discription 查询一个公告
     */
    public List<Map<String, Object>> selectAnotice(String nid){
        return ad.selectAnotice(nid);
    }
    /**
     * @discription 删除一个公告
     */
    public boolean deleteAnotice(String nid){
        return ad.deleteAnotice(nid);
    }
    /**
     * @discription 查询用户
     */
    public List<Map<String, Object>> selectUsers(){
        return ad.selectUsers();
    }
    /**
     * @discription 删除用户
     */
    public String deleteUser(String bid){
        if(!ad.isDeleteUser(bid)){
            return "nodelete";
        }else{
            ad.deleteUser(bid);
            return "ok";
        }
    }
    /**
     * @discription 查询用户
     */
    public List<Map<String, Object>> selectOrders(){
        return ad.selectOrders();
    }
    /**
```

```
     * @discription 删除订单
     */
    public boolean deleteOrder(String ordersn){
        if(ad.deleteOrderdetail(ordersn)){
            if(ad.deleteOrderbasetable(ordersn)){
                return true;
            }else{
                return false;
            }
        }else{
            return false;
        }
    }
}
```

❷ **BeforeUserService.java 的代码**

```
package service;
import java.util.List;
import java.util.Map;
import dao.BeforeUserDao;
import dto.UserDTO;
public class BeforeUserService {
    BeforeUserDao bd = new BeforeUserDao();
    /**
     * @discription 获得最新商品列表
     */
    public List<Map<String, Object>> getLastedGoods(String typeid){
        return bd.getLastedGoods(typeid);
    }
    /**
     * @discription 获得销售排行商品
     */
    public List<Map<String, Object>> getSaleOrder(){
        return bd.getSaleOrder();
    }
    /**
     * @discription 获得人气排行商品
     */
    public List<Map<String, Object>> getFocusOrder(){
        return bd.getFocusOrder();
    }
    /**
     * @discription 获得公告
     */
    public List<Map<String, Object>> getNotice(){
```

```java
        return bd.getNotice();
    }
    /**
     * @discription 注册
     */
    public boolean register(UserDTO ud){
        return bd.register(ud);
    }
    /**
     * 查询 E-mail
     */
    public List<Map<String, Object>> isExit(UserDTO ud){
        return bd.isExit(ud);
    }
    /**
     * @discription 登录判定
     */
    public List<Map<String, Object>> isLogin(UserDTO ud){
        return bd.isLogin(ud);
    }
    /**
     * @discription 关注商品
     */
    public String focusGoods(String gno, int bid){
        //已关注
        if(bd.isFocus(gno, bid).size() > 0){
            return "focused";
        }else{//未关注
            if(bd.focusGoods(gno, bid)){
                return "focusSuccess";
            }else{
                return "focusFail";
            }
        }
    }
    /**
     * @discription 放入购物车
     */
    public boolean putCart(String gno, int bid, String mknum){
        return bd.putCart(gno, bid, mknum);
    }
    /**
     * 修改购物车
     */
    public boolean updateCart(String shoppingnum,String gno, int bid) {
```

```java
        return bd.updateCart(shoppingnum, gno, bid);
    }
    /**
     * @discription 查询购物车
     */
    public List<Map<String, Object>> selectCart(int bid){
        return bd.selectCart(bid);
    }
    /**
     * @discription 删除购物车
     */
    public boolean deleteCart(int bid, String gno){
        return bd.deleteCart(bid, gno);
    }
    /**
     * @discription 提交订单
     */
    public int orderSubmit(int bid, String amount){
        //向订单表插入记录
        int id = bd.addOrder(bid, amount);
        //向订单详情表插入订单详情
        bd.addOrderDetail(id, bid);
        //清空购物车
        bd.deleteCart(bid, null);
        //更新商品库存
        bd.updateGoodsStore(id);
        return id;
    }
    /**
     * @discription 订单支付
     */
    public boolean pay(String ordersn){
        return bd.pay(ordersn);
    }
    /**
     * @discription 头部搜索框
     */
    public List<Map<String, Object>> search(String keyword){
        return bd.search(keyword);
    }
    /**
     * @discription 我的订单
     */
    public List<Map<String, Object>> myOrder(int bid){
        return bd.myOrder(bid);
```

```
    }
    /**
     * @discription 我的关注
     */
    public List<Map<String, Object>> myFocus(int bid){
        return bd.myFocus(bid);
    }
    /**
     * @discription 我的订单详情
     */
    public List<Map<String, Object>> myOrderDetail(String ordersn){
        return bd.myOrderDetail(ordersn);
    }
}
```

13.4.6　工具类

本系统使用的工具类 MyUtil 的代码如下：

```
package util;
import java.text.SimpleDateFormat;
import java.util.Date;
import javax.servlet.http.Part;
public class MyUtil {
    /**
     * @discription 获取一个时间串
     */
    public static String getStringID(){
        String id=null;
        Date date=new Date();
        SimpleDateFormat sdf=new SimpleDateFormat("yyyyMMddHHmmssSSS");
        id=sdf.format(date);
        return id;
    }
    /**
     * @discription 从 Part 中获得原始文件名
     */
    public static String getFileName(Part part){
        if(part == null)
            return null;
        //fileName 形式为: form-data; name="resPath"; filename="20140920_110531.jpg"
        String fileName = part.getHeader("content-disposition");
        //没有选择文件
```

```
            if(fileName.lastIndexOf("=") + 2 == fileName.length() - 1)
                return null;
            return fileName.substring(fileName.lastIndexOf("=") + 2, fileName.
            length() - 1);
        }
    }
```

13.5 后台管理子系统的实现

与后台管理子系统相关的 JSP 页面、CSS、JavaScript 和图片位于 WebContent/admin 目录下。在 13.4 节中已经介绍了系统的实体模型、数据库操作以及业务模型，本节只是介绍 JSP 页面和 Servlet 的核心实现。

13.5.1 管理员登录

管理员输入用户名和密码后，系统将对管理员的用户名和密码进行验证。如果用户名和密码同时正确，则成功登录，进入系统管理主页面（main.jsp）；如果用户名或密码有误，则提示错误。

❶ 视图

login.jsp 页面提供登录信息输入的界面。login.jsp 页面效果如图 13.7 所示。

图 13.7 管理员登录界面

login.jsp 的代码略。

❷ 控制器

该控制器 Servlet 对象的 urlPatterns 是{ "/admin_login" }，类名为 AdminLoginServlet。控制器获取视图的请求后，将视图中的信息封装在实体模型 AdminDTO 中，然后调用业务模型 AdminService（见 13.4.5 节）中的 adminLogin()方法执行登录的业务处理。登录成功后，首先将登录人信息存入 session；然后获得商品的类型并存入 session，用于添加商品和修改商品；最后进入 main.jsp 页面。登录失败回到 loginerror.jsp 页面。

AdminLoginServlet.java 的核心代码如下：

```
...
//将管理员信息保存到实体模型中
AdminDTO adt = new AdminDTO();
adt.setAname(request.getParameter("aname"));
adt.setApwd(request.getParameter("apwd"));
//业务层
AdminService as = new AdminService();
//登录成功
if(as.adminLogin(adt)){
    //将登录存储到session中
    HttpSession session = request.getSession(true);
    session.setAttribute("admin", adt);
    //获得商品类型
    List<Map<String, Object>>list = as.getGoodsType();
    session.setAttribute("goodsType", list);
    //跳转到后台主页面
    RequestDispatcher rds = request.getRequestDispatcher("admin/main.jsp");
    rds.forward(request, response);
}else{//登录失败
    response.sendRedirect("admin/loginerror.jsp");
}
...
```

13.5.2 添加商品

管理员输入商品的名称、原价、折扣价、数量,并选择图片与类型等信息后,该模块首先检查输入的信息是否合法。如果合法,则实现添加;如果不合法,则回到添加页面。

❶ 视图

addGoods.jsp 页面是添加商品信息的输入界面。该页面中商品类型从 session 中获取。添加商品页面如图 13.8 所示。

图 13.8　添加商品页面

addGoods.jsp 的核心代码如下:

```html
...
<body>
    <form action="admin_addGoods?act=add" method="post" enctype="multipart/form-data">
        <table border=1 style="border-collapse: collapse">
            <caption>
                <font size=4 face=华文新魏>添加商品</font>
            </caption>
            <tr>
                <td>商品名称<font color="red">*</font></td>
                <td><input type="text" name="gname" />
                </td>
            </tr>
            <tr>
                <td>商品原价<font color="red">*</font></td>
                <td><input type="text" name="goprice" /></td>
            </tr>
            <tr>
                <td>商品折扣价<font color="red">*</font></td>
                <td><input type="text" name="grprice" /></td>
            </tr>
            <tr>
                <td>商品数量<font color="red">*</font></td>
                <td><input type="text" name="gstore" /></td>
            </tr>
            <tr>
                <td>商品图片<font color="red">*</font></td>
                <td>
                    <input type="file" id="gpicture" name="gpicture"/></td>
            </tr>
            <tr>
                <td>商品类型<font color="red">*</font></td>
                <td><select name="typeid">
                    <c:forEach var="gtp" items="${goodsType}">
                        <option value="${gtp.id}">${gtp.typename}</option>
                    </c:forEach>
                </select></td>
            </tr>
            <tr>
                <td align="center"><input type="submit" value="提交" /></td>
                <td align="left"><input type="reset" value="重置" /></td>
            </tr>
        </table>
```

```
            </form>
    </body>
...
```

❷ 控制器

该控制器 Servlet 对象的 urlPatterns 是 { "/admin_addGoods" }，类名为 AddGoodsServlet。控制器获取视图的请求后，将视图中的信息封装在实体模型 GoodsDTO 中，并将商品图片上传到服务器；然后根据动作类型调用业务模型中的 addGoods() 方法执行添加的业务处理。添加成功，进入查询商品页面；添加失败，回到 addGoods.jsp 页面。

AddGoodsServlet.java 的核心代码如下：

```
...
GoodsDTO gd = new GoodsDTO();//实体模型
//获得动作类型，不是通过表单域提交，所以可以通过 request 获取
String act = request.getParameter("act");
//获得 Part 对象
Part part = request.getPart("gpicture");
//指定上传的文件保存到服务器的 uploadFile 目录中
File uploadFileDir = new File(getServletContext().getRealPath("/uploadFile"));
if(!uploadFileDir.exists()){
    uploadFileDir.mkdir();
}
//获得原始文件名
String oldName = MyUtil.getFileName(part);
try{
    String gname = request.getParameter("gname");
    String gpicture = null;
    if("add".equals(act)){
        if(oldName != null){
            //上传时的新文件名
            gpicture = MyUtil.getStringID() + oldName.substring(oldName.lastIndexOf("."));
            //上传图片
            part.write(uploadFileDir + File.separator + gpicture);
        }
    }else if("update".equals(act)){
        gd.setId(Integer.parseInt(request.getParameter("gno")));
        //修改时没有选择图片，使用旧图片
        if(oldName == null){
            gpicture = request.getParameter("oldgpicture");
        }else{
            //上传时的新文件名
            gpicture = MyUtil.getStringID() + oldName.substring(oldName.lastIndexOf("."));
```

```java
            //上传图片
            part.write(uploadFileDir + File.separator + gpicture);
        }
    }
    double goprice = Double.parseDouble(request.getParameter("goprice"));
    double grprice = Double.parseDouble(request.getParameter("grprice"));
    int gstore = Integer.parseInt(request.getParameter("gstore"));
    int typeid = Integer.parseInt(request.getParameter("typeid"));
    gd.setGname(gname);
    gd.setGpicture(gpicture);
    gd.setGoprice(goprice);
    gd.setGrprice(grprice);
    gd.setGoodstype_id(typeid);
    gd.setGstore(gstore);
}catch(Exception e){
    //数据输入有误
    response.sendRedirect("admin/addGoods.jsp");
    return ;
}
//业务层
AdminService as = new AdminService();
//添加
if("add".equals(act)){
    //成功
    if(as.addGoods(gd)){
        //跳转到后台主页面
        response.sendRedirect("admin_selectGoods?act=select");
    }else{
        response.sendRedirect("admin/addGoods.jsp");
    }
}
//修改
else if("update".equals(act)){
    //成功
    if(as.updateGoods(gd)){
        //跳转到后台主页面
        response.sendRedirect("admin_selectGoods?act=select");
    }else{
        //失败跳到修改页面
        response.sendRedirect("admin_detail?act=updateAgoods&gno=" +
        gd.getId());
    }
}
...
```

13.5.3 查询商品

管理员登录成功后，进入后台管理子系统的主页面，在主页面中初始显示查询商品页面 selectAllGoods.jsp。查询页面运行效果如图 13.9 所示。

商品ID	商品名称	商品类型	查看详情
12	上衣X	服装	详情
13	苹果	水果	详情
14	樱桃	水果	详情
15	微波炉	家电	详情
16	电视99	家电	详情
17	孕装55	孕婴	详情
18	牛奶11	食品	详情
19	红酒99	酒类	详情
20	化妆品22	化妆品	详情
22	上衣xx	服装	详情

共17条记录 共2页 第1页 下一页

图 13.9 查询页面

单击主页面中的"查询商品"超链接，打开查询页面 selectAllGoods.jsp。"查询商品"超链接的目标地址是一个 Servlet，Servlet 对象的 urlPatterns 是{ "/admin_selectGoods" }，类名为 SelectGoodsServlet。在该 Servlet 中，根据动作类型分为"条件查询""修改查询""查询"以及"删除查询"。调用业务模型中的 selectGoods()方法执行"查询"的业务处理。根据动作类型，将查询结果转发到 selectAllGoods.jsp 页面。

在 selectAllGoods.jsp 页面中单击"详情"超链接，打开商品详细信息页面 detail.jsp。"详情"超链接的目标地址是个 Servlet，Servlet 对象的 urlPatterns 是{ "/admin_detail" }，类名为 DetailServlet。在该 Servlet 中，根据商品编号调用业务模型中的 selectAGoods()方法执行查询一个商品信息的业务处理。根据动作类型，将查询结果转发给详细信息页面 detail.jsp。商品详细信息页面如图 13.10 所示。

图 13.10 商品详情

❶ 视图

查询商品模块涉及两个 JSP 页面：selectAllGoods.jsp 和 detail.jsp。
selectAllGoods.jsp 的核心代码如下：

```jsp
...
<body>
    <form action="admin_selectGoods" method="post">
    <center><br>
    <table>
        <tr>
            <td width="250px">
                商品名称:<input type="text" name="gname" value="${gname1 }">
            </td>
            <td width="200px">
                商品类型:<select name="typeid">
                    <option value="all">全部</option>
                    <c:forEach var="gtp" items="${goodsType}">
                    <option value="${gtp.id}"
                            <c:if test="${typeid!='all'&&typeid==gtp.id}">
                            selected </c:if>>${gtp.typename}</option>
                    </c:forEach>
                </select>
                <input type="hidden" name="act" value="condiction"/>
                <input type="submit" value="条件查询"/>
            </td>
        </tr>
    </table>
    <br></center>
    </form>
    <table border="1" bordercolor="PaleGreen">
        <tr>
            <th width="250px">商品ID</th>
            <th width="200px">商品名称</th>
            <th width="200px">商品类型</th>
            <th width="200px">查看详情</th>
        </tr>
        <c:forEach var="g" items="${requestScope.allGoods}">
        <tr onmousemove="changeColor(this)" onmouseout="changeColor1(this)">
            <td>${g.id}</td>
            <td>${g.gname}</td>
            <td>${g.typename}</td>
            <td><a href="admin_detail?gno=${g.id}&act=detail" target="_blank">详情</a></td>
        </tr>
        </c:forEach>
```

```jsp
            <tr>
                <td colspan="4" align="right">
                       
           共${totalCount}条记录  共${totalPage}
        页  
                    第${pageCur}页  
                    <%
                        //解决URL不能传中文的问题，条件查询
                        String gname1 =(String)request.getAttribute("gname1");
                        if(gname1 != null){
                            gname1 = java.net.URLEncoder.encode(gname1,"utf-8");
                            request.setAttribute("gname1",gname1);
                        }
                    %>
<c:url var="url_pre" value="admin_selectGoods?act=${act }&&gname1=${gname1}
&&typeid=${typeid}">
                        <c:param name="pageCur" value="${pageCur - 1 }"/>
                    </c:url>
<c:url var="url_next" value="admin_selectGoods?act=${act }&&gname1=${gname1}
&&typeid=${typeid}">
                        <c:param name="pageCur" value="${pageCur + 1 }"/>
                    </c:url>
                    <!-- 第一页，没有上一页 -->
                    <c:if test="${pageCur != 1 }">
                        <a href="${url_pre}">上一页</a>    
                    </c:if>
                    <!-- 最后一页，没有下一页 -->
                    <c:if test="${pageCur != totalPage && totalPage !=0 }">
                        <a href="${url_next}">下一页</a>
                    </c:if>
                </td>
            </tr>
    </table>
</body>
...
```

detail.jsp 的代码略。

❷ 控制器

查询商品模块有两个 Servlet：SelectGoodsServlet.java 和 DetailServlet.java。SelectGoodsServlet.java 的核心代码如下：

```java
...
AdminService as = new AdminService();
List<Map<String, Object>> list = null;
//获得动作类型
String act = request.getParameter("act");
```

```java
//条件查询
String gname = request.getParameter("gname");
String typeid = request.getParameter("typeid");
//在条件查询分页 URL 中使用
String gname1 = request.getParameter("gname1");
//条件查询首页
if(gname1 == null) {
    gname1 = gname;
}else {
    //解码 URL 中的中文
    gname1 = java.net.URLDecoder.decode(gname1,"utf-8");
}
//条件查询非首页
if(gname == null) {
    gname = gname1;
}
//条件查询
if("condiction".equals(act)){
    list = as.conditionSelectGoods(gname, typeid);
}else{
    list = as.selectGoods();
}
RequestDispatcher rds = null;
//分页查询
int temp = list.size();
request.setAttribute("totalCount", temp);
int totalPage = 0;
if (temp == 0) {
    totalPage = 0;//总页数
} else {
    //返回大于或者等于指定表达式的最小整数
    totalPage = (int) Math.ceil((double) temp / 10);
}
String pageCur11 = request.getParameter("pageCur");
if (pageCur11 == null) {
    pageCur11 = "1";
}
int  pageCur = Integer.parseInt(pageCur11);
if ((pageCur - 1) * 10 > temp) {
    pageCur = pageCur - 1;
}
//分页查询
int startIndex = (pageCur - 1) * 10;//起始位置
int perPageSize = 10;//每页 10 个
if("condiction".equals(act)) {
```

```java
            list = as.conditionSelectGoodsByPage(gname, typeid, startIndex, perPageSize);
    }else {
        list = as.selectAllGoodsByPage(startIndex, perPageSize);
    }
    //回显查询条件
    request.setAttribute("gname1", gname1);
    request.setAttribute("act", act);
    request.setAttribute("typeid", typeid);
    request.setAttribute("allGoods", list);
    request.setAttribute("totalPage", totalPage);
    request.setAttribute("pageCur", pageCur);
    //查询页面
    if("select".equals(act) || "condiction".equals(act)){
        rds = request.getRequestDispatcher("admin/selectAllGoods.jsp");
    }
    //修改查询页面
    else if("updateSelect".equals(act)){
        rds = request.getRequestDispatcher("admin/updateSelect.jsp");
    }
    //删除查询页面
    else if("deleteSelect".equals(act)){
        rds = request.getRequestDispatcher("admin/deleteSelect.jsp");
    }
    rds.forward(request, response);
    ...
```

DetailServlet.java 的核心代码如下：

```java
...
//获得商品编号
String gno = request.getParameter("gno");
AdminService as = new AdminService();
List<Map<String, Object>> list = as.selectAGoods(gno);
//将一个商品详细信息存到 request 中
request.setAttribute("agoods", list.get(0));
RequestDispatcher rds = null;
//详情页面
if("detail".equals(request.getParameter("act"))){
    rds = request.getRequestDispatcher("admin/detail.jsp");
}
//修改一个商品页面
else if("updateAgoods".equals(request.getParameter("act"))){
    rds = request.getRequestDispatcher("admin/updateAgoods.jsp");
}
rds.forward(request, response);
...
```

13.5.4 修改商品

单击管理信息主页面中的"修改商品"超链接，打开修改查询页面 updateSelect.jsp。"修改商品"超链接的目标地址是个 Servlet，该 Servlet 与 13.5.3 节查询商品的 SelectGoodsServlet.java 类同。在 SelectGoodsServlet 中，根据动作类型，将查询结果转发给修改查询页面。

单击 updateSelect.jsp 页面中的"修改"超链接，打开修改商品信息页面 updateAgoods.jsp。"修改"超链接的目标地址是个 Servlet，该 Servlet 与 13.5.3 节查询商品的 DetailServlet.java 类同。在 DetailServlet 中，根据动作类型，将查询结果转发给 updateAgoods.jsp 页面显示。

输入要修改的信息后，单击"提交"按钮，将商品信息提交给控制器 Servlet，该 Servlet 与 13.5.2 节添加商品的 AddGoodsServlet.java 类同。在 AddGoodsServlet 中，将提交的信息封装在实体模型 GoodsDTO 中，并将商品图片上传到服务器（若修改图片）；然后根据动作类型调用业务模型中的 updateGoods () 方法执行修改的业务处理。修改成功，进入查询商品页面；修改失败，回到 updateAgoods.jsp 页面。

updateSelect.jsp 页面的运行效果如图 13.11 所示，updateAgoods.jsp 页面的运行效果如图 13.12 所示。

商品ID	商品名称	商品类型	操作类型
12	上衣X	服装	修改
13	苹果	水果	修改
14	樱桃	水果	修改
15	微波炉	家电	修改
16	电视99	家电	修改
17	孕装55	孕婴	修改
18	牛奶11	食品	修改
19	红酒99	酒类	修改
20	化妆品22	化妆品	修改
22	上衣xx	服装	修改

共17条记录 共2页 第1页 下一页

图 13.11　updateSelect.jsp 页面

图 13.12　updateAgoods.jsp 页面

updateSelect.jsp 和 updateAgoods.jsp 的代码略。

13.5.5 删除商品

单击后台管理主页面中的"删除商品"超链接,打开删除查询页面 deleteSelect.jsp。"删除商品"超链接的目标地址是个 Servlet,该 Servlet 与 13.5.3 节查询商品的 SelectGoodsServlet.java 类同。在 SelectGoodsServlet 中,根据动作类型,将查询结果转发给 deleteSelect.jsp 页面。deleteSelect.jsp 页面的运行效果如图 13.13 所示。

商品ID	商品名称	商品类型	查看详情	操作类型
☐ 12	上衣X	服装	详情	删除
☐ 13	苹果	水果	详情	删除
☐ 14	樱桃	水果	详情	删除
☐ 15	微波炉	家电	详情	删除
☐ 16	电视99	家电	详情	删除
☐ 17	孕装55	孕婴	详情	删除
☐ 18	牛奶11	食品	详情	删除
☐ 19	红酒99	酒类	详情	删除
☐ 20	化妆品22	化妆品	详情	删除
☐ 22	上衣xx	服装	详情	删除

共17条记录 共2页 第1页 下一页

删除

图 13.13 deleteSelect.jsp 页面

在图 13.13 的复选框中选择要删除的商品,单击"删除"按钮,将要删除商品的 ID 提交给控制器 Servlet,Servlet 对象的 urlPatterns 是{ "/admin_deleteGoods" },类名为 DeleteGoodsServlet。在该 Servlet 中,根据动作类型调用业务模型中的 deleteManygoods() 方法执行批量删除的业务处理。

单击图 13.13 中的"删除"超链接,将当前行的商品 ID 提交给控制器 Servlet,Servlet 的类名为 DeleteGoodsServlet。在该 Servlet 中,根据动作类型调用业务模型中的 deleteAgoods()方法执行单个删除的业务处理。

删除成功后,进入删除查询页面。

❶ 视图

deleteSelect.jsp 的核心代码如下:

```
...
<body>
    <br>
    <form action="admin_deleteGoods" name="deleteForm">
    <table border="1" bordercolor="PaleGreen">
        <tr>
            <th width="250px">商品 ID</th>
            <th width="200px">商品名称</th>
            <th width="200px">商品类型</th>
            <th width="200px">查看详情</th>
            <th width="200px">操作类型</th>
        </tr>
        <c:forEach var="g" items="${requestScope.allGoods}">
```

```html
            <tr onmousemove="changeColor(this)" onmouseout="changeColor1(this)">
                <td>
                    <input type="checkbox" name="gno" value="${g.id}">
                    ${g.id}
                </td>
                <td>${g.gname}</td>
                <td>${g.typename}</td>
            <td><a href="admin_detail?gno=${g.id}&act=detail" target="_blank">详情</a></td>
                <td>
                    <a href="javascript:checkDel('${g.id}')">删除</a>
                </td>
            </tr>
</c:forEach>
  <tr>
        <td colspan="5" align="right">
               
   共${totalCount}条记录  共${totalPage}页  
            第${pageCur}页  
            <c:url var="url_pre" value="admin_selectGoods?act=${act }">
                <c:param name="pageCur" value="${pageCur - 1 }"/>
            </c:url>
            <c:url var="url_next" value="admin_selectGoods?act=${act }">
                <c:param name="pageCur" value="${pageCur + 1 }"/>
            </c:url>
            <!-- 第一页，没有上一页 -->
            <c:if test="${pageCur != 1 }">
                <a href="${url_pre}">上一页</a>   
                 
            </c:if>
            <!-- 最后一页，没有下一页 -->
            <c:if test="${pageCur != totalPage && totalPage !=0 }">
                <a href="${url_next}">下一页</a>
            </c:if>
        </td>
    </tr>
<tr>
    <td colspan="5">
        <input type="hidden" name="act" value="button">
        <input type="button" value="删除" onclick="confirmDelete()">
    </td>
</tr>
```

```
        </table>
    </form>
</body>
...
```

❷ 控制器

DeleteGoodsServlet.java 的核心代码如下：

```
...
AdminService as = new AdminService();
PrintWriter out = response.getWriter();
//获得动作类型
String act = request.getParameter("act");
String rs = null;
if("link".equals(act)){//单击链接删除
    String gno = request.getParameter("gno");
    //单个删除
    rs = as.deleteAgoods(gno);
}else if("button".equals(act)){//单击按钮删除
    //获取页面上选中的商品ID
    String gno[] = request.getParameterValues("gno");
    //批量删除
    rs = as.deleteManygoods(gno);
}
if("ok".equals(rs)){//删除成功
    //返回查询页面
    out.println("删除成功！3秒钟返回！");
}else{
    out.println("有子记录，删除失败！3秒钟返回！");
}
response.setHeader("refresh", "3; url=admin_selectGoods?act=deleteSelect");
return;
...
```

13.5.6 订单管理

单击后台管理主页面中"订单管理"的"删除订单"超链接，打开订单管理页面 orderManager.jsp。"订单管理"超链接的目标地址是个 Servlet，该 Servlet 的 urlPatterns 是 {"/admin_orderManager"}，类名为 OrderManagerServlet。在 OrderManagerServlet 中，调用业务模型中的 selectOrders()方法查询订单，将查询结果转发给 orderManager.jsp 页面。orderManager.jsp 页面的运行效果如图 13.14 所示。

订单编号	用户E-mail	订单金额	订单状态	订单日期	操作
8	1@126.com	39600.0	已付款	2018-02-12 05:04:33.0	
10	1@126.com	324.0	已付款	2018-02-13 14:54:29.0	
14	1@126.com	5.0	已付款	2018-02-26 08:22:12.0	
16	1@126.com	19.0	未付款	2018-02-26 15:32:40.0	删除

图 13.14 订单管理页面

在订单管理页面，管理员可以删除未支付的订单。单击"删除"超链接，可以删除当前行对应的订单。"删除"超链接的目标地址是一个 Servlet，该 Servlet 的类名是 OrderManagerServlet。在该 Servlet 中，根据动作类型调用业务模型中的 deleteOrder ()方法执行订单删除的业务处理。成功删除订单后回到订单管理页面。

❶ 视图

orderManager.jsp 的核心代码如下：

```jsp
...
<body>
    <br>
    <table border="1" bordercolor="PaleGreen">
        <tr>
            <th width="150px">订单编号</th>
            <th width="150px">用户 E-mail</th>
            <th width="100px">订单金额</th>
            <th width="100px">订单状态</th>
            <th width="150px">订单日期</th>
            <th width="100px">操作</th>
        </tr>
        <c:forEach var="n" items="${orderList}">
        <tr onmousemove="changeColor(this)" onmouseout="changeColor1(this)">
            <td>${n.id}</td>
            <td>${n.bemail}</td>
            <td>${n.amount}</td>
<td><c:if test="${n.status == 0}" >未付款</c:if><c:if test="${n.status == 1}" >已付款</c:if></td>
            <td>${n.orderdate}</td>
            <td>
                <c:if test="${n.status == 0}" >
                    <a href="javaScript:checkDel('${n.id}')">删除</a>
                </c:if>
                 
            </td>
        </tr>
        </c:forEach>
    </table>
</body>
...
```

❷ 控制器

OrderManagerServlet.java 的核心代码如下：

```java
...
String act = request.getParameter("act");//获得动作类型
AdminService as = new AdminService();
```

```
PrintWriter out = response.getWriter();
List<Map<String, Object>> list = as.selectOrders();
request.setAttribute("orderList", list);
RequestDispatcher rds = null;
//查询
if("manager".equals(act)){
    rds = request.getRequestDispatcher("admin/orderManager.jsp");
    rds.forward(request, response);
}else if("delete".equals(act)){//删除
    String ordersn = request.getParameter("ordersn");
    if(as.deleteOrder(ordersn)){
        out.println("删除成功！3秒钟返回！");
    }else{
        out.println("删除失败！3秒钟返回！");
    }
    response.setHeader("refresh", "3; url=admin_orderManager?act=manager");
    return;
}
...
```

13.5.7 用户管理

单击后台管理主页面中"用户管理"的"删除用户"超链接，打开用户管理页面 userManager.jsp。"用户管理"超链接的目标地址是一个 Servlet，该 Servlet 的 urlPatterns 是{ "/admin_userManager" }，类名为 UserManagerServlet。在 UserManagerServlet 中，根据动作类型调用业务模型中的 selectUsers()方法查询用户，将查询结果转发给 userManager.jsp 页面。userManager.jsp 页面的运行效果如图 13.15 所示。

用户ID	用户E-mail	用户密码	删除
5	1@126.com	●●●●●	删除
6	2@126.com	●●●●●	删除

图 13.15 用户管理页面

在用户管理页面，管理员可以删除某个用户（未有关联记录的用户）。单击"删除"超链接，可以删除当前行对应的用户。"删除"超链接的目标地址是个 Servlet，该 Servlet 的类名是 UserManagerServlet。在该 Servlet 中，根据动作类型调用业务模型中的 deleteUser()方法执行删除用户的业务处理。成功删除用户后回到用户管理页面。

❶ 视图

userManager.jsp 的代码略。

❷ 控制器

UserManagerServlet.java 的核心代码如下：

...

```
String act = request.getParameter("act");//获得动作类型
AdminService as = new AdminService();
PrintWriter out = response.getWriter();
List<Map<String, Object>> list = as.selectUsers();
request.setAttribute("userList", list);
RequestDispatcher rds = null;
//查询
if("manager".equals(act)){
    rds = request.getRequestDispatcher("admin/userManager.jsp");
    rds.forward(request, response);
}else if("delete".equals(act)){
    String bid = request.getParameter("bid");
    String rs = as.deleteUser(bid);
    if("nodelete".equals(rs)){
        out.println("有子记录，删除失败！3秒钟返回！");
    }else{
        out.println("删除成功！3秒钟返回！");
    }
    response.setHeader("refresh", "3; url=admin_userManager?act=manager");
    return;
}
...
```

13.5.8 公告管理

单击后台管理主页面中"公告管理"的"添加公告"超链接，打开公告管理页面 noticeManager.jsp。"公告管理"超链接的目标地址是一个 Servlet，该 Servlet 的 urlPatterns 是{ "/admin_noticeManager" }，类名为 NoticeManagerServlet。在 NoticeManagerServlet 中，根据动作类型调用业务模型中的 selectNotice()方法查询公告，将查询结果转发给 noticeManager.jsp 页面。noticeManager.jsp 页面的运行效果如图 13.16 所示。

公告ID	公告标题	公告时间	查看编辑
5	2018年大减价	2018-02-11 07:06:25.0	详情 删除
6	大减价	2018-02-11 07:11:41.0	详情 删除
7	春节不打烊	2018-02-11 07:11:59.0	详情 删除
8	春节不打烊春节不打烊	2018-02-11 07:12:07.0	详情 删除
9	欢度春节	2018-02-11 07:12:27.0	详情 删除
10	欢度春节欢度春节	2018-02-11 07:12:39.0	详情 删除
11	低价大放送	2018-02-11 07:13:10.0	详情 删除
12	低价大放送低价大放送	2018-02-11 07:13:19.0	详情 删除
13	勇往直前	2018-02-11 07:14:29.0	详情 删除
14	勇往直前勇往直前	2018-02-11 07:14:38.0	详情 删除
16	2018再出发	2018-02-26 15:11:24.0	详情 删除
17	2018整装待发	2018-02-26 15:11:42.0	详情 删除
18	2018一路发发	2018-02-26 15:11:56.0	详情 删除
添加公告			

图 13.16 公告管理页面

在公告管理页面中单击"添加公告"超链接，打开添加公告页面 addNotice.jsp。

addNotice.jsp 页面的运行效果如图 13.17 所示。

图 13.17　添加公告页面

在添加公告页面中输入公告信息后，单击"提交"按钮，将公告信息提交给控制器 Servlet，该 Servlet 的类名为 NoticeManagerServlet。在 NoticeManagerServlet 中，根据动作类型调用业务模型中的 addNotice ()方法添加公告，成功后回到公告管理页面。

在公告管理页面中单击"详情"超链接，打开公告详情页面 noticeDetail.jsp。"详情"超链接的目标地址是一个 Servlet，该 Servlet 的 urlPatterns 是 { "/admin_noticeDetail" }，类名为 NoticeDetailServlet。在 NoticeDetailServlet 中，根据动作类型调用业务模型中的 selectAnotice()方法查询公告信息，将查询结果转发给 noticeDetail.jsp 页面。noticeDetail.jsp 页面的运行效果如图 13.18 所示。

图 13.18　公告详情页面

在公告管理页面中单击"删除"超链接，将该行对应的公告删除。"删除"超链接的目标地址是一个 Servlet，该 Servlet 的类名为 NoticeDetailServlet。在 NoticeDetailServlet 中，根据动作类型调用业务模型中的 deleteAnotice()方法删除公告。成功删除后，回到公告管理页面。

❶ 视图

该模块的视图共涉及 3 个 JSP 页面：noticeManager.jsp、addNotice.jsp 和 noticeDetail.jsp。noticeManager.jsp、addNotice.jsp 和 noticeDetail.jsp 代码略。

❷ 控制器

该模块的控制器共涉及两个 Servlet：NoticeManagerServlet.java 和 NoticeDetailServlet.java。
NoticeManagerServlet.java 的核心代码如下：

```java
...
String act = request.getParameter("act");//获得动作类型
AdminService as = new AdminService();
List<Map<String, Object>> list = as.selectNotice();//查询公告
request.setAttribute("noticeList", list);
RequestDispatcher rds = null;
//查询
if("manager".equals(act)){
    rds = request.getRequestDispatcher("admin/noticeManager.jsp");
}else if("add".equals(act)){
    NoticeDTO nd = new NoticeDTO();
    String ntitle = request.getParameter("ntitle");
    String ncontent = request.getParameter("ncontent");
    nd.setNtitle(ntitle);
    nd.setNcontent(ncontent);
    as.addNotice(nd);//添加
    rds = request.getRequestDispatcher("admin_noticeManager?act=manager");
}
rds.forward(request, response);
...
```

NoticeDetailServlet.java 的核心代码如下：

```java
...
AdminService as = new AdminService();
String act = request.getParameter("act");//获得动作类型
String nid = request.getParameter("nid");
List<Map<String, Object>> list = as.selectAnotice(nid);
request.setAttribute("anotice", list.get(0));
RequestDispatcher rds = null;
//查看详情
if("detail".equals(act)){
    rds = request.getRequestDispatcher("admin/noticeDetail.jsp");
}
//删除
```

```
else if("delete".equals(act)){
    as.deleteAnotice(nid);
    //删除后回到查询
    rds = request.getRequestDispatcher("admin_noticeManager?act=manager");
}
rds.forward(request, response);
...
```

13.5.9 退出系统

在后台管理主页面中，单击"退出系统"超链接，将返回后台登录页面。"退出系统"超链接的目标地址是一个 Servlet，该 Servlet 的 urlPatterns 是{ "/admin_loginOut" }，类名为 LoginOutServlet。在 LoginOutServlet 中执行 session.invalidate()，session 将失效，并返回后台登录页面。

LoginOutServlet.java 的核心代码如下：

```
...
HttpSession session = request.getSession(true);
session.invalidate();
String act = request.getParameter("act");
if("admin".equals(act)){//后台退出系统
    response.sendRedirect("admin/login.jsp");
}
...
```

13.6 前台电子商务子系统的实现

与前台电子商务子系统相关的 JSP 页面、CSS、JavaScript 和图片位于 WebContent/beforeUser 目录下。在 13.4 节中已经介绍了系统的实体模型、数据库操作以及业务模型，本节只介绍 JSP 页面和 Servlet 的核心实现。

13.6.1 导航栏

在前台每个 JSP 页面中，都引入了一个名为 head.jsp 的页面，引入代码如下：

```
<jsp:include page="head.jsp"></jsp:include>
```

head.jsp 中的商品类型是从 session 对象获得的，该 session 对象是用户请求电子商务子系统的首页时初始化的。在请求首页时，执行了类名为 FirstServlet 的 Servlet，在该 Servlet 中调用业务模型 AdminService 中的 getGoodsType()方法获得商品类型，并将商品类型列表保存到 session 对象中；调用 BeforeUserService 业务模型中的 getLastedGoods()、getSaleOrder()、getFocusOrder()和 getNotice()方法分别获得最新商品列表、销售排行列表、人气排行列表和公告列表，并将这些列表保存到 request 对象中。最后转发到 beforeUser/index.jsp 页面。index.jsp 页面的实现过程见 13.3.2 节。head.jsp 页面的运行效果如图 13.19 所示。

图 13.19　导航栏

在导航栏的搜索框中输入信息，单击"搜索"按钮，将搜索信息提交给一个 Servlet，该 Servlet 的 urlPatterns 为 { "/before_search" }，类名是 SearchServlet。在 SearchServlet 中，调用业务模型中的 search() 方法对商品名称以模糊查询的方式搜索，并将搜索到的商品信息转发给 searchResult.jsp。searchResult.jsp 页面的运行效果如图 13.20 所示。

图 13.20　搜索结果

❶ 视图

导航栏模块的视图涉及两个 JSP 页面：head.jsp 和 searchResult.jsp。
head.jsp 的核心代码如下：

```
...
<p id="content">
    <c:if test="${sessionScope.user != null }">欢迎 ${sessionScope.user.bemail}</c:if>
    <c:if test="${sessionScope.user == null }"><a href="beforeUser/userLogin.jsp">登录</a></c:if>
</p>
...
<form action="before_search" name="myForm" method="post">
<div class="div2">
```

```
<input type="text" name="mykey" class="txt" value="请输入您要查询的内容"
    onfocus="clearValue()" />
    </div>
    <div class="div1">
        <input type="submit" class="an" value="搜索" />
    </div>
</form>
...
<div class="front_daohang">
    <ul>
        <li class="backbj"><a href="before_first?typeid=0">首页</a></li>
        <!-- 显示商品类型 -->
        <c:forEach var="g" items="${sessionScope.goodsType}">
            <li><a href="before_first?typeid=${g.id }">${g.typename }
            </a></li>
        </c:forEach>
        <li class="buy">
        <p class="car">
        <a href="before_selectCart">购物车</a>
        </p>
        </li>
    </ul>
</div>
...
```

searchResult.jsp 的核心代码如下：

```
...
<c:forEach var="mf" items="${requestScope.searchlist}">
    <tr>
        <td bgcolor="#ffffff" align="center">${mf.id}</td>
        <td bgcolor="#ffffff" align="center">${mf.gname}</td>
        <td align="center" bgcolor="#ffffff" height="60px">
            <imgstyle="width: 50px; height: 50px;" src="uploadFile/${mf.gpicture}"
            border="0" title="${mf.gname}" />
        </td>
        <td align="center" bgcolor="#ffffff">${mf.grprice}</td>
        <td align="center" bgcolor="#ffffff"><a style="text-decoration:
        none;" class="f6"
            href="before_detail?gno=${mf.id}">去看看</a></td>
    </tr>
</c:forEach>
...
```

❷ 控制器

该模块的控制器涉及两个 Servlet：FirstServlet.java 和 SearchServlet.java。
FirstServlet.java 的核心代码如下：

```
...
String typeid = request.getParameter("typeid");//获得商品类型
AdminService as = new AdminService();
BeforeUserService bs = new BeforeUserService();
//将登录存储到session中
HttpSession session = request.getSession(true);
//获得商品类型
List<Map<String, Object>> list = as.getGoodsType();
session.setAttribute("goodsType", list);
//获得最新商品
List<Map<String, Object>> lastedlist = bs.getLastedGoods(typeid);
//获得销售排行商品
List<Map<String, Object>> salelist = bs.getSaleOrder();
//获得人气排行
List<Map<String, Object>> focuslist = bs.getFocusOrder();
//获得公告
List<Map<String, Object>> noticelist = bs.getNotice();
request.setAttribute("lastedlist", lastedlist);
request.setAttribute("salelist", salelist);
request.setAttribute("focuslist", focuslist);
request.setAttribute("noticelist", noticelist);
RequestDispatcher rds = request.getRequestDispatcher("beforeUser/index.jsp");
rds.forward(request, response);
...
```

SearchServlet.java 的核心代码如下：

```
...
String mykey = request.getParameter("mykey");
BeforeUserService bs = new BeforeUserService();
List<Map<String, Object>> list = bs.search(mykey);
request.setAttribute("searchlist", list);
request.getRequestDispatcher("beforeUser/searchResult.jsp").forward(request, response);
...
```

13.6.2 销售排行

销售排行是以订单详情表中每种商品的销量总和排序的，具体 SQL 语句请参考 BeforeUserDao 的 getSaleOrder()方法。

13.6.3 人气排行

人气排行是以关注表中每种商品的关注次数总和排序的，具体 SQL 语句请参考 BeforeUserDao 的 getFocusOrder ()方法。

13.6.4　最新商品

最新商品是以商品 ID 排序的，因为商品 ID 是递增的。具体 SQL 语句请参考 BeforeUserDao 的 getLastedGoods ()方法。

13.6.5　公告栏

公告栏的 SQL 语句请参考 BeforeUserDao 的 getNotice ()方法。

13.6.6　用户注册

单击导航栏的"注册"超链接，将打开注册页面 register.jsp。register.jsp 页面的运行效果如图 13.21 所示。

图 13.21　注册页面

输入用户信息，单击"注册账号"按钮，将用户信息提交给一个 urlPatterns 为 {"/before_register"}的 Servlet，Servlet 的类名为 RegisterServlet。在 RegisterServlet 中，调用业务模型中的 register()方法实现注册。在注册前使用 Ajax 异步提交请求，调用 isExit() 方法判断 E-mail 是否可用。

❶ 视图

register.jsp 的代码略。

❷ 控制器

RegisterServlet.java 的核心代码如下：

```
…
String bemail = request.getParameter("bemail");
```

```java
String bpwd = request.getParameter("bpwd");
String code1 = request.getParameter("code");
String flag = request.getParameter("flag");
HttpSession session = request.getSession(true);
//获取验证码
String code2 = (String)session.getAttribute("rand");
//实体
UserDTO ud = new UserDTO();
ud.setBemail(bemail);
ud.setBpwd(bpwd);
//业务
BeforeUserService bs = new BeforeUserService();
RequestDispatcher rds = null;
if("0".equals(flag)) {//注册按钮
    //验证码输入正确
    if(code2.equalsIgnoreCase(code1)){
        if(bs.register(ud)){
            rds = request.getRequestDispatcher("beforeUser/userLogin.jsp");
        }else{
            rds = request.getRequestDispatcher("beforeUser/register.jsp");
        }
    }else{//验证码输入有误
        request.setAttribute("codeError", "验证码输入错误！");
        rds = request.getRequestDispatcher("beforeUser/register.jsp");
    }
    rds.forward(request, response);
}else {//验证邮箱
    //查询 E-mail
    List<Map<String, Object>> list = bs.isExit(ud);
    PrintWriter out = response.getWriter();
    //E-mail 已注册
    if(list.size() > 0){
        out.print("<font color=red size=5>该 E-mail 已注册！</font>");
    }else {
        out.print("<font color=green size=5>该 E-mail 可注册！</font>");
    }
}
...
```

13.6.7 用户登录

用户注册成功后，进入登录页面 userLogin.jsp。userLogin.jsp 页面的运行效果如图 13.22 所示。

图 13.22　登录页面

输入 E-mail、密码以及验证码，单击"确定"按钮，将用户输入的这些信息提交给一个类名为 UserLoginServlet 的 Servlet。在 UserLoginServlet 中，首先调用业务模型中的 isExit() 方法判断用户的邮箱是否正确；然后调用业务模型中的 isLogin ()方法判断密码是否正确。登录成功后，将用户的登录信息保存在 session 对象中，然后回到网站的首页。

❶ 视图

userLogin.jsp 的代码略。

❷ 控制器

UserLoginServlet.java 的核心代码如下：

```
...
String bemail = request.getParameter("bemail");
String bpwd = request.getParameter("bpwd");
String code1 = request.getParameter("code");
HttpSession session = request.getSession(true);
//实体
UserDTO ud = new UserDTO();
ud.setBemail(bemail);
ud.setBpwd(bpwd);
//业务
BeforeUserService bs = new BeforeUserService();
//获取验证码
String code2 = (String)session.getAttribute("rand");
RequestDispatcher rds = null;
//验证码输入正确
if(code2.equalsIgnoreCase(code1)){
    //邮箱输入错误
    if(bs.isExit(ud).size() < 1 ){
        request.setAttribute("bemailError", "邮箱输入错误！");
        rds = request.getRequestDispatcher("beforeUser/userLogin.jsp");
    }else if(bs.isLogin(ud).size() < 1){//邮箱正确，密码错误
        request.setAttribute("bpwdError", "密码输入错误！");
        rds = request.getRequestDispatcher("beforeUser/userLogin.jsp");
    }else{
```

```
            //将 id 存到 ud 中
            ud.setId((Integer)(bs.isLogin(ud).get(0)).get("id"));
            session.setAttribute("user", ud);
            //登录成功,转到首页
            rds = request.getRequestDispatcher("before_first?typeid=0");
        }
    }else{
        request.setAttribute("codeError", "验证码输入错误!");
        rds = request.getRequestDispatcher("beforeUser/userLogin.jsp");
    }
    rds.forward(request, response);
    ...
```

13.6.8　商品详情

可以从销售排行、人气排行以及最新商品等位置处,单击商品图片或商品名称进入商品详情页面 goodsdetail.jsp。goodsdetail.jsp 页面的运行效果如图 13.23 所示。

图 13.23　商品详情页面

在进入商品详情页面之前,程序先执行一个类名为 BeforeDetailServlet 的 Servlet。在该 Servlet 中,根据商品编号调用业务模型 AdminService 中的 selectAGoods()方法获得商品详情,并转发给 goodsdetail.jsp 页面显示。

❶ 视图

goodsdetail.jsp 的核心代码如下:

```
...
<form action="before_putCart" name="putcartform" method="post">
    <div id="goodsInfo">
        <div class="imgInfo">
            <input type="hidden" name="gno" value="${agoods.id}"/>
            <imgsrc="uploadFile/${agoods.gpicture}"width="230px"height="230px"/>
```

```html
                </div>
            </div>
            <div class="goods_desc">
                <div class="bt">
                    ${agoods.gname}
                </div>
                <div class="goods_show">
                    <ul>
                        <li><span>价格:</span> <strong class="yj">
                            ${agoods.goprice}元</strong></li>
                        <li><span>折扣价:</span><strong
                            class="xj">${agoods.grprice}元</strong> </li>
                        <li><span>类型:</span> ${agoods.typename}</li>
                        <li><span>购买数量:</span>${agoods.gname}
                        <input type="text" name="mknum"
                            class="good_txt" value="1" /> (库存${agoods.gstore}件)</li>
                    </ul>
                </div>
                <p class="bottom10 top5">
                    <img src="beforeUser/images/goods_ann2.gif" style="cursor: pointer"
                        onclick="goCart()" />  <input type="button"
                        style="cursor: pointer" class="sh_bnt2"
                        onclick="gofocus('${agoods.id}')" value="关注" />
                </p>
            </div>
</form>
...
```

❷ 控制器

BeforeDetailServlet.java 的核心代码如下：

```
...
//获得商品编号
String gno = request.getParameter("gno");
AdminService as = new AdminService();
List<Map<String, Object>> list = as.selectAGoods(gno);
//将一个商品详细信息存到 request 中
request.setAttribute("agoods", list.get(0));
RequestDispatcher rds =request.getRequestDispatcher("beforeUser/goodsdetail.jsp");
rds.forward(request, response);
...
```

13.6.9 关注商品

登录成功的用户可以在商品详情页面单击"关注"按钮关注该商品。此时，程序调用一个类名为 FocusServlet 的 Servlet。在该 Servlet 中，根据商品编号和用户 ID 调用业务模型中的 focusGoods()方法完成关注。关注成功后，回到商品详情页面。

FocusServlet.java 的核心代码如下：

```
...
String gno = request.getParameter("gno");
HttpSession session = request.getSession(true);
int bid = ((UserDTO)session.getAttribute("user")).getId();
//业务层
BeforeUserService bs = new BeforeUserService();
String rs = bs.focusGoods(gno, bid);
PrintWriter out = response.getWriter();
//已关注
if("focused".equals(rs)){
    out.println("已关注该商品！3 秒钟返回！");
}else{
    if("focusSuccess".equals(rs)){
        out.println("成功关注该商品！3 秒钟返回！");
    }else{
        out.println("关注失败！3 秒钟返回！");
    }
}
response.setHeader("refresh", "3; url=before_detail?gno=" + gno);
return;
...
```

13.6.10　购物车

单击商品详情页面中的"加入购物车"按钮或导航栏中的"购物车"超链接，打开购物车页面 cart.jsp。在该页面可以修改购买数量，运行效果如图 13.24 所示。

图 13.24　购物车

与购物车有关的 Servlet 有 PutCartServlet、SelectCartServlet 和 DeleteCartServlet。在 PutCartServlet 中，根据用户 ID 和商品编号调用业务模型中的 putCart()方法，将商品加入购物车；在 SelectCartServlet 中，根据用户 ID 调用业务模型中的 selectCart()方法查询购物车；在 DeleteCartServlet 中，根据用户 ID 和商品编号调用业务模型中的 deleteCart()方法，从购物车中删除商品。

❶ 视图

cart.jsp 的核心代码如下：

```
...
<c:forEach var="ce" items="${cartlist}">
    <tr>
        <td bgcolor="#ffffff" align="center"><a href="before_detail?gno=
        ${ce.id}"> <img
            style="width: 100px; height: 100px;"
            src="uploadFile/${ce.gpicture}" border="0"
            title="${ce.gname}" />
</a><br/><a style="text-decoration: none;" href="before_detail?gno=${ce.id}"
    class="f6">${ce.gname}</a></td>
        <td bgcolor="#ffffff" width="110px" align="center"><span>${ce.grprice}
        </span></td>
        <td align="center" bgcolor="#ffffff" width="115px"
            valign="middle"><input type="text" name="goods_number"
            value="${ce.shoppingnum}" size="4" onblur="updateNum(this.value,
            '${ce.id}')"
            class="inputBg"
            style="text-align: center; width: 36px; color: #999999" /></td>
    <td align="center" bgcolor="#ffffff" width="115px">¥ <span>${ce.smallsum}
    </span> 
        </td>
        <td align="center" bgcolor="#ffffff" width="185px"><a
            style="text-decoration: none;" href="javaScript:deleteAgoods
            ('${ce.id}')"
            class="f6" title="删除"><img src="beforeUser/images/sc.png" />
        </a></td>
    </tr>
</c:forEach>
<tr>
    <td align="right" bgcolor="#ffffff" colspan="4" height="41px;"
        style="border-left: 0 none;"><font
        style="color: #a60401; font-size: 13px; font-weight: bold; letter-
        spacing: 0px;">
            购物金额总计(不含运费) ¥ <span id="stotal"></span>
        ${total}元
    </font></td>
```

```html
        <td align="center" bgcolor="#ffffff"><input type="button"
            value="清空购物车" onclick="godelete()" class="bnt_blue_1" id="bnt11" />
        </td>
</tr>
```
...

❷ 控制器

本模块涉及 4 个 Servlet：PutCartServlet、UpdateCartServlet、SelectCartServlet 和 DeleteCartServlet。

PutCartServlet.java 的核心代码如下：

```java
...
String gno = request.getParameter("gno");//商品ID
String mknum = request.getParameter("mknum");//购买数量
HttpSession session = request.getSession(true);
int bid = ((UserDTO)session.getAttribute("user")).getId();
//业务层
BeforeUserService bs = new BeforeUserService();
RequestDispatcher rds = null;
//成功到购物车页面
if(bs.putCart(gno, bid, mknum)){
    rds = request.getRequestDispatcher("before_selectCart");
}else{
    rds = request.getRequestDispatcher("beforeUser/operateerror.jsp");
}
rds.forward(request, response);
...
```

UpdateCartServlet.java 的核心代码如下：

```java
...
String shoppingnum = request.getParameter("shoppingnum");
String gno = request.getParameter("gno");
HttpSession session = request.getSession(true);
int bid = ((UserDTO)session.getAttribute("user")).getId();
//业务层
BeforeUserService bs = new BeforeUserService();
RequestDispatcher rds = null;
if(bs.updateCart(shoppingnum,gno, bid)){
    rds = request.getRequestDispatcher("before_selectCart");
}else{
    rds = request.getRequestDispatcher("beforeUser/operateerror.jsp");
}
rds.forward(request, response);
...
```

SelectCartServlet.java 的核心代码如下：

```
...
//业务层
BeforeUserService bs = new BeforeUserService();
HttpSession session = request.getSession(true);
int bid = ((UserDTO)session.getAttribute("user")).getId();
List<Map<String, Object>> list = bs.selectCart(bid);
double sum = 0;
for (Map<String, Object> map : list) {
    sum = sum + (Double)map.get("smallsum");
}
request.setAttribute("total", sum);
request.setAttribute("cartlist", list);
String act = request.getParameter("act");
//购物确认
if("confirm".equals(act)){
    request.getRequestDispatcher("beforeUser/orderconfirm.jsp").forward
    (request, response);
}else{
    request.getRequestDispatcher("beforeUser/cart.jsp").forward(request,
    response);
}
...
```

DeleteCartServlet.java 的核心代码如下：

```
...
String act = request.getParameter("act");
HttpSession session = request.getSession(true);
int bid = ((UserDTO)session.getAttribute("user")).getId();
//业务层
BeforeUserService bs = new BeforeUserService();
//清空购物车
if("all".equals(act)){
    bs.deleteCart(bid, null);
}else{//删除一个商品
    String gno = request.getParameter("gno");
    bs.deleteCart(bid, gno);
}
request.getRequestDispatcher("before_selectCart").forward(request, response);
...
```

13.6.11 下单

在购物车页面单击"去结算"按钮，进入订单确认页面 orderconfirm.jsp。orderconfirm.jsp 页面的运行效果如图 13.25 所示。

第 13 章 电子商务平台的设计与实现

图 13.25 订单确认

在订单确认页面单击"提交订单"按钮,完成订单。订单完成时,提示页面效果如图 13.26 所示。

图 13.26 订单完成

❶ 视图

"下单"模块的视图涉及两个 JSP 页面:orderconfirm.jsp 和 orderdone.jsp。orderconfirm.jsp 与 orderdone.jsp 的代码略。

❷ 控制器

下单时,使用的控制器是一个类名为 OrderSubmitServlet 的 Servlet。该 Servlet 的核心代码如下:

```
...
String amount = request.getParameter("amount");
HttpSession session = request.getSession(true);
int bid = ((UserDTO)session.getAttribute("user")).getId();
//业务层
BeforeUserService bs = new BeforeUserService();
//订单提交
int id = bs.orderSubmit(bid, amount);
if(id > 0){
    request.setAttribute("ordersn", id);
    request.getRequestDispatcher("beforeUser/orderdone.jsp").forward
```

```
        (request, response);
    }else{
        request.getRequestDispatcher("beforeUser/operateerror.jsp").forward
        (request, response);
    }
...
```

13.6.12 用户中心

在导航栏的上方,单击"用户中心"超链接,进入用户中心页面 userCenter.jsp。userCenter.jsp 页面的运行效果如图 13.27 所示。

图 13.27 用户中心

在用户中心页面,单击"去看看"超链接,打开该行商品的详情页面;单击"详情"超链接,打开该行订单的详情页面 userOrderDetail.jsp。userOrderDetail.jsp 页面的运行效果如图 13.28 所示。

图 13.28 用户订单详情

❶ 视图

"用户中心"模块的视图涉及两个 JSP 页面:userCenter.jsp 和 userOrderDetail.jsp。

userCenter.jsp 和 userOrderDetail.jsp 的代码略。

❷ 控制器

该模块的控制器涉及一个类名为 UserCenterServlet 的 Servlet。该 Servlet 的核心代码如下：

```java
...
String act = request.getParameter("act");
HttpSession session = request.getSession(true);
int bid = ((UserDTO)session.getAttribute("user")).getId();
//业务层
BeforeUserService bs = new BeforeUserService();
if("index".equals(act)){//用户中心首页
    //我的订单
    List<Map<String, Object>> myOrder = bs.myOrder(bid);
    //我的关注
    List<Map<String, Object>> myFocus = bs.myFocus(bid);
    request.setAttribute("myOrder", myOrder);
    request.setAttribute("myFocus", myFocus);
    request.getRequestDispatcher("beforeUser/userCenter.jsp").forward
    (request, response);
}else if("detail".equals(act)){
    //订单详情
    String ordersn = request.getParameter("ordersn");
    List<Map<String, Object>> myOrderDetail = bs.myOrderDetail(ordersn);
    request.setAttribute("myOrderDetail", myOrderDetail);
    request.getRequestDispatcher("beforeUser/userOrderDetail.jsp").
    forward(request, response);
}
...
```

13.7 本章小结

本章讲述了电子商务平台通用功能的设计与实现。通过本章的学习，读者不仅掌握了 Servlet MVC 模式应用开发的流程、方法和技术，还应该熟悉电子商务平台的业务需求、设计以及实现。

第 14 章 Struts 2 框架基础

学习目的与要求

本章对 Struts 2 框架进行简要介绍。通过本章的学习，用户能够掌握使用 Eclipse 平台开发 Struts 2 应用的基本步骤，了解 Struts 2 的基本流程。

本章主要内容

- Struts 2 的体系结构
- Struts 2 的工作环境
- Struts 2 的 Action
- 基于 Struts 2 框架的注册与登录

Struts 2 是一款优秀的、基于 MVC 思想的应用框架，是 Apache Struts 和 WebWork 组合产生的新产品，是非常灵活、简单的 MVC 组件。

14.1 Struts 2 的体系结构

视频讲解

Struts 2 体系与 Struts 1 体系的差别非常大，因为 Struts 2 使用了 WebWork 的设计核心，而不是 Struts 1 的设计核心。Struts 2 大量使用拦截器来处理用户的请求，从而允许用户的业务逻辑控制器和 Servlet API 分离。在处理请求的过程中以用户的业务逻辑控制器为目标，创建一个控制器代理。控制器代理回调业务控制器中的 execute 方法来处理用户的请求。该方法的返回值决定了 Struts 2 以怎样的视图资源呈现给用户，如图 14.1 所示。

从图 14.1 中可以发现 Struts 2 的大致处理流程如下：

① 浏览器发送请求，例如 "admin/login.action"。

② 核心控制器 StrutsPrepareAndExecuteFilter 根据请求调用对应的 Action。

③ Struts 2 的拦截器链自动对请求进行相关应用的拦截，例如验证、数据类型转换或文件上传下载等。

④ 回调 Action 的 execute 方法。该方法先获取参数，然后执行某种业务操作，既可以

将数据保存到数据库中，也可以将数据从数据库中提取出来。实际上，Action 只是一个控制器，它会调用业务逻辑组件来处理用户的请求。

图 14.1　Struts 2 的体系结构

⑤ 将 Action 中 execute 方法的处理结果存入 Stack Context 中，并返回一个字符串。核心控制器 StrutsPrepareAndExecuteFilter 将根据返回的字符串跳转到指定的视图资源，该视图资源将会读取 Stack Context 中的信息，并在浏览器中显示信息。

14.2　Struts 2 的工作环境

在第 2 章 Java Web 开发环境的基础上，导入 Struts 2 的相关 JAR 包，即可开发 Struts 2 应用。

视频讲解

14.2.1　Struts 2 的下载

在浏览器中打开网址"http://struts.apache.org/download.cgi"，下载 Struts 2（本书采用的是 struts-2.5.18-all.zip）。

建议读者下载 Struts 2 的完整版（Full Distribution），将下载到的 ZIP 文件解压缩，解压缩后的目录结构如图 14.2 所示。图 14.2 中，各目录解释如下。

- apps：该文件夹下包含了基于 Struts 2 的示例应用（可以使用 Eclipse Import 这些 WAR 文件，具体步骤为：Import → Web →WAR file），这些示例应用对于读者是非常有用的。
- docs：该文件夹下包含了 Struts 2 的相关文档，包括 Struts 2 的快速入门、Struts 2 的文档以及 API 文件等内容。

名称	修改日期
apps	2018/9/18 13:51
docs	2018/9/18 13:51
lib	2018/9/18 13:51
src	2018/9/18 13:51
CLASSWORLDS-LICENSE.txt	2018/9/18 13:36
FREEMARKER-LICENSE.txt	2018/9/18 13:36
LICENSE.txt	2018/9/18 13:36
NOTICE.txt	2018/9/18 13:36
OGNL-LICENSE.txt	2018/9/18 13:36
OVAL-LICENSE.txt	2018/9/18 13:36
SITEMESH-LICENSE.txt	2018/9/18 13:36
XPP3-LICENSE.txt	2018/9/18 13:36
XSTREAM-LICENSE.txt	2018/9/18 13:36

图 14.2　Struts 2 的目录结构

- lib：该文件夹下包含了 Struts 2 框架的核心类库，以及 Struts 2 的第三方插件类库。
- src：该文件夹下包含了 Struts 2 框架的全部源代码。

14.2.2　使用 Eclipse 开发 Struts 2 的 Web 应用

使用 Eclipse 开发 Struts 2 的 Web 应用的具体步骤如下。

❶ 创建 Java Web 应用（Dynamic Web Project）

创建 Java Web 应用时，首先在对话框的 Project Name 文本框中输入项目名称 ch14；然后两次单击 Next 按钮后，选中 Generate web.xml deployment descriptor 复选框，如图 14.3 所示；最后单击 Finish 按钮，完成项目 ch14 的创建。

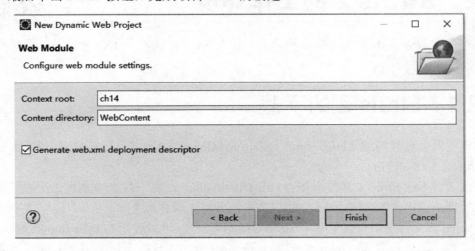

图 14.3　选择生成 web.xml

❷ 为 Web 应用添加 Struts 2 的 JAR 包

将 Struts 2 解压缩后的 lib 目录下的 commons-fileupload-1.3.3.jar、commons-io-2.5.jar、 commons-lang3-3.6.jar、freemarker-2.3.26-incubating.jar、javassist-3.20.0-GA.jar、log4j-api-2.10.0.jar、ognl-3.1.15.jar、struts2-core-2.5.18.jar、struts2-json-plugin-2.5.18.jar 等必需 JAR 包复制到 Web 应用的 WEB-INF/lib 目录下。如果需要使用 Struts 2 的更多特性，则将相应的 JAR 包复制到 Web 应用的 WEB-INF/lib 目录下即可。

❸ 在 web.xml 文件中配置 Struts 2 的核心控制器

要想在开发 Web 应用时使用 Struts 2 框架，还需要在 web.xml 中配置 Struts 2 的核心过滤器，代码如下：

```xml
<?xml version="1.0" encoding="UTF-8"?>
<web-app xmlns:xsi="http://www.w3.org/2001/XMLSchema-instance" xmlns="http:
//xmlns.jcp.org/xml/ns/javaee" xsi:schemaLocation="http://xmlns.jcp.org/
xml/ns/javaee http://xmlns.jcp.org/xml/ns/javaee/web-app_3_1.xsd" id="WebApp_ID"
version="3.1">
  <display-name>struts2cardManage</display-name>
  <welcome-file-list>
    <welcome-file>index.html</welcome-file>
    <welcome-file>index.htm</welcome-file>
    <welcome-file>index.jsp</welcome-file>
    <welcome-file>default.html</welcome-file>
    <welcome-file>default.htm</welcome-file>
    <welcome-file>default.jsp</welcome-file>
  </welcome-file-list>
  <filter>
    <filter-name>struts2</filter-name>
    <filter-class>org.apache.struts2.dispatcher.filter.StrutsPrepareAndExecuteFilter
    </filter-class>
  </filter>
  <filter-mapping>
    <filter-name>struts2</filter-name>
    <url-pattern>/*</url-pattern>
  </filter-mapping>
</web-app>
```

经过上面 3 个步骤，已经可以在 "ch14" Web 应用中使用 Struts 2 的基本功能了。

14.2.3　第一个 Struts 2 应用

本节在 14.2.2 节的基础上，讲解第一个 Struts 2 应用的具体开发步骤。

❶ 创建用户请求页面

在 ch14 应用中，需要用户提交请求。submit.jsp 页面的代码如下：

```
<%@ page language="java" contentType="text/html; charset=UTF-8" pageEncoding="UTF-8"%>
<!DOCTYPE html>
```

```html
<html>
<head>
<meta charset="UTF-8">
<title>Insert title here</title>
</head>
<body>
    <form action="hello/world.action" method="post">
        <input type="submit" value="提交"/>
    </form>
</body>
</html>
```

❷ **Action 实现**

在 ch14 应用的 src 目录下创建 HelloWorldAction 类来处理用户提交的请求。HelloWorldAction.java 的代码如下：

```java
package action;
public class HelloWorldAction {
    public String execute(){
        return "success";
    }
}
```

❸ **配置 Action**

定义好 HelloWorldAction 后，还需要在 src 目录下提供一个 struts.xml 文件，并且需要在该 xml 文件中配置 HelloWorldAction。在 struts.xml 文件中主要配置处理请求的名称以及所对应的 Action 类，还需要配置 Action 处理结果和资源之间的映射关系。下面是 HelloWorldAction 应用中的 struts.xml 的代码：

```xml
<?xml version="1.0" encoding="UTF-8" ?>
<!DOCTYPE struts PUBLIC "-//Apache Software Foundation//DTD Struts Configuration 2.5//EN" "http://struts.apache.org/dtds/struts-2.5.dtd">
<struts>
    <package name="hello" namespace="/hello" extends="struts-default">
        <action name="world" class="action.HelloWorldAction">
            <result name="success">/hello.jsp</result>
        </action>
    </package>
</struts>
```

上面的映射文件定义了 name 为 world 的 Action，Action 将调用自身的 execute 方法处理用户请求。如果 execute 方法返回的结果为 success 字符串，那么请求将转发到 hello.jsp。hello.jsp 的代码如下：

```jsp
<%@ page language="java" contentType="text/html; charset=UTF-8" pageEncoding="UTF-8"%>
<!DOCTYPE html>
```

```
<html>
<head>
<meta charset="UTF-8">
<title>Insert title here</title>
</head>
<body>
    <h1>Hello World!</h1>
</body>
</html>
```

❹ 在 Eclipse 中发布并运行 Web 应用

在 Eclipse 中第一次运行 Web 应用 ch14 时，需要将 Web 应用发布到 Tomcat。例如，运行 HelloWorld 应用时，可以右击 jsp 文件 submit.jsp，选择 Run As | Run on Server 命令，在对话框中单击 Finish 按钮完成发布并运行。

14.3 Struts 2 的 Action

视频讲解

在使用 Struts 2 进行 Web 应用开发时，Action 是应用的核心，Action 类包含了对用户请求的处理逻辑，因此被称为业务控制器。业务控制器 Action 是用户请求和业务逻辑之间的"桥梁"，是 Struts 2 框架最核心的部分，负责具体的业务逻辑处理。

14.3.1 Action 的创建与配置

❶ **Action 的创建**

在 Struts 2 应用中创建 Action 有 3 种方式：一个普通的类，不需要实现任何接口或继承任何类；实现 Action 接口；继承 ActionSupport 类，这是最常用的一种方式，好处在于可以直接使用 Struts 2 封装好的方法。在这里以 14.2.3 节的 Action 类为例介绍 Action 的 3 种创建方式。

1）一个普通的类

Struts 2 采用低侵入式设计，Struts 2 的 Action 可以不用继承或实现任何类或接口，代码重用性较好。14.2.3 节中的 HelloWorldAction 类就是一个普通的类，示例代码如下：

```
package action;
public class HelloWorldAction {
    public String execute(){
        return "success";
    }
}
```

2）实现 Action 接口

为了使开发者开发的 Action 更加规范，Struts 2 提供了一个 Action 接口，该接口定义了 Struts 2 的 Action 处理类应该实现的规范。标准 Action 接口的代码如下：

```java
package com.opensymphony.xwork2;
public interface Action {
    //定义Action接口中包含的一些结果字符串
    public static final String SUCCESS = "success";
    public static final String NONE = "none";
    public static final String ERROR = "error";
    public static final String INPUT = "input";
    public static final String LOGIN = "login";
    //定义处理用户请求的execute方法
    public String execute() throws Exception;
}
```

在上面的 Action 接口代码中，只定义了一个 execute 方法，该方法返回一个字符串。除此之外，该接口还定义了 5 个字符串常量，它们的作用是统一 execute 方法的返回值。例如，当 Action 类处理用户请求成功后，有人喜欢返回 success 字符串，有人喜欢返回 doSuccess 字符串……这样不利于项目的统一管理。可以将 14.2.3 节中的 HelloWorldAction 修改为实现 Action 接口，代码如下所示：

```java
package action;
import com.opensymphony.xwork2.Action;
public class HelloWorldAction implements Action{
    public String execute() throws Exception{
        return SUCCESS;
    }
}
```

在上面的 HelloWorldAction 代码中，execute 方法处理成功返回一个 SUCCESS 常量，不用直接返回一个 success 字符串，这样统一了 execute 方法处理成功的返回值。

3）继承 ActionSupport 类

Struts 2 还提供了 Action 接口的一个实现类 ActionSupport。有关 ActionSupport 类的实现代码可以去"struts-2.x.x\src\core\src\main\java\com\opensymphony\xwork2"目录中查阅。ActionSupport 类是 Struts 2 默认的 Action 处理类，该类已经提供了许多默认方法，包括获取国际化信息方法、数据校验方法、默认的处理用户请求的方法等。如果让开发者的 Action 类继承该 ActionSupport 类，则会大大简化 Action 的开发。继承 ActionSupport 类创建 Action 的方式是最常用的一种方式。可以将 14.2.3 节中的 HelloWorldAction 修改为继承 ActionSupport 类，代码如下所示：

```java
package action;
import com.opensymphony.xwork2.ActionSupport;
public class HelloWorldAction extends ActionSupport{
    public String execute(){
        return SUCCESS;
    }
}
```

❷ **Action 的配置**

在 Struts 2 框架中每一个 Action 是一个工作单元。Struts 2 框架负责将一个用户请求对应一个 Action，当一个 Action 类匹配一个请求的时候，这个 Action 类就会被 Struts 2 框架调用。struts.xml 中的每一对 action 元素就是一个 Action 映射。

Struts 2 框架提供多种 Action 配置方法，使得开发者有更多的选择。但在实际的项目开发过程中，开发团队必须采用统一的方法来开发，方便以后的维护。

例如，在 Struts 2 的 Web 应用 struts.xml 中，有如下配置：

```xml
<package name="user" namespace="/" extends="struts-default">
    <action name="login" class="action.LoginAction">
        <result>/loginSuccess.jsp</result>
        <result name="fail">/login.jsp</result>
    </action>
</package>
```

从上述的配置中，读者可以得知：类 LoginAction 提供一个 execute 方法，用来处理用户登录（login）的请求。Action 的名字是 login，用户请求此 Action 时需要将 URL 的访问路径设置为 http://localhost:8080/ch2/login.action。现在的问题是类 LoginAction 只能配置一个 Action 吗？如果将注册（register）也放在类 LoginAction 中，应该怎样配置呢？

Struts 2 配置文件中的 action 元素提供了一个名为 method 的属性，该属性指定执行 Action 时调用 Action 类中的任意公有、非静态的方法。那么登录和注册功能可以映射为：

```xml
<package name="user" namespace="/" extends="struts-default">
    <action name="login" class="action.LoginAction">
        <result>/loginSuccess.jsp</result><!--与execute方法的return SUCCESS 对应 -->
        <result name="loginfail">/login.jsp</result><!--与execute方法的return "loginfail"对应-->
    </action>
    <action name="register" class="action.LoginAction" method="register">
        <result>/login.jsp</result><!--与register方法的return SUCCESS 对应 -->
        <result name="registerfail">/register.jsp</result><!-- return "registerfail"对应-->
    </action>
</package>
```

从上述的配置中，读者可以得知：类 LoginAction 提供一个 register 方法，用来处理用户注册的请求。Action 的名字是 register，用户请求此 Action 时需要将 URL 的访问路径设置为 http://localhost:8080/ch2/register.action。

类 LoginAction 的伪代码如下：

```java
public class LoginAction extends ActionSupport{
    /**
     * 实现登录功能
     */
```

```
    public String execute(){
        if(登录成功){
            return SUCCESS;
        }
        return "loginfail";//与配置文件中result的name属性对应
    }
    /**
     * 实现注册功能
     */
    public String register(){
        if(注册成功){
            return SUCCESS;
        }
        return "registerfail";//与配置文件中result的name属性对应
    }
}
```

14.3.2　Action 接收请求参数

如何将用户的请求信息提交给 Action 是 Struts 2 框架的重点知识，因为 Action 只有获取了用户的请求信息后，才能对请求做出处理。在 Struts 2 的 Action 中，有 3 种方式接收用户的请求参数，分别是使用 Action 属性接收参数、使用 DomainModel（实体）接收参数和使用 ModelDriven<T>接口接收参数。本节使用同一个完整示例讲解 Action 接收请求参数的 3 种方式。Action 的配置统一如下：

```
<package name="test" namespace="/test" extends="struts-default">
    <action name="input" class="action.ReceiveAction">
        <result>/hello.jsp</result>
    </action>
</package>
```

❶ 使用 Action 属性接收参数

使用 Action 属性接收参数就是在 Action 类中定义与请求参数同名的属性，并为属性提供 setter 和 getter 方法，Struts 2 便能自动接收请求参数并赋值给同名属性。下面通过一个完整的示例来讲解 Action 接收请求参数。

首先，提供一个 input.jsp 页面，代码如下：

```
<body>
    <form action=" test/input.action" method="post">
        <h3>信息提交</h3>
        用户名：<input type="text" name="uname"><br>
        年龄：<input type="text" name="uage"><br>
        <input type="submit" value="提交">
    </form>
</body>
```

其次，需要提供一个 Action 类，该 Action 是通过属性来保存用户请求的参数，要保证 input.jsp 页面 form 表单中元素的 name 属性与 Action 类中的属性名称一致。ReceiveAction 代码如下：

```java
package action;
import com.opensymphony.xwork2.ActionSupport;
public class ReceiveAction extends ActionSupport{
    //用来封装用户请求中用户名的信息
    private String uname;
    //用来封装用户请求中年龄的信息
    private int uage;
    public String getUname() {
        return uname;
    }
    public void setUname(String uname) {
        this.uname = uname;
    }
    public int getUage() {
        return uage;
    }
    public void setUage(int uage) {
        this.uage = uage;
    }
    /**
     * 处理用户请求的方法
     */
    public String execute(){
        System.out.println("uname:"+uname);
        System.out.println("uage:"+uage);
        return SUCCESS;
    }
}
```

❷ 使用 DomainModel（实体）接收参数

使用 DomainModel 接收参数就是将用户请求的参数封装到 JavaBean 对象中，所以首先提供一个 JavaBean 类 Information，代码如下：

```java
package model;
public class Information {
    //用来封装用户请求中用户名的信息，请求参数 uname 与属性名相同
    private String uname;
    //用来封装用户请求中年龄的信息，请求参数 uage 与属性名相同
    private int uage;
    public String getUname() {
        return uname;
```

```java
    }
    public void setUname(String uname) {
        this.uname = uname;
    }
    public int getUage() {
        return uage;
    }
    public void setUage(int uage) {
        this.uage = uage;
    }
}
```

其次,提供处理请求的 ReceiveAction,代码如下:

```java
package action;
import model.Information;
import com.opensymphony.xwork2.ActionSupport;
public class ReceiveAction extends ActionSupport{
    //声明实体(JavaBean)对象 info,提供 setter 和 getter 方法
    private Information info;
    public Information getInfo() {
        return info;
    }
    public void setInfo(Information info) {
        this.info = info;
    }
    /**
     * 处理用户请求的方法
     */
    public String execute(){
        System.out.println("uname:"+info.getUname());
        System.out.println("uage:"+info.getUage());
        return SUCCESS;
    }
}
```

最后,input.jsp 页面 form 表单中元素的 name 属性值需要加上实体对象名 "info",代码如下:

```html
<body>
    <form action=" test/input.action" method="post">
        <h3>信息提交</h3>
        用户名:<input type="text" name="info.uname"><br>
        年龄:<input type="text" name="info.uage"><br>
        <input type="submit" value="提交">
    </form>
</body>
```

❸ 使用 **ModelDriven<T>接口接收参数**

使用 ModelDriven<T>接口接收参数时，Action 类需要实现 ModelDriven<T>接口，T 为 JavaBean 类。ReceiveAction 的代码如下：

```java
package action;
import model.Information;
import com.opensymphony.xwork2.ActionSupport;
import com.opensymphony.xwork2.ModelDriven;
public class ReceiveAction extends ActionSupport implements ModelDriven<Information>{
    //定义info对象，需要创建该对象，不会主动帮你创建
    private Information info = new Information();
    /**
     * 处理用户请求的方法
     */
    public String execute(){
        System.out.println("uname:"+info.getUname());
        System.out.println("uage:"+info.getUage());
        return SUCCESS;
    }
    @Override
    public Information getModel() {
        return info;
    }
}
```

input.jsp 页面 form 表单中元素的 name 属性值与 JavaBean 类中的属性名称一致，代码如下：

```html
<body>
    <form action=" test/input.action" method="post">
        <h3>信息提交</h3>
        用户名：<input type="text" name="uname"><br>
        年龄：<input type="text" name="uage"><br>
        <input type="submit" value="提交">
    </form>
</body>
```

通过学习 Action 接收请求参数的 3 种方式，可以发现各有优点，开发者可根据实际情况灵活使用。

14.3.3 Action 访问 Servlet API

在 Struts 2 中，Action 已经与 Servlet API 分离，使得 Action 具有更加灵活和低耦合的特性。Action 作为 Web 应用的控制器，不访问 Servlet API 几乎是不可能的，例如跟踪 HTTP Session 状态等。Web 应用中通常需要访问的 Servlet API 是 HttpServletRequest、HttpSession 和 ServletContext，这 3 个类分别代表 request、session 和 application 等 JSP 内置对象。

Action 访问 Servlet API 的常用方式有通过 ActionContext，实现 RequestAware、SessionAware 和 ApplicationAware 接口，通过 ServletActionContext，实现 ServletRequestAware 接口等。

❶ 通过 ActionContext

Struts 2 提供了一个 ActionContext 类，在 Action 中可以通过该类访问 Servlet API。ActionContext 类中包含的几个常用方法如表 14.1 所示。

表 14.1 ActionContext 类的常用方法

序号	方　法	功　能　说　明
1	Object get(Object key)	通过参数 key 获取当前 ActionContext 中的值
2	Map getApplication( )	返回一个 Map 类型的 Application 对象
3	static ActionContext getContext( )	返回当前线程的 ActionContext 对象
4	Map getParameters( )	返回一个包含所有 HttpServletRequest 参数信息的 Map 对象
5	Map getSession( )	返回一个 Map 类型的 HttpSession 对象

下面的 Web 应用将在 Action 中通过 ActionContext 访问 Servlet API，以如图 14.4 所示的登录页面为示例。yourLogin.jsp 的代码如下：

```
<body>
    <h3>登录页面</h3>
    <form action="test/yourLogin.action" method="post">
    <table>
        <tr>
            <td>用户名：</td>
            <td><input type="text" name="userName"/></td>
        </tr>
        <tr>
            <td>密　码：</td>
            <td><input type="password" name="userpwd"/></td>
        </tr>
        <tr>
            <td><input type="submit" value="提交"/></td>
            <td><input type="reset" value="重置"/></td>
        </tr>
    </table>
    </form>
</body>
```

图 14.4 简单的登录页面

当单击如图 14.4 所示页面中的"提交"按钮时，系统将提交到 yourLogin.action，该 Action 对应的处理类 YourLoginAction 的代码如下：

```java
package action;
import java.util.Map;
import model.MyUser;
import com.opensymphony.xwork2.ActionContext;
import com.opensymphony.xwork2.ActionSupport;
import com.opensymphony.xwork2.ModelDriven;
public class YourLoginAction extends ActionSupport implements ModelDriven<MyUser>{
    private static final long serialVersionUID = 1L;
    //MyUser 为 JavaBean 类
    private MyUser u = new MyUser();
     //声明 Map 对象
    private Map<String,Object> request;
    private Map<String,Object> session;
    private Map<String,Object> application;
     //构造方法，初始化 Map 对象
    public YourLoginAction(){
        //获取 ActionContext 实例，通过该实例访问 Servlet API
        ActionContext ctx = ActionContext.getContext();
        request = (Map<String,Object>)ctx.get("request");
        session = ctx.getSession();
        application = ctx.getApplication();
    }
    public String execute(){
        if("陈恒".equals(u.getUserName())
                && "123456".equals(u.getUserpwd())){
            session.put("usersession", u);
            application.put("userapplication", u);
            return SUCCESS;//跳转到 yourLoginSuccess.jsp 页面
        }else{
            //类似于 request.setAtrribute("userrequest",u);
            request.put("userrequest", u);
            return "fail";//跳转到 yourLoginFail.jsp 页面
        }
    }
    @Override
    public MyUser getModel() {
        //TODO Auto-generated method stub
        return u;
    }
}
```

上面的 Action 访问了 HttpServletRequest、HttpSession 和 ServletContext。将该 Action

配置在 struts.xml 文件中，具体代码如下：

```
...
<action name="yourLogin" class="action.YourLoginAction">
    <result>/yourLoginSuccess.jsp</result>
    <result name="fail">/yourLoginFail.jsp</result>
</action>
...
```

在图 14.4 所示的页面中输入用户名为"陈恒"，密码为"123456"时，单击"提交"按钮跳转到 yourLoginSuccess.jsp 页面，否则跳转到 yourLoginFail.jsp 页面。

yourLoginSuccess.jsp 页面的代码如下：

```
<body>
      从 session 中取值<br>
      用户名：${sessionScope.usersession.userName }<br>
      密码：${sessionScope.usersession.userpwd }<br>
      从 application 中取值<br>
      用户名：${applicationScope.userapplication.userName }<br>
      密码：${applicationScope.userapplication.userpwd }
</body>
```

yourLoginFail.jsp 页面的代码如下：

```
<body>
      从 request 中取值<br>
      用户名：${requestScope.userrequest.userName }<br>
      密码：${requestScope.userrequest.userpwd }
</body>
```

❷ 实现 RequestAware、SessionAware 和 ApplicationAware 接口

在 Struts 2 的 Action 中还可以通过实现一系列接口 RequestAware、SessionAware 和 ApplicationAware 来访问 Servlet API。现在只需要将 Action 类的 YourLoginAction 修改为如下代码：

```
package action;
import java.util.Map;
import model.MyUser;
import org.apache.struts2.interceptor.ApplicationAware;
import org.apache.struts2.interceptor.RequestAware;
import org.apache.struts2.interceptor.SessionAware;
import com.opensymphony.xwork2.ActionSupport;
import com.opensymphony.xwork2.ModelDriven;
public class YourLoginAction extends ActionSupport implements ModelDriven<MyUser>,RequestAware,SessionAware,ApplicationAware{
    private static final long serialVersionUID = 1L;
    private MyUser u = new MyUser();
    //声明 Map 对象
```

```
    private Map<String,Object> request;
    private Map<String,Object> session;
    private Map<String,Object> application;
    public String execute(){
        if("陈恒".equals(u.getUserName())
                && "123456".equals(u.getUserpwd())){
            session.put("usersession", u);
            application.put("userapplication", u);
            return SUCCESS;
        }else{
            //类似request.setAtrribute("userrequest",u);
            request.put("userrequest", u);
            return "fail";
        }
    }
    @Override
    public MyUser getModel() {
        return u;
    }
    @Override
    public void setSession(Map<String, Object> arg0) {
        session = arg0;
    }
    @Override
    public void setRequest(Map<String, Object> arg0) {
        request = arg0;
    }
    @Override
    public void setApplication(Map<String, Object> arg0) {
        application = arg0;
    }
}
```

❸ 通过 ServletActionContext

在 Struts 2 的 Action 中，除了前面两种间接方式访问 Servlet API 之外，还可以直接访问 Servlet API。直接访问 Servlet API 有两种方式：通过 ServletActionContext 和实现 ServletRequestAware 接口。

通过 ServletActionContext 直接访问 Servlet API，只需要将 Action 类的 YourLoginAction 修改为如下代码：

```
package action;
import javax.servlet.ServletContext;
import javax.servlet.http.HttpServletRequest;
import javax.servlet.http.HttpSession;
import org.apache.struts2.ServletActionContext;
import model.MyUser;
```

```java
import com.opensymphony.xwork2.ActionSupport;
import com.opensymphony.xwork2.ModelDriven;
public class YourLoginAction extends ActionSupport implements ModelDriven<MyUser> {
    private static final long serialVersionUID = 1L;
    private MyUser u = new MyUser();
    private HttpServletRequest request;
    private HttpSession session;
    private ServletContext application;
    public YourLoginAction() {
        request = ServletActionContext.getRequest();
        session = request.getSession();
        application = ServletActionContext.getServletContext();
    }
    public String execute() {
        if ("陈恒".equals(u.getUserName()) && "123456".equals(u.getUserpwd())) {
            session.setAttribute("usersession", u);
            application.setAttribute("userapplication", u);
            return SUCCESS;
        } else {
            request.setAttribute("userrequest", u);
            return "fail";
        }
    }
    @Override
    public MyUser getModel() {
        return u;
    }
}
```

❹ 实现 ServletRequestAware 接口

实现 ServletRequestAware 接口直接访问 Servlet API，只需要将 Action 类的 YourLoginAction 修改为如下代码：

```java
package action;
import javax.servlet.ServletContext;
import javax.servlet.http.HttpServletRequest;
import javax.servlet.http.HttpSession;
import model.MyUser;
import org.apache.struts2.interceptor.ServletRequestAware;
import com.opensymphony.xwork2.ActionSupport;
import com.opensymphony.xwork2.ModelDriven;
public class YourLoginAction extends ActionSupport implements ModelDriven<MyUser>,ServletRequestAware{
    private static final long serialVersionUID = 1L;
    private MyUser u = new MyUser();
```

```
    private HttpServletRequest request;
    private HttpSession session;
    private ServletContext application;
    public String execute(){
        if("陈恒".equals(u.getUserName())
              && "123456".equals(u.getUserpwd())){
            session.setAttribute("usersession", u);
            application.setAttribute("userapplication", u);
            return SUCCESS;
        }else{
            request.setAttribute("userrequest", u);
            return "fail";
        }
    }
    @Override
    public MyUser getModel() {
        return u;
    }
    @Override
    public void setServletRequest(HttpServletRequest arg0) {
        request = arg0;
        session = arg0.getSession();
        application = arg0.getServletContext();
    }
}
```

在 Action 中访问 Servlet API，本节介绍了 4 种常用方式。其中，"通过 ActionContext" 与 "实现 RequestAware、SessionAware 和 ApplicationAware 接口" 这两种方式与 Servlet 进行了解耦。在实际编程中，具体使用哪种方式，根据程序需要进行选择。本书推荐使用 "实现 RequestAware、SessionAware 和 ApplicationAware 接口" 的方式访问 Servlet API。

14.3.4　Action 中常见的结果类型

Struts 2 的 Action 处理用户请求结束后，返回一个普通字符串——逻辑视图名，必须在配置文件中完成逻辑视图和物理视图资源的映射，才可让系统转到实际的视图资源。视图资源告诉用户系统执行的情况。但有的时候，返回的仅仅是个 XML 文件、JSON 串或输出流之类的，于是原有的代码很可能会因不同的返回结果而修改。Struts 2 提供了可配置的方式来帮助用户完成上述需求，用户不需要手动处理各种返回结果的细节，需要做的仅仅是在配置文件中配置所希望的返回结果类型。

❶ chain 类型

chain 结果类型是 Action 链式处理的结果类型，其两个 Action 的配置如下：

```
...
<action name="addUser" class="action.UserAction" method="add">
```

```xml
        <result type="chain">queryUser</result>
        <result name="fail">/User/addUser.jsp</result>
</action>
<action name="queryUser" class="action.UserAction" method="query">
        <result>/User/queryUser.jsp</result>
</action>
...
```

上面的配置可以这样理解，成功执行名为 addUser 的 Action 后，直接使用 chain 类型跳转到同一个包中名为 queryUser 的 Action 中。这里的包是配置文件中的包。

如果需要从一个 Action 中跳转到不同包的 Action 中，该如何配置呢？示例代码如下所示：

```xml
<package name="test1" namespace="/test1" extends="struts-default">
    <action name="addUser" class="action.UserAction" method="add">
        <result type="chain">
            <!--actionName 为固定写法, queryUser 为 Action 的名字 -->
            <param name="actionName">queryUser</param>
            <!-- namespace 为固定写法, test2 为包的 namespace-->
            <param name="namespace">/test2</param>
        </result>
        <result name="fail">/User/addUser.jsp</result>
    </action>
    ...
</package>
...
<package name="test2" namespace="/test2" extends="struts-default">
    <action name="queryUser" class="action.UserAction" method="query">
        <result>/User/queryUser.jsp</result>
    </action>
    ...
</package>
```

需要注意的是，chain 类型会将第一个 Action 执行完的状态复制到第二个 Action 中，供其使用。这样，在第二个 Action 中可以直接使用或修改第一个 Action 中的值，所以在最终页面显示的时候，很可能结果是不正确的。另外，chain 类型也不能实现跳转时通过"?"传递参数。因此，不推荐使用 chain 结果类型。

如果需要从一个 Action 中跳转到不同包的 Action 中，并通过"?"传递参数，又该如何配置呢？这时可以使用 redirectAction 类型或 redirect 类型。

❷ **redirectAction 类型**

当需要让一个 Action 处理结束后，直接将请求重定向（请求参数、属性和处理结果都将丢失）到另一个 Action 时，应该使用 redirectAction 类型。

配置 redirectAction 结果类型时，可以指定如下两个参数。

- actionName：该参数指定重定向的 Action 名。

- namespace：该参数指定重定向的 Action 所在包的命名空间。

例如，需要从一个 Action 中跳转到不同包的 Action 中，并通过"?"传递参数，示例代码如下：

```xml
<package name="test1" namespace="/test1" extends="struts-default">
    <action name="addUser" class="action.UserAction" method="add">
        <result type="redirectAction">
            <!--由于是 XML 文档，传参时必须将 "&" 写成 "&amp;"，id 和 uname 是参数名，
            idv 和 unamev 是 Action 中的属性名，${ } 为 EL 表达式。-->
            <param name="actionName">queryUser?id=${idv}&amp;uname=${unamev}
            </param>
            <param name="namespace">/test2</param>
        </result>
        <result name="fail">/User/addUser.jsp</result>
    </action>
    ...
</package>
...
<package name="test2" namespace="/test2" extends="struts-default">
    <action name="queryUser" class="action.UserAction" method="query">
        <result>/User/queryUser.jsp</result>
    </action>
    ...
</package>
```

❸ redirect 类型

redirect 类型与 redirectAction 类型不同的是，既可以重定向到 JSP 页面，也可以重定向到 Action，甚至重定向到其他网站，这种配置比较常用。

配置 redirect 结果类型时，可以指定参数 location，该参数指定 Action 处理完用户请求后跳转的 Action 地址。

示例代码如下：

```xml
<package name="test1" namespace="/test1" extends="struts-default">
    <action name="addUser" class="action.UserAction" method="add">
    <result type="redirect">
        <!-- 同一个包时，location 的参数值直接写 xxx.action；不同包时，加上另一个
        包的 namespace，如：/xxx/yyy.action -->
        <param name="location">queryUser.action</param>
        <!-- 直接设置参数，'query'为字符串常量，请求 addUser 成功后，地址栏显示：
        http://localhost:8080/ch14/queryUser.action?act=query -->
        <param name="act">${'query'}</param>
    </result>
    <!--上面的 result 还可以这样简单配置（推荐）
```

```xml
            <result type="redirect">queryUser.action?act=${'query'}</result>
            -->
            <!-- 也可以通过?设置参数，请求 addUser 失败后，地址栏显示：
                http://localhost:8080/ch14/User/addUser.jsp?act=add-->
            <result type="redirect" name="addfail">/User/addUser.jsp?act=${'add'}
            </result>
        </action>
        ...
        <action name="queryUser" class="action.UserAction" method="query">
            <result>/User/queryUser.jsp</result>
        </action>
    ...
    </package>
```

❹ dispatcher 类型

dispatcher 结果类型是将请求转发到指定的 JSP 资源，在目标页面中可以访问 request 作用域的数据。默认结果类型就是 dispatcher 类型，即 result 没有指定 type 属性时，该 result 类型就是 dispatcher 类型。

❺ 全局 result

一般情况下，<result>节点都配置在<action>中，作为 Action 的返回结果。但有些情况下，这种做法却会造成冗余的配置。例如，系统只有在用户合法登录的情况下才能继续操作，否则应该跳转到登录页面。这时候，就可以使用全局 result 了。

Struts 2 允许在<package>节点中使用<global-results>定义全局结果。当 Action 执行完成后，首先会从自己的<result>节点中匹配，如果匹配不成功，再从定义好的全局结果中继续查找。示例代码如下：

```xml
<package name="test1" namespace="/test1" extends="struts-default">
    ...
    <!-- 定义全局结果 -->
    <global-results>
        <!-- 在 Action 中，返回 noLogin 时，跳转到 login.jsp -->
        <result name="noLogin">/login.jsp</result>
        <!-- 对应异常 -->
        <result name="exception">/exception.jsp</result>
        <result name="sqlException">/sqlException.jsp</result>
    </global-results>
    <!-- 定义异常 -->
    <global-exception-mappings>
        <exception-mapping result="exception" exception="java.lang.Exception">
        </exception-mapping>
        <exception-mapping result="sqlException" exception="java.sql.SQLException"> </exception-mapping>
```

```xml
    </global-exception-mappings>
    ...
</package>
```

对于庞大而复杂的系统，模块化设计是一种非常好的思想，对于配置文件也是如此。开发者开发 Web 应用时可按照业务功能模块进行划分，然后为各个业务功能模块分别提供一个 XML 配置文件，最后再将这些 XML 配置文件通过 include 元素统一配置在 struts.xml 文件中。Struts 在解析时，会自动加载这些 XML 配置文件。例如，在 struts.xml 中有如下配置：

```xml
<struts >
    <include file="conf/struts-user.xml"/ >
    <include file="conf/struts-admin.xml"/ >
    <include file="conf/struts-common.xml"/ >
    ...
</struts>
```

上述 struts.xml 配置文件说明了，在与 struts.xml 同目录下有目录 conf，conf 目录下有 struts-user.xml、struts-admin.xml 和 struts-common.xml 等子配置文件，这些子配置文件的结构应该与 struts.xml 一样，只是分别处理不同的逻辑模块。这对于多人协同开发来说非常重要。因此从实际开发来说，struts.xml 就如同上面的配置那样，只将其他子配置文件包含进来，然后再定义一些全局共享的配置，struts.xml 只起到一个统领全局配置文件的作用，更具体的配置都应该放在子配置文件中。

在 Struts 2 的配置文件中，<package>节点提供了类似于 Java package 的概念，具有继承的特征。子<package>节点既可以继承父<package>节点中的配置，也可以覆盖。一般将一些公共的配置放到父<package>节点中配置，例如拦截器、全局结果等。<package>继承示例代码如下：

```xml
<package name="my-default" extends="struts-default">
    <!-- 拦截器配置-->
    <interceptors>
        <interceptor name="myinter" class="intercepter.FirstIntercepter"/>
        <interceptor-stack name="myStack">
            <interceptor-ref name="defaultStack"></interceptor-ref>
            <interceptor-ref name="myinter"></interceptor-ref>
        </interceptor-stack>
    </interceptors>
    <global-results>
        <result name="myyy">/index.jsp</result>
        <result name="exception">/exception.jsp</result>
    </global-results>
    <global-exception-mappings>
    <exception-mapping  result="exception"  exception="java.lang.Exception">
    </exception-mapping>
```

```xml
        </global-exception-mappings>
        ...
</package>
<!-- 包 hello 继承包 my-default -->
<package name="hello" namespace="/hello" extends="my-default">
    <action name="world" class="action.HelloWorldAction">
        <interceptor-ref name="myStack"></interceptor-ref>
        <result name="success">/hello.jsp</result>
        <!-- 从一个包下，转到另一个包下的 action，并传参数 -->
        <result name="fail" type="redirectAction">
            <param name="actionName">world1?id=${id}&amp;uname=${uname}
            </param>
            <param name="namespace">/hello1</param>
        </result>
    </action>
</package>
```

上述配置文件中，包 my-default 继承了包 struts-default，而包 hello 又继承了包 my-default。需要注意的是，Struts 2 的配置文件是顺序解析的，如果子包 hello 定义在父包 my-default 之前，是错误的。

14.4 基于 Struts 2 框架的注册与登录

视频讲解

创建 Struts 2 应用 ch14SS，实现以下功能：
- 未注册用户请运行注册页面进行注册，注册时检查用户名是否已注册（要求使用 Ajax 异步提交检查）。
- 已注册的用户运行登录页面进行登录，登录成功的用户可以查看所有用户姓名。

实现应用 ch14SS 的步骤如下：

❶ 创建数据表 usertable

在 test 数据库中创建数据表 usertable，具体代码如下：

```sql
DROP TABLE IF EXISTS 'usertable';
CREATE TABLE 'usertable' (
  'uname' varchar(100) NOT NULL DEFAULT '',
  'upwd' varchar(100) DEFAULT NULL,
  PRIMARY KEY ('uname')
) ENGINE=InnoDB DEFAULT CHARSET=latin1;
```

❷ 创建应用 ch14SS，并导入相关 JAR 文件

除了导入 Struts 2 框架所需要的 JAR 文件，还需要导入 JSTL、MySQL 数据库驱动和 JSON 需要的 JAR 文件。JSON 需要的 JAR 文件 struts2-json-plugin-2.5.18.jar，请从 Struts 2 解压缩后的 lib 目录下复制。具体如图 14.5 所示。

第 14 章 Struts 2 框架基础

```
            WEB-INF
              lib
                commons-fileupload-1.3.3.jar
                commons-io-2.5.jar
                commons-lang3-3.6.jar
                freemarker-2.3.26-incubating.jar
                javassist-3.20.0-GA.jar
                log4j-api-2.10.0.jar
                mysql-connector-java-5.1.45-bin.jar
                ognl-3.1.15.jar
                struts2-core-2.5.18.jar
                struts2-json-plugin-2.5.18.jar
                taglibs-standard-impl-1.2.5.jar
                taglibs-standard-spec-1.2.5.jar
```

图 14.5 相关 JAR 文件

❸ 创建相关 JSP 文件

应用中需要的 js 和图片，请参见提供的源代码，这里只说明 JSP 相关代码。

1）注册页面 register.jsp 的代码

```jsp
<%@ page language="java" contentType="text/html; charset=UTF-8" pageEncoding="UTF-8"%>
<%@ taglib prefix="c" uri="http://java.sun.com/jsp/jstl/core"%>
<%
String path = request.getContextPath();
String basePath = request.getScheme()+"://"+request.getServerName()+":"+request.getServerPort()+path+"/";
%>
<!DOCTYPE html>
<html>
<base href="<%=basePath%>">
<head>
<meta charset="UTF-8">
<style type="text/css">
    .textSize{
        width: 100pt;
        height: 15pt
    }
</style>
<title>注册页面</title>
<script type="text/javascript" src="js/jquery-3.2.1.min.js"></script>
<script type="text/javascript">
    //输入姓名后，调用该方法，判断用户名是否可用
    function nameIsNull(){
        $.ajax({//提交检查用户名是否可用请求
            //请求路径
            url : "jsontest/isUse.action",
```

```
                //请求类型
                type : "post",
                //data 表示发送的数据
                data : {
                    uname : $("#uname").val()
                },
                //成功响应的结果
                success : function(obj){
                    if(obj.isResult == "no"){
                        $("#isExit").html("<font color=red size=5>×</font>");
                        alert("用户已存在，请修改！");
                    }else{
                        $("#isExit").html("<font color=green size=5>√</font>");
                        alert("用户可用");
                    }
                },
                error : function() {
                    alert("处理异常！");
                }
            });
        }
    </script>
</head>
<body style="background-image: url('images/bb.jpg');">
    <form action="user/regist.action" method="post" name="registForm">
        <table>
            <tr>
                <td>姓名：</td>
                <td>
                    <input class="textSize" type="text" id="uname" name="uname" onblur="nameIsNull()" />
                    <span id="isExit"></span>
                </td>
            </tr>
            <tr>
                <td>密码：</td>
                <td><input class="textSize" type="password" name="upass"/></td>
            </tr>
            <tr>
                <td>确认密码：</td>
                <td><input class="textSize" type="password" name="reupass"/></td>
            </tr>
            <tr>
                <td colspan="2" align="center"><input type="button" value="注册" onclick=
```

```
            "allIsNull()"/></td>
          </tr>
        </table>
      </form>
 </body>
</html>
```

2)登录页面 login.jsp 的代码

```jsp
<%@ page language="java" contentType="text/html; charset=UTF-8" pageEncoding="UTF-8"%>
<%
String path = request.getContextPath();
String basePath = request.getScheme()+"://"+request.getServerName()+":"+request.getServerPort()+path+"/";
%>
<!DOCTYPE html>
<html>
<head>
<base href="<%=basePath%>">
<meta charset="UTF-8">
<title>后台登录</title>
<script type="text/javascript">
    //确定按钮
    function gogo(){
        document.forms[0].submit();
    }
    //取消按钮
    function cancel(){
        document.forms[0].action = "";
    }
</script>
  </head>
  <body>
    <form action="user/login.action" method="post">
    <table>
      <tr>
          <td colspan="2"><img src="images/login.gif"></td>
      </tr>
      <tr>
          <td>姓名：</td>
          <td><input type="text" name="uname" class="textSize"/></td>
      </tr>
      <tr>
          <td>密码：</td>
          <td><input type="password" name="upass" class="textSize"/></td>
```

```
            </tr>
            <tr>
                <td colspan="2">
                    <input type="image" src="images/ok.gif" onclick="gogo()" >
                    <input type="image" src="images/cancel.gif" onclick="cancel()">
                </td>
            </tr>
        </table>
    </form>
  </body>
</html>
```

3）主页面 main.jsp 的代码

```
<%@ page language="java" contentType="text/html; charset=UTF-8" pageEncoding="UTF-8"%>
<%@ taglib prefix="c" uri="http://java.sun.com/jsp/jstl/core"%>
<%
String path = request.getContextPath();
String basePath = request.getScheme()+"://"+request.getServerName()+":"+request.getServerPort()+path+"/";
%>
<!DOCTYPE html>
<html>
<head>
<base href="<%=basePath%>">
<meta charset="UTF-8">
<title>后台主页面</title>
</head>
<body>
    查询出所有用户：
    <br>
    <c:forEach items="${allUsers}" var="u">
        ${u}<br>
    </c:forEach>
</body>
</html>
```

❹ 创建相关配置文件

这里只提供 struts.xml 的代码，web.xml 的内容请读者参考 ch14 应用。

struts.xml 的代码如下：

```
<?xml version="1.0" encoding="UTF-8" ?>
<!DOCTYPE struts PUBLIC "-//Apache Software Foundation//DTD Struts Configuration 2.5//EN" "http://struts.apache.org/dtds/struts-2.5.dtd">
<struts>
    <package name="user" namespace="/user" extends="struts-default">
```

```xml
        <global-results>
            <!-- 全局转发，程序出错时，也可使用全局异常 -->
            <result name="error">/error.jsp</result>
        </global-results>
        <action name="regist" class="action.UserAction" method="register">
            <result name="register">/register.jsp</result>
            <result>/login.jsp</result>
        </action>
        <action name="login" class="action.UserAction" method="login">
            <result name="loginFail">/login.jsp</result>
            <result type="redirect">getAllUser.action</result>
        </action>
        <action name="getAllUser" class="action.UserAction" method="getAllUser">
            <result>/main.jsp</result>
        </action>
    </package>
<!-- package extends 为 json-default,
只有在 json-default 包中使用 result 的 type 为 json。result 类型需显式指明为 json,
result 标签内，无须指明视图指向的界面。 -->
    <package name="jsontest" extends="json-default" namespace="/jsontest">
        <action name="isUse" class="action.UserAction" method="isUse">
            <result name="success" type="json">
                <!-- 返回的 json 对象的属性 isResult -->
                <param name="includeProperties">isResult</param>
            </result>
        </action>
    </package>
</struts>
```

❺ 创建 Action 类

UserAction.java 的代码如下：

```java
package action;
import java.util.ArrayList;
import java.util.Map;
import org.apache.struts2.interceptor.RequestAware;
import org.apache.struts2.interceptor.SessionAware;
import com.opensymphony.xwork2.ActionSupport;
import dao.UserDao;
public class UserAction extends ActionSupport implements SessionAware, RequestAware{
    private static final long serialVersionUID = 1L;
    private String uname;
    private String upass;
    private Map<String, Object> session;
    private Map<String, Object> request;
```

```java
private ArrayList<String> allUsers;
private String isResult;
UserDao ud = new UserDao();
/**
 * 检查用户名是否可用
 */
public String isUse() {
    try {
        if(ud.isExit(uname)){//用户名存在
            isResult = "no";
        }else{
            isResult = "ok";
        }
        return SUCCESS;
    } catch (Exception e) {
        e.printStackTrace();
        return ERROR;
    }
}
/**
 * 注册
 */
public String register(){
    try {
        if(ud.isRegist(uname, upass)){
            return SUCCESS;
        }else{
            return "register";
        }
    } catch (Exception e) {
        e.printStackTrace();
        return ERROR;
    }
}
/**
 * 登录
 */
public String login(){
    try {
        if(ud.isLogin(uname, upass)){
            session.put("userName", uname);
            session.put("userPWD", upass);
            return SUCCESS;
        }
        request.put("errorMsg", "用户名或密码错误!");
```

```java
            return "loginFail";
        } catch (Exception e) {
            //TODO Auto-generated catch block
            e.printStackTrace();
            return ERROR;
        }
    }
    /**
     * 查询所有用户
     */
    public String getAllUser(){
        try {
            allUsers = ud.getAllUsers();
        } catch (Exception e) {
            e.printStackTrace();
            return ERROR;
        }
        return SUCCESS;
    }
    public String getUname() {
        return uname;
    }
    public void setUname(String uname) {
        this.uname = uname;
    }
    public String getUpass() {
        return upass;
    }
    public void setUpass(String upass) {
        this.upass = upass;
    }
    @Override
    public void setSession(Map<String, Object> arg0) {
        session = arg0;
    }
    public String getIsResult() {
        return isResult;
    }
    public void setIsResult(String isResult) {
        this.isResult = isResult;
    }
    @Override
    public void setRequest(Map<String, Object> arg0) {
        request = arg0;
    }
```

```java
    public ArrayList<String> getAllUsers() {
        return allUsers;
    }
    public void setAllUsers(ArrayList<String> allUsers) {
        this.allUsers = allUsers;
    }
}
```

❻ 创建数据访问层

该层有两个 dao 类，一个是 BaseDao，一个是 UserDao。还有一个数据库信息文件 jdbc.properties。

BaseDao.java 的代码如下：

```java
package dao;
import java.sql.Connection;
import java.sql.DriverManager;
import java.sql.PreparedStatement;
import java.sql.ResultSet;
import java.util.ArrayList;
import java.util.Properties;
public class BaseDao {
    //list 连接池
    static ArrayList<Connection> list = new ArrayList<Connection>();
    /**
     * 从连接池中获得一个连接
     */
    public synchronized static Connection getConnection() throws Exception{
        Connection con = null;
        //如果连接池有连接
        if (list.size() > 0) {
            return list.remove(0);
        }
        //连接池中没有连接
        else {
            Properties p = new Properties();
            //加载配置文件
            p.load(BaseDao.class.getClassLoader().getResourceAsStream
            ("dao/jdbc.properties"));
            String driverClass = p.getProperty("jdbc.driverClass");
            String jdbcUrl = p.getProperty("jdbc.jdbcUrl");
            String username = p.getProperty("jdbc.username");
            String password = p.getProperty("jdbc.password");
            //加载驱动
            Class.forName(driverClass);
            //和指定的数据库建立连接
            for (int i = 0; i < 10; i++) {
```

```java
            con = DriverManager.getConnection(jdbcUrl, username,password);
            list.add(con);
        }
    }
    return list.remove(0);
}
/**
 * 关闭结果集
 * @param rs 代表结果集
 */
public static void close(ResultSet rs) throws Exception{
    if (rs != null)
        rs.close();
}
/**
 * 关闭预处理对象
 * @param pst 代表预处理
 */
public static void close(PreparedStatement pst) throws Exception{
    if (pst != null)
        pst.close();
}
/**
 * 关闭连接对象
 * @param con 代表连接对象
 */
public synchronized static void close(Connection con) {
    if (con != null)
        list.add(con);
}
/**
 * 关闭结果集、预处理、连接等对象
 * @param rs 结果集
 * @param ps 预处理
 * @param con 连接
 */
public static void close(ResultSet rs, PreparedStatement ps, Connection con) throws Exception{
    close(rs);
    close(ps);
    close(con);
}
}
```

UserDao.java 的代码如下：

```java
package dao;
import java.sql.Connection;
import java.sql.PreparedStatement;
import java.sql.ResultSet;
import java.util.ArrayList;
public class UserDao extends BaseDao{
    /**
     * 判断用户名是否存在
     * @throws Exception
     */
    public boolean isExit(String uname) throws Exception{
        boolean b = false;//用户名不存在
        Connection con = getConnection();
        String sql = "select * from usertable where uname=?";
        PreparedStatement ps = con.prepareStatement(sql);
        ps.setString(1, uname);
        ResultSet rs = ps.executeQuery();
        if(rs.next()){//用户名存在
            b = true;
        }
        close(rs, ps, con);
        return b;
    }
    /**
     * 注册
     * @throws Exception
     */
    public boolean isRegist(String uname, String upass) throws Exception{
        boolean b = false;
        Connection con = getConnection();
        String sql = "insert into usertable values(?,?)";
        PreparedStatement ps = con.prepareStatement(sql);
        ps.setString(1, uname);
        ps.setString(2, upass);
        int i = ps.executeUpdate();
        if(i > 0){//注册成功
            b = true;
        }
        close(null, ps, con);
        return b;
    }
    /**
     * 登录
     * @throws Exception
     */
```

```java
public boolean isLogin(String uname, String upass) throws Exception{
    boolean b = false;//用户名不存在
    Connection con = getConnection();
    String sql = "select * from usertable where uname=? and upwd=?";
    PreparedStatement ps = con.prepareStatement(sql);
    ps.setString(1, uname);
    ps.setString(2, upass);
    ResultSet rs = ps.executeQuery();
    if(rs.next()){//登录成功
        b = true;
    }
    close(rs, ps, con);
    return b;
}
/**
 * 查询所有用户
 */
public ArrayList<String> getAllUsers() throws Exception{
    Connection con = getConnection();
    String sql = "select * from usertable ";
    PreparedStatement ps = con.prepareStatement(sql);
    ResultSet rs = ps.executeQuery();
    ArrayList<String> allUser = new ArrayList<String>();
    while(rs.next()){//登录成功
        allUser.add(rs.getString(1));
    }
    close(rs, ps, con);
    return allUser;
}
}
```

jdbc.properties 的代码如下：

```
jdbc.driverClass=com.mysql.jdbc.Driver
jdbc.jdbcUrl=jdbc:mysql://localhost:3306/test?characterEncoding=utf-8
jdbc.username=root
jdbc.password=root
```

14.5 本章小结

本章主要讲解了 Struts 2 的工作环境和 Struts 2 的 Action。通过本章的学习，务必掌握如何编写 Action 类以及如何配置 Action。

习题 14

1. result 的类型 redirectAction 表示（　　）。
   A．处理 action 链，跳转到下一个 action　　B．转发到一个 JSP
   C．重定向到一个 action　　D．重定向到一个 JSP

2. result 的类型 redirect 表示（　　）。
   A．处理 action 链，跳转到下一个 action　　B．转发到一个 JSP
   C．重定向到一个 action 或一个 JSP　　D．重定向到一个 JSP

3. Struts 2 的默认配置文件是（　　）。
   A．xeb.xml　　B．struts.xml　　C．user.xml　　D．server.xml

4. 某 action 的配置如下：

```
<action name="hello" class="action.HelloAction">
    <result name="error">/error.jsp</result>
</action>
```

当执行该 action 时会调用对应类中的（　　）方法。
   A．execute()　　B．doPost()　　C．doGet()　　D．service()

5. 在 Struts 2 的 Web 应用 hello 中 struts.xml 的配置如下：

```
<package name="user" extends="struts-default" namespace="/my">
    <action name="hello" class="action.HelloAction">
        <result>/success.jsp</result>
    </action>
</package>
```

需要访问该 action 时，输入的 URL 地址是（　　）。
   A．http://localhost:8080/hello/user/hello.action
   B．http://localhost:8080/hello/my/hello.jsp
   C．http://localhost:8080/hello/my/hello.action
   D．http://localhost:8080/hello/user/hello.jsp

6. result 的类型 dispatcher 表示（　　）。
   A．处理 action 链，跳转到下一个 action　　B．转发到一个 JSP
   C．重定向到一个 action　　D．重定向到一个 JSP

7. 某 action 的配置如下：

```
<action name="hello" class="action.HelloAction">
    <result>/main.jsp</result>
    <result name="sussecc">/success.jsp</result>
    <result name="error">/error.jsp</result>
    <result name="input">/login.jsp</result>
</action>
```

调用该 action 后返回值是"success"，则显示的页面是（　　）。

A. main.jsp　　　　B. success.jsp　　　　C. error.jsp　　　　D. login.jsp

8. 某 action 的配置如下：

```
<action name="hello" >
    <result >/success.jsp</result>
</action>
```

访问该 action 时会调用（　　）类。

A. Action　　　　B. ActionSupport　　　　C. HelloAction　　　　D. UserAction

9. 某 action 的配置如下：

```
<action name="hello" class="action.HelloAction" method="update">
    <result >/success.jsp</result>
</action>
```

访问该 action 时会执行 HelloAction 类中的（　　）方法。

A. execute()　　　　B. delete()　　　　C. select()　　　　D. update()

10. 在 Struts 2 的 Action 类中访问 Servlet API 有哪几种方法？其中与 HttpServlet 耦合的方法是哪几种？

11. 结果类型 redirect 与 redirectAction 的区别是什么？

12. 在 Action 中接收请求参数有哪几种方式？你认为哪种方式比较方便？

第 15 章 Spring MVC 框架基础

学习目的与要求

本章重点讲解 Spring MVC 的工作原理、控制器、数据绑定以及数据库编程。通过本章的学习，用户能够了解 Spring MVC 的工作原理，掌握 Spring MVC 应用的开发步骤。

本章主要内容

- Spring MVC 的工作原理
- Spring MVC 的工作环境
- 基于注解的控制器
- 表单标签库与数据绑定
- Spring 的数据库编程
- 基于 Spring MVC 框架的注册与登录

MVC 的思想将一个应用分成 3 个基本部分：Model（模型）、View（视图）和 Controller（控制器），让这 3 个部分以最低的耦合进行协同工作，从而提高应用的可扩展性及可维护性。Spring MVC 是一款优秀的、基于 MVC 思想的应用框架，它是 Spring 提供的一个实现了 Web MVC 设计模式的轻量级 Web 框架。

15.1 Spring MVC 的工作原理

视频讲解

Spring MVC 框架是高度可配置的，包含多种视图技术，如 JSP 技术、Velocity、Tiles、iText 和 POI。Spring MVC 框架并不关心使用的视图技术，也不会强迫开发者只使用 JSP 技术，但本书使用的视图是 JSP。

Spring MVC 框架主要由 DispatcherServlet、处理器映射、控制器、视图解析器、视图组成，其工作原理如图 15.1 所示。

从图 15.1 可总结出 Spring MVC 的工作流程如下：

第 15 章　Spring MVC 框架基础

图 15.1　Spring MVC 工作原理图

① 客户端请求提交到 DispatcherServlet。
② 由 DispatcherServlet 控制器寻找一个或多个 HandlerMapping，找到处理请求的 Controller。
③ DispatcherServlet 将请求提交到 Controller。
④ Controller 调用业务逻辑处理后，返回 ModelAndView。
⑤ DispatcherServlet 寻找一个或多个 ViewResolver 视图解析器，找到 ModelAndView 指定的视图。
⑥ 视图负责将结果显示到客户端。

图 15.1 中包含 4 个 Spring MVC 接口：DispatcherServlet、HandlerMapping、Controller 和 ViewResolver。

Spring MVC 所有的请求都经过 DispatcherServlet 来统一分发。DispatcherServlet 将请求分发给 Controller 之前，需要借助于 Spring MVC 提供的 HandlerMapping 定位到具体的 Controller。

HandlerMapping 接口负责完成客户请求到 Controller 映射。

Controller 接口将处理用户请求，这和 Java Servlet 扮演的角色是一致的。一旦 Controller 处理完用户请求，则返回 ModelAndView 对象给 DispatcherServlet 前端控制器，ModelAndView 中包含了模型（Model）和视图（View）。从宏观角度考虑，DispatcherServlet 是整个 Web 应用的控制器；从微观角度考虑，Controller 是单个 Http 请求处理过程中的控制器，而 ModelAndView 是 Http 请求过程中返回的模型（Model）和视图（View）。

ViewResolver 接口（视图解析器）在 Web 应用中负责查找 View 对象，从而将相应结果渲染给客户。

15.2 Spring MVC 的工作环境

视频讲解

在第 2 章 Java Web 开发环境的基础上，导入 Spring MVC 的相关 JAR 包，即可开发 Spring MVC 应用。

15.2.1 Spring 的下载与安装

使用 Spring 框架开发应用程序时，除了引用 Spring 自身的 JAR 包外，还需要引用 commons.logging 的 JAR 包。

❶ Spring 的 JAR 包

Spring 官方网站升级后，建议通过 Maven 和 Gradle 下载。对于不使用 Maven 和 Gradle 下载的开发者，本书给出一个 Spring Framework jar 官方直接下载路径 http://repo.springsource.org/libs-release-local/org/springframework/spring/ 。本书采用的是 spring-framework-5.0.2.RELEASE-dist.zip。将下载到的 ZIP 文件解压缩，解压缩后的目录结构如图 15.2 所示。

图 15.2 spring-framework-5.0.2 的目录结构

其中，docs 目录包含 Spring 的 API 文档和开发规范。

libs 目录包含开发 Spring 应用所需要的 JAR 包和源代码。该目录下有三类 JAR 文件，其中，以 RELEASE.jar 结尾的文件是 Spring 框架 class 的 JAR 包，即开发 Spring 应用所需要的 JAR 包；以 RELEASE-javadoc.jar 结尾的文件是 Spring 框架 API 文档的压缩包；以 RELEASE-sources.jar 结尾的文件是 Spring 框架源文件的压缩包。该目录下有 4 个基础包：spring-core-5.0.2.RELEASE.jar、spring-beans-5.0.2.RELEASE.jar、spring-context-5.0.2.RELEASE.jar 和 spring-expression-5.0.2.RELEASE.jar，分别对应 Spring 核心容器的 4 个模块：Spring-core、Spring-beans、Spring-context 和 Spring-expression。

schema 目录包含开发 Spring 应用所需要的 schema 文件，这些 schema 文件定义了 Spring 相关配置文件的约束。

❷ commons.logging 的 JAR 包

Spring 框架依赖于 Apache Commons Logging 组件，该组件的 JAR 包可以通过网址 http://commons.apache.org/proper/commons-logging/download_logging.cgi 下载，本书下载的是 commons-logging-1.2-bin.zip，解压缩后，即可找到 commons-logging-1.2.jar。

❸ **Spring MVC 所需的 JAR 包**

对于 Spring MVC 框架的初学者，开发 Spring MVC 应用时，只需要将 Spring 的 4 个基础包、commons-logging-1.2.jar、注解时需要的 JAR 包 spring-aop-5.0.2.RELEASE.jar 和 Spring MVC 相关的 JAR 包（spring-web-5.0.2.RELEASE.jar 和 spring-webmvc-5.0.2.RELEASE.jar）复制到 Web 应用的 WEB-INF/lib 目录下即可。添加后的 JAR 包如图 15.3 所示。

图 15.3　Spring MVC 所需要的 JAR 包

15.2.2　使用 Eclipse 开发 Spring MVC 的 Web 应用

本节通过一个简单的 Web 应用 ch15，来演示 Spring MVC 入门程序的实现过程，具体实现步骤如下。

❶ **创建 Web 应用并导入 JAR 包**

创建 Web 应用 ch15，导入如图 15.3 所示的 JAR 包。

❷ **在 web.xml 文件中部署 DispatcherServlet**

在开发 Spring MVC 应用时，需要在 web.xml 中部署 DispatcherServlet，代码如下：

```
<?xml version="1.0" encoding="UTF-8"?>
<web-app
xmlns:xsi="http://www.w3.org/2001/XMLSchema-instance"
xmlns="http://xmlns.jcp.org/xml/ns/javaee"
xsi:schemaLocation="http://xmlns.jcp.org/xml/ns/javaee
http://xmlns.jcp.org/xml/ns/javaee/web-app_3_1.xsd"
id="WebApp_ID" version="3.1">
<!--配置 springmvcDispatcherServlet-->
<servlet>
    <servlet-name>springmvc</servlet-name>
    <servlet-class>org.springframework.web.servlet.DispatcherServlet
    </servlet-class>
    <load-on-startup>1</load-on-startup>
</servlet>
<servlet-mapping>
```

```
        <servlet-name>springmvc</servlet-name>
        <url-pattern>/</url-pattern>
    </servlet-mapping>
</web-app>
```

上述 DispatcherServlet 的 servlet 对象 springmvc 初始化时,将在应用程序的 WEB-INF 目录下查找一个配置文件,该配置文件的命名规则是 servletName-servlet.xml,例如 springmvc-servlet.xml。

❸ 创建 Web 应用首页

在 ch15 应用的 WebContent 目录下,有个应用首页 index.jsp。index.jsp 的代码如下:

```
<%@ page language="java" contentType="text/html; charset=UTF-8" pageEncoding="UTF-8"%>
<!DOCTYPE html>
<html>
<head>
<meta charset="UTF-8">
<title>Insert title here</title>
</head>
<body>
    没注册的用户,请<a href="index/register">注册</a>! <br>
    已注册的用户,去<a href="index/login">登录</a>!
</body>
</html>
```

❹ 创建 Controller 类

在 src 目录下创建包 controller,并在该包中创建基于注解的名为 IndexController 的控制器类,该类中有两个处理请求方法,分别处理首页中"注册"和"登录"超链接请求。

```
package controller;
import org.springframework.stereotype.Controller;
import org.springframework.web.bind.annotation.RequestMapping;
/**"@Controller"表示 IndexController 的实例是一个控制器
 * @Controller 相当于@Controller("indexController")
 * 或@Controller(value = "indexController")
 */
@Controller
@RequestMapping("/index")
public class IndexController {
    @RequestMapping("/login")
    public String login() {
        /**login 代表逻辑视图名称,需要根据 Spring MVC 配置文件中
         * internalResourceViewResolver 的前缀和后缀找到对应的物理视图
         */
        return "login";
    }
```

```
    @RequestMapping("/register")
    public String register() {
        return "register";
    }
}
```

❺ 创建 Spring MVC 的配置文件

在 Spring MVC 中，使用扫描机制找到应用中所有基于注解的控制器类。所以，为了让控制器类被 Spring MVC 框架扫描到，需要在配置文件中声明 spring-context，并使用 <context:component-scan/> 元素指定控制器类的基本包（请确保所有控制器类都在基本包及其子包下）。另外，需要在配置文件中定义 Spring MVC 的视图解析器（ViewResolver），示例代码如下：

```
<bean class="org.springframework.web.servlet.view.InternalResourceViewResolver"
        id="internalResourceViewResolver">
    <!-- 前缀 -->
    <property name="prefix" value="/WEB-INF/jsp/" />
    <!-- 后缀 -->
    <property name="suffix" value=".jsp" />
</bean>
```

上述视图解析器配置了前缀和后缀两个属性。因此，控制器类中视图路径仅需提供 register 和 login，视图解析器将会自动添加前缀和后缀。

因此，在 ch15 应用程序的 WEB-INF 目录下创建名为 springmvc-servlet.xml 的配置文件，其代码如下：

```
<?xml version="1.0" encoding="UTF-8"?>
<beans xmlns="http://www.springframework.org/schema/beans"
    xmlns:xsi="http://www.w3.org/2001/XMLSchema-instance"
    xmlns:context="http://www.springframework.org/schema/context"
    xmlns:mvc="http://www.springframework.org/schema/mvc"
    xsi:schemaLocation=
        "http://www.springframework.org/schema/beans
        http://www.springframework.org/schema/beans/spring-beans.xsd
        http://www.springframework.org/schema/context
        http://www.springframework.org/schema/context/spring-context.xsd
        http://www.springframework.org/schema/mvc
        http://www.springframework.org/schema/mvc/spring-mvc.xsd">
    <!-- 使用扫描机制，扫描控制器类 -->
    <context:component-scan base-package="controller"/>
    <mvc:annotation-driven />
    <!-- 使用 resources 过滤掉不需要 dispatcher servlet 的资源（即静态资源，如 css、
    js、html、images）。使用 resources 时,必须使用 annotation-driven,不然 resources
    元素会阻止任意控制器被调用。
    -->
```

```xml
<!-- 允许 css 目录下所有文件可见 -->
<mvc:resources location="/css/" mapping="/css/**"></mvc:resources>
<!-- 配置视图解析器 -->
<bean class="org.springframework.web.servlet.view.InternalResourceViewResolver"
      id="internalResourceViewResolver">
    <!-- 前缀 -->
    <property name="prefix" value="/WEB-INF/jsp/" />
    <!-- 后缀 -->
    <property name="suffix" value=".jsp" />
</bean>
</beans>
```

❻ 应用的其他页面

IndexController 控制器的 register 方法处理成功后，跳转到/WEB-INF/jsp/register.jsp 视图；IndexController 控制器的 login 方法处理成功后，跳转到/WEB-INF/jsp/login.jsp 视图。因此，应用的/WEB-INF/jsp 目录下应有 register.jsp 和 login.jsp 页面，这两个 JSP 页面代码略。

❼ 发布并运行 Spring MVC 应用

在 Eclipse 中第一次运行 Spring MVC 应用时，需要将应用发布到 Tomcat。例如，运行 ch15 应用时，可以右击应用名称 ch15，选择 Run As | Run on Server 命令，打开如图 15.4 所示的对话框，在对话框中单击 Finish 按钮即可完成发布并运行。

图 15.4　在 Eclipse 中发布并运行 Spring MVC 应用

15.3　基于注解的控制器

视频讲解

在使用 Spring MVC 进行 Web 应用开发时，Controller 是 Web 应用的核心。Controller 实现类包含了对用户请求的处理逻辑，是用户请求和业务逻辑之间的"桥梁"，是 Spring MVC 框架的核心部分，负责具体的业务逻辑处理。

15.3.1　Controller 注解类型

在 Spring MVC 中，使用 org.springframework.stereotype.Controller 注解类型声明某类的实例是一个控制器。例如，15.2.2 节中的 IndexController 控制器类。别忘了在 Spring MVC 的配置文件中使用<context:component-scan/>元素指定控制器类的基本包，进而扫描所有注解的控制器类。

15.3.2　RequestMapping 注解类型

在基于注解的控制器类中，可以为每个请求编写对应的处理方法。如何将请求与处理方法一一对应呢？需要使用 org.springframework.web.bind.annotation.RequestMapping 注解类型。

❶ 方法级别注解

方法级别注解示例代码如下：

```
package controller;
import org.springframework.stereotype.Controller;
import org.springframework.web.bind.annotation.RequestMapping;
@Controller
public class IndexController {
    @RequestMapping(value = "/index/login")
    public String login() {
        /**login 代表逻辑视图名称，需要根据 Spring MVC 配置文件中
         * internalResourceViewResolver 的前缀和后缀找到对应的物理视图
         */
        return "login";
    }
    @RequestMapping(value = "/index/register")
    public String register() {
        return "register";
    }
}
```

上述示例中有两个 RequestMapping 注解语句，它们都作用在处理方法上。注解的 value

属性将请求 URI 映射到方法，value 属性是 RequestMapping 注解的默认属性，如果就一个 value 属性，则可省略该属性。可以使用如下 URL 访问 login 方法（请求处理方法），在访问 login 方法之前，需要事先在/ WEB-INF/jsp/目录下创建 login.jsp：

```
http://localhost:8080/ch10/index/login
```

❷ 类级别注解

类级别注解示例代码如下：

```
package controller;
import org.springframework.stereotype.Controller;
import org.springframework.web.bind.annotation.RequestMapping;
@Controller
@RequestMapping("/index")
public class IndexController {
    @RequestMapping("/login")
    public String login() {
        return "login";
    }
    @RequestMapping("/register")
    public String register() {
        return "register";
    }
}
```

在类级别注解的情况下，控制器类中的所有方法都将映射为类级别的请求。可以使用如下 URL 访问 login 方法：

```
http://localhost:8080/ch10/index/login
```

为了方便程序维护，建议开发者采用类级别注解，将相关处理放在同一个控制器类中，例如，对商品的增、删、改、查处理方法都可以放在 GoodsOperate 控制类中。

15.3.3　编写请求处理方法

在控制器类中每个请求处理方法可以有多个不同类型的参数，以及一个多种类型的返回结果。

❶ 请求处理方法中常出现的参数类型

如果需要在请求处理方法中使用 Servlet API 类型，那么可以将这些类型作为请求处理方法的参数类型。Servlet API 参数类型示例代码如下：

```
package controller;
import javax.servlet.http.HttpServletRequest;
import javax.servlet.http.HttpSession;
import org.springframework.stereotype.Controller;
import org.springframework.web.bind.annotation.RequestMapping;
```

```
@Controller
@RequestMapping("/index")
public class IndexController {
    @RequestMapping("/login")
    public String login(HttpSession session, HttpServletRequest request) {
        session.setAttribute("skey", "session 范围的值");
        request.setAttribute("rkey", "request 范围的值");
        return "login";
    }
}
```

除了 Servlet API 参数类型外，还有输入输出流、表单实体类、注解类型、与 Spring 框架相关的类型等，这些类型在后续章节中使用时再详细介绍。但特别重要的类型是 org.springframework.ui.Model 类型，该类型是一个包含 Map 的 Spring 框架类型。每次调用请求处理方法时，Spring MVC 都将创建 org.springframework.ui.Model 对象。Model 参数类型示例代码如下：

```
package controller;
import org.springframework.stereotype.Controller;
import org.springframework.ui.Model;
import org.springframework.web.bind.annotation.RequestMapping;
@Controller
@RequestMapping("/index")
public class IndexController {
    @RequestMapping("/register")
    public String register(Model model) {
        /*在视图中可以使用 EL 表达式${success}取出 model 中的值，有关 EL 相关知识，请
        参考本书有关内容。*/
        model.addAttribute("success", "注册成功");
        return "register";
    }
}
```

❷ 请求处理方法常见的返回类型

最常见的返回类型，就是代表逻辑视图名称的 String 类型，如前面章节中的请求处理方法。除了 String 类型外，还有 Model、View 以及其他任意的 Java 类型。

15.3.4　Controller 接收请求参数的常见方式

Controller 接收请求参数的方式有很多种，有的适合 get 请求方式，有的适合 post 请求方式，有的二者都适合。下面介绍几个常用的方式，读者可根据实际情况选择合适的接收方式。

❶ 通过实体 bean 接收请求参数

通过一个实体 bean 来接收请求参数，适用于 get 和 post 提交请求方式。需要注意的是，

bean 的属性名称必须与请求参数名称相同。下面通过完善应用 ch15，讲解"通过实体 bean 接收请求参数"。

1）完善视图文件

在应用 ch15 的 / WEB-INF/jsp/ 目录下有 register.jsp、login.jsp 和 main.jsp 文件。main.jsp 的代码略。

register.jsp 的代码如下：

```jsp
<%@ page language="java" contentType="text/html; charset=UTF-8" pageEncoding="UTF-8"%>
<!DOCTYPE html>
<html>
<head>
<meta charset="UTF-8">
<title>Insert title here</title>
</head>
<body>
<form action="${pageContext.request.contextPath }/user/register" method="post" name="registForm">
    <table border=1>
        <tr>
            <td>姓名：</td>
            <td>
                <input type="text" name="uname" value="${user.uname }"/>
            </td>
        </tr>
        <tr>
            <td>密码：</td>
            <td><input type="password" name="upass"/></td>
        </tr>
        <tr>
            <td>确认密码：</td>
            <td><input type="password" name="reupass"/></td>
        </tr>
        <tr>
            <td colspan="2" align="center">
                <input type="submit" value="注册" />
            </td>
        </tr>
    </table>
</form>
</body>
</html>
```

login.jsp 的代码如下：

```jsp
<%@ page language="java" contentType="text/html; charset=UTF-8" pageEncoding=
```

```jsp
"UTF-8"%>
<!DOCTYPE html>
<html>
<head>
<meta charset="UTF-8">
<title>Insert title here</title>
</head>
<body>
    <form action="${pageContext.request.contextPath }/user/login" method="post">
    <table>
        <tr>
            <td align="center" colspan="2">登录</td>
        </tr>
        <tr>
            <td>姓名：</td>
            <td><input type="text" name="uname"></td>
        </tr>
        <tr>
            <td>密码：</td>
            <td><input type="password" name="upass"></td>
        </tr>
        <tr>
            <td colspan="2">
                <input type="submit" value="提交" >
                <input type="reset" value="重置" >
            </td>
        </tr>
    </table>
    ${messageError }
    </form>
</body>
</html>
```

2）创建 POJO 实体类

在应用 ch15 的 src 目录下创建包 pojo，并在该包中创建实体类 UserForm，代码如下：

```java
package pojo;
public class UserForm {
    private String uname;//与请求参数名称相同
    private String upass;
    private String reupass;
    //省略 getter 和 setter 方法
}
```

3）创建控制器类

在应用 ch15 的 controller 包中，创建控制器类 UserController。

UserController 的代码如下：

```java
package controller;
import javax.servlet.http.HttpSession;
import org.apache.commons.logging.Log;
import org.apache.commons.logging.LogFactory;
import org.springframework.stereotype.Controller;
import org.springframework.ui.Model;
import org.springframework.web.bind.annotation.RequestMapping;
import pojo.UserForm;
@Controller
@RequestMapping("/user")
public class UserController {
    //得到一个用来记录日志的对象，这样打印信息的时候能够标记打印的是哪个类的信息
    private static final Log logger = LogFactory.getLog(UserController.class);
    /**
     * 处理登录
     * 使用 UserForm 对象（实体 bean）user 接收登录页面提交的请求参数
     */
    @RequestMapping("/login")
    public String login(UserForm user, HttpSession session, Model model) {
        if("zhangsan".equals(user.getUname())
                && "123456".equals(user.getUpass())) {
            session.setAttribute("u", user);
            logger.info("成功");
            return "main";//登录成功，跳转到 main.jsp
        }else{
            logger.info("失败");
            model.addAttribute("messageError", "用户名或密码错误");
            return "login";
        }
    }

    /**
     *处理注册
     *使用 UserForm 对象（实体 bean）user 接收注册页面提交的请求参数
     */
    @RequestMapping("/register")
    public String register(UserForm user, Model model) {
        if("zhangsan".equals(user.getUname())
                && "123456".equals(user.getUpass())) {
            logger.info("成功");
            return "login";//注册成功，跳转到 login.jsp
        }else{
            logger.info("失败");
```

```
            //在register.jsp页面上可以使用EL表达式取出model的uname值
            model.addAttribute("uname", user.getUname());
            return "register";//返回register.jsp
        }
    }
}
```

4）测试应用

运行应用 ch15 的首页面 index.jsp，进行程序测试。

❷ **通过处理方法的形参接收请求参数**

通过处理方法的形参接收请求参数，也就是直接把表单参数写在控制器类相应方法的形参中，即形参名称与请求参数名称完全相同。该接收参数方式适用于 get 和 post 提交请求方式。可以将控制器类 UserController 中 register 方法的代码修改如下：

```
@RequestMapping("/register")
/**
 * 通过形参接收请求参数, 形参名称与请求参数名称完全相同
 */
public String register(String uname, String upass, Model model) {
    if("zhangsan".equals(uname)
            && "123456".equals(upass)) {
        logger.info("成功");
        return "login";//注册成功，跳转到login.jsp
    }else{
        logger.info("失败");
        //在register.jsp页面上可以使用EL表达式取出model的uname值
        model.addAttribute("uname", uname);
        return "register";//返回register.jsp
    }
}
```

❸ **通过@RequestParam 接收请求参数**

通过@RequestParam 接收请求参数，适用于 get 和 post 提交请求方式。可以将控制器类 UserController 中 register 方法的代码修改如下：

```
@RequestMapping("/register")
/**
 * 通过@RequestParam 接收请求参数
 */
public String register(@RequestParam String uname, @RequestParam String upass, Model model) {
    if("zhangsan".equals(uname)
            && "123456".equals(upass)) {
        logger.info("成功");
        return "login";//注册成功，跳转到login.jsp
    }else{
```

```
        logger.info("失败");
        //在register.jsp页面上可以使用EL表达式取出model的uname值
        model.addAttribute("uname", uname);
        return "register";//返回register.jsp
    }
}
```

"通过@RequestParam接收请求参数"与"通过处理方法的形参接收请求参数"的区别是：当请求参数与接收参数名不一致时，"通过处理方法的形参接收请求参数"不会报400错误，而"通过@RequestParam接收请求参数"会报400错误。

❹ 通过@ModelAttribute接收请求参数

@ModelAttribute注解放在处理方法的形参上时，用于将多个请求参数封装到一个实体对象，从而简化数据绑定流程，而且自动暴露为模型数据用于视图页面展示时使用。而"通过实体bean接收请求参数"只是将多个请求参数封装到一个实体对象，并不能暴露为模型数据（需要使用model.addAttribute语句才能暴露为模型数据，数据绑定与模型数据展示可参考第15.4节的内容）。

通过@ModelAttribute注解接收请求参数，适用于get和post提交请求方式。可以将控制器类UserController中register方法的代码修改如下：

```
@RequestMapping("/register")
public String register(@ModelAttribute("user") UserForm user) {
    if("zhangsan".equals(user.getUname())
            && "123456".equals(user.getUpass())){
        logger.info("成功");
        return "login";//注册成功，跳转到login.jsp
    }else{
        logger.info("失败");
        //使用@ModelAttribute("user")与model.addAttribute("user", user)功
        //能相同
    //在register.jsp页面上可以使用EL表达式${user.uname}取出ModelAttribute
    //的uname值
        return "register";//返回register.jsp
    }
}
```

15.3.5 Spring MVC框架中的重定向与转发

重定向是将用户从当前处理请求定向到另一个视图（如JSP）或处理请求，以前的请求（request）中存放的信息全部失效，并进入一个新的request作用域；转发是将用户对当前处理的请求转发给另一个视图或处理请求，以前的request中存放的信息不会失效。

在Spring MVC框架中，控制器类中处理方法的return语句默认就是转发实现，只不过实现的是转发到视图。示例代码如下：

```
@RequestMapping("/register")
```

```
public String register() {
    return "register";//转发到 register.jsp
}
```

在 Spring MVC 框架中，重定向与转发的示例代码如下：

```
package controller;
import org.springframework.stereotype.Controller;
import org.springframework.web.bind.annotation.RequestMapping;
@Controller
@RequestMapping("/index")
public class IndexController {
    @RequestMapping("/login")
    public String login() {
        //转发到一个请求方法（同一个控制器类里，可省略/index/）
        return "forward:/index/isLogin";
    }
    @RequestMapping("/isLogin")
    public String isLogin() {
        //重定向到一个请求方法
        return "redirect:/index/isRegister";
    }
    @RequestMapping("/isRegister")
    public String isRegister() {
        //转发到一个视图
        return "register";
    }
}
```

在 Spring MVC 框架中，不管重定向还是转发，都需要符合视图解析器的配置，如果直接转发到一个不需要 DispatcherServlet 的资源，例如：

```
return "forward:/html/my.html";
```

需要使用 mvc:resources 配置：

```
<mvc:resources location="/html/" mapping="/html/**"></mvc:resources>
```

15.3.6　应用@Autowired 进行依赖注入

在前面学习的控制器中，并没有体现 MVC 的 M 层，这是因为控制器既充当 C 层，又充当 M 层。这样设计程序的系统结构很不合理，应该将 M 层从控制器中分离出来。Spring MVC 框架本身就是一个非常优秀的 MVC 框架，它具有一个依赖注入的优点。可以通过 org.springframework.beans.factory.annotation.Autowired 注解类型将依赖注入一个属性（成员变量）或方法，例如：

```
@Autowired
public UserService userService;
```

在 Spring MVC 中，为了能被作为依赖注入，服务层的类必须使用 org.springframework.stereotype.Service 注解类型并注明为@Service（一个服务）。另外，还需要在配置文件中使用<context:component-scan base-package="基本包"/>元素来扫描依赖基本包。下面将 ch15 应用的"登录"和"注册"的业务逻辑处理分离出来，使用 Service 层实现。

首先，创建 service 包，在包中创建 UserService 接口和 UserServiceImpl 实现类。UserService 接口的具体代码如下：

```java
package service;
import pojo.UserForm;
public interface UserService {
    boolean login(UserForm user);
    boolean register(UserForm user);
}
```

UserServiceImpl 实现类的具体代码如下：

```java
package service;
import org.springframework.stereotype.Service;
import pojo.UserForm;
//注解为一个服务
@Service
public class UserServiceImpl implements UserService{
    @Override
    public boolean login(UserForm user) {
        if("zhangsan".equals(user.getUname())
            && "123456".equals(user.getUpass()))
            return true;
        return false;
    }
    @Override
    public boolean register(UserForm user) {
        if("zhangsan".equals(user.getUname())
            && "123456".equals(user.getUpass()))
            return true;
        return false;
    }
}
```

其次，在配置文件中添加一个<context:component-scan base-package="基本包"/>元素，具体代码如下：

```xml
<context:component-scan base-package="service"/>
```

最后，修改控制器类 UserController，具体代码如下：

```java
package controller;
```

```java
import javax.servlet.http.HttpSession;
import org.apache.commons.logging.Log;
import org.apache.commons.logging.LogFactory;
import org.springframework.beans.factory.annotation.Autowired;
import org.springframework.stereotype.Controller;
import org.springframework.ui.Model;
import org.springframework.web.bind.annotation.RequestMapping;
import pojo.UserForm;
import service.UserService;
@Controller
@RequestMapping("/user")
public class UserController {
    //得到一个用来记录日志的对象，这样打印信息的时候能够标记打印的是哪个类的信息
    private static final Log logger = LogFactory.getLog(UserController.class);
    //将服务层依赖注入属性userService
    @Autowired
     public UserService userService;
    /**
     * 处理登录
     */
    @RequestMapping("/login")
    public String login(UserForm user, HttpSession session, Model model) {
        if(userService.login(user)){
            session.setAttribute("u", user);
            logger.info("成功");
            return "main";//登录成功，跳转到main.jsp
        }else{
            logger.info("失败");
            model.addAttribute("messageError", "用户名或密码错误");
            return "login";
        }
    }
    /**
     *处理注册
     */
    @RequestMapping("/register")
    public String register(@ModelAttribute("user") UserForm user) {
        if(userService.register(user)){
            logger.info("成功");
            return "login";//注册成功，跳转到login.jsp
        }else{
            logger.info("失败");
            //使用@ModelAttribute("user")与model.addAttribute("user", user)
            //功能相同
            //在register.jsp页面上可以使用EL表达式${user.uname}取出ModelAttribute
```

```
            //的 uname 值
        return "register";//返回 register.jsp
    }
  }
}
```

15.3.7 @ModelAttribute

通过 org.springframework.web.bind.annotation.ModelAttribute 注解类型，可经常实现如下两个功能。

❶ 绑定请求参数到实体对象（表单的命令对象）

该用法与 15.3.4 节的"通过@ModelAttribute 接收请求参数"一样：

```
@RequestMapping("/register")
public String register(@ModelAttribute("user") UserForm user) {
    if("zhangsan".equals(user.getUname())
            && "123456".equals(user.getUpass())){
        return "login";
    }else{
        return "register";
    }
}
```

上述代码中，"@ModelAttribute("user") UserForm user"语句的功能有两个：一是将请求参数的输入封装到 user 对象中；一是创建 UserForm 实例，以 user 为键值存储在 Model 对象中，与"model.addAttribute("user", user)"语句功能一样。如果没有指定键值，即"@ModelAttribute UserForm user"，那么创建 UserForm 实例时，以 userForm 为键值存储在 Model 对象中，与"model.addAttribute("userForm", user)"语句功能一样。

❷ 注解一个非请求处理方法

被@ModelAttribute 注解的方法，将在每次调用该控制器类的请求处理方法前被调用。这种特性可以用来控制登录权限，当然控制登录权限的方法很多，例如拦截器、过滤器等。

使用该特性控制登录权限的示例代码如下：

```
package controller;
import javax.servlet.http.HttpSession;
import org.springframework.web.bind.annotation.ModelAttribute;
public class BaseController {
    @ModelAttribute
    public void isLogin(HttpSession session) throws Exception {
        if(session.getAttribute("user") == null){
            throw new Exception("没有权限");
        }
    }
}
```

```
}
package controller;
import org.springframework.stereotype.Controller;
import org.springframework.web.bind.annotation.RequestMapping;
@Controller
@RequestMapping("/admin")
public class ModelAttributeController extends BaseController{
    @RequestMapping("/add")
    public String add(){
        return "addSuccess";
    }
    @RequestMapping("/update")
    public String update(){
        return "updateSuccess";
    }
    @RequestMapping("/delete")
    public String delete(){
        return "deleteSuccess";
    }
}
```

上述 ModelAttributeController 类中的 add、update、delete 请求处理方法执行时，首先执行父类 BaseController 中的 isLogin 方法判断登录权限。可以通过地址 http://localhost:8080/ch15/admin/add 测试登录权限。

15.4 表单标签库与数据绑定

视频讲解

数据绑定是将用户参数输入值绑定到领域模型的一种特性。在 Spring MVC 的 Controller 和 View 参数数据传递中，所有 HTTP 请求参数的类型均为字符串。如果模型需要绑定的类型为 double 或 int，则需要手动进行类型转换；而有了数据绑定后，就不再需要手动将 HTTP 请求中的 String 类型转换为模型需要的类型。数据绑定的另一个好处是，当输入验证失败时，会重新生成一个 HTML 表单，无须重新填写输入字段。在 Spring MVC 中，为了方便、高效地使用数据绑定，还需要学习表单标签库。

15.4.1 表单标签库

表单标签库中包含了可以用在 JSP 页面中渲染 HTML 元素的标签。JSP 页面使用 Spring 表单标签库时，必须在 JSP 页面开头处声明 taglib 指令，指令代码如下：

```
<%@ taglib prefix="form" uri="http://www.springframework.org/tags/form" %>
```

表单标签库中有 form、input、password、hidden、textarea、checkbox、checkboxes、radiobutton、radiobuttons、select、option、options、errors。

form：渲染表单元素。

input：渲染<input type="text"/>元素。

password：渲染<input type="password"/>元素。

hidden：渲染<input type="hidden"/>元素。

textarea：渲染 textarea 元素。

checkbox：渲染一个<input type="checkbox"/>元素。

checkboxes：渲染多个<input type="checkbox"/>元素。

radiobutton：渲染一个<input type="radio"/>元素。

radiobuttons：渲染多个<input type="radio"/>元素。

select：渲染一个选择元素。

option：渲染一个选项元素。

options：渲染多个选项元素。

errors：在 span 元素中渲染字段错误。

❶ 表单标签

表单标签的语法格式如下：

```
<form:form modelAttribute="xxx" method="post" action="xxx">
   …
</form:form>
```

除了具有 HTML 表单元素属性外，表单标签还具有 acceptCharset、commandName、cssClass、cssStyle、htmlEscape 和 modelAttribute 等属性。各属性含义如下所示。

acceptCharset：定义服务器接受的字符编码列表。

commandName：暴露表单对象的模型属性名称，默认为 command。

cssClass：定义应用到 form 元素的 CSS 类。

cssStyle：定义应用到 form 元素的 CSS 样式。

htmlEscape：true 或 false，表示是否进行 HTML 转义。

modelAttribute：暴露 form backing object 的模型属性名称，默认为 command。

其中，commandName 和 modelAttribute 属性功能基本一致，属性值绑定一个 JavaBean 对象。假设控制器类 UserController 的方法 inputUser 是返回 userAdd.jsp 的请求处理方法。

inputUser 方法的代码如下：

```
@RequestMapping(value = "/input")
public String inputUser(Model model) {
    …
    model.addAttribute("user", new User());
    return "userAdd";
}
```

userAdd.jsp 的表单标签代码如下：

```
<form:form modelAttribute="user" method="post" action="user/save">
   …
</form:form>
```

注意：在 inputUser 方法中，如果没有 Model 属性 user，userAdd.jsp 页面就会抛出异常，因为表单标签无法找到在其 modelAttribute 属性中指定的 form backing object。

❷ input 标签

input 标签的语法格式如下：

```
<form:input path="xxx"/>
```

该标签除了 cssClass、cssStyle、htmlEscape 属性外，还有一个最重要的属性 path。path 属性将文本框输入值绑定到 form backing object 的一个属性。示例代码如下：

```
<form:form modelAttribute="user" method="post" action="user/save">
    <form:input path="userName"/>
</form:form>
```

上述代码将输入值绑定到 user 对象的 userName 属性。

❸ password 标签

password 标签的语法格式如下：

```
<form:password path="xxx"/>
```

该标签与 input 标签用法完全一致，不再赘述。

❹ hidden 标签

hidden 标签的语法格式如下：

```
<form:hidden path="xxx"/>
```

该标签与 input 标签用法基本一致，只不过它不可显示，不支持 cssClass 和 cssStyle 属性。

❺ textarea 标签

textarea 基本上就是一个支持多行输入的 input 元素，其语法格式如下：

```
<form:textarea path="xxx"/>
```

该标签与 input 标签用法完全一致，不再赘述。

❻ checkbox 标签

checkbox 标签的语法格式如下：

```
<form:checkbox path="xxx" value="xxx"/>
```

多个 path 相同的 checkbox 标签是一个选项组，允许多选。选项值绑定到一个数组属性。示例代码如下：

```
<form:checkbox path="friends" value="张三"/>张三
<form:checkbox path="friends" value="李四"/>李四
<form:checkbox path="friends" value="王五"/>王五
<form:checkbox path="friends" value="赵六"/>赵六
```

上述示例代码中复选框的值绑定到一个字符串数组属性 friends（String[] friends）。该标签的其他用法与 input 标签基本一致，不再赘述。

❼ **checkboxes 标签**

checkboxes 标签渲染多个复选框，是一个选项组，等价于多个 path 相同的 checkbox 标签。它有 3 个非常重要的属性：items、itemLabel 和 itemValue。

items：用于生成 input 元素的 Collection、Map 或 Array。

itemLabel：items 属性中指定的集合对象的属性，为每个 input 元素提供 label。

itemValue：items 属性中指定的集合对象的属性，为每个 input 元素提供 value。

checkboxes 标签的语法格式如下：

```
<form:checkboxes items="xxx" path="xxx"/>
```

示例代码如下：

```
<form:checkboxes items="${hobbys}" path="hobby" />
```

上述示例代码是将 model 属性 hobbys 的内容（集合元素）渲染为复选框。itemLabel 和 itemValue 默认的情况下，如果集合是数组，复选框的 label 和 value 相同；如果集合是 Map，复选框的 label 是 Map 的值（value），复选框的 value 是 Map 的关键字（key）。

❽ **radiobutton 标签**

radiobutton 标签的语法格式如下：

```
<form:radiobutton path="xxx" value="xxx"/>
```

多个 path 相同的 radiobutton 标签是一个选项组，只允许单选。

❾ **radiobuttons 标签**

radiobuttons 标签渲染多个 radio，是一个选项组，等价于多个 path 相同的 radiobutton 标签。radiobuttons 标签的语法格式如下：

```
<form:radiobuttons path="xxx" items="xxx"/>
```

该标签的 itemLabel 和 itemValue 属性与 checkboxes 标签的 itemLabel 和 itemValue 属性完全一样，但只允许单选。

❿ **select 标签**

select 标签的选项可能来自其属性 items 指定的集合，或者来自一个嵌套的 option 标签或 options 标签。其语法格式如下：

```
<form:select path="xxx" items="xxx" />
```

或

```
<form:select path="xxx" items="xxx" >
    <option value="xxx">xxx</option>
</ form:select>
```

或

```
<form:select path="xxx">
    <form:options items="xxx"/>
</form:select>
```

该标签的 itemLabel 和 itemValue 属性与 checkboxes 标签的 itemLabel 和 itemValue 属性完全一样。

⑪ options 标签

options 标签生成一个 select 标签的选项列表。因此，需要与 select 标签一同使用，具体用法参见 select 标签。

⑫ errors 标签

errors 标签渲染一个或者多个 span 元素，每个 span 元素包含一个错误消息。它可以用于显示一个特定的错误消息，也可以显示所有错误消息。其语法如下：

```
<form:errors path="*"/>
```

或

```
<form:errors path="xxx"/>
```

其中，"*"表示显示所有错误消息；"xxx"表示显示由"xxx"指定的特定错误消息。

15.4.2 数据绑定

为了让读者进一步学习数据绑定和表单标签，本节给出了一个应用范例 ch15SS。应用中实现了 User 类属性和 JSP 页面中表单参数的绑定，同时在 JSP 页面中分别展示了 input、textarea、checkbox、checkboxs、select 等标签。

❶ 应用的相关配置

在 ch15SS 应用中需要使用 JSTL，因此，不仅需要将 Spring MVC 相关 jar 包复制到应用的 WEN-INF/lib 目录下，还需要从 Tomcat 的 webapps\examples\WEB-INF\lib 目录下，将 JSTL 相关 jar 包复制到应用的 WEN-INF/lib 目录下。ch15SS 的 JAR 包如图 15.5 所示。

图 15.5 ch15SS 的 JAR 包

为了避免中文乱码问题，需要在 web.xml 文件中增加编码过滤器，同时 JSP 页面编码设置为 UTF-8，form 表单的提交方式必须为 post。

web.xml 的代码如下：

```
<?xml version="1.0" encoding="UTF-8"?>
<web-app xmlns:xsi="http://www.w3.org/2001/XMLSchema-instance"
```

```xml
     xmlns="http://xmlns.jcp.org/xml/ns/javaee"
     xsi:schemaLocation="http://xmlns.jcp.org/xml/ns/javaee http://xmlns.
jcp.org/xml/ns/javaee/web-app_3_1.xsd"
     id="WebApp_ID" version="3.1">
<!--配置DispatcherServlet -->
<servlet>
    <servlet-name>springmvc</servlet-name>
    <servlet-class>org.springframework.web.servlet.DispatcherServlet
    </servlet-class>
    <load-on-startup>1</load-on-startup>
</servlet>
<servlet-mapping>
    <servlet-name>springmvc</servlet-name>
    <url-pattern>/</url-pattern>
</servlet-mapping>
<!-- 避免中文乱码 -->
<filter>
    <filter-name>characterEncodingFilter</filter-name>
    <filter-class>org.springframework.web.filter.CharacterEncodingFilter
    </filter-class>
    <init-param>
        <param-name>encoding</param-name>
        <param-value>UTF-8</param-value>
    </init-param>
    <init-param>
        <param-name>forceEncoding</param-name>
        <param-value>true</param-value>
    </init-param>
</filter>
<filter-mapping>
    <filter-name>characterEncodingFilter</filter-name>
    <url-pattern>/*</url-pattern>
</filter-mapping>
</web-app>
```

配置文件 springmvc-servlet.xml 与 ch15 一样，不再赘述。

❷ View 层

View 层包含两个 JSP 页面，一个是信息输入页面 userAdd.jsp，一个是信息显示页面 userList.jsp。在 ch15SS 应用的 WEB-INF/jsp/目录下，创建这两个 JSP 页面。

在 userAdd.jsp 页面中将 Map 类型的 hobbys 绑定到了 checkboxes 上，将 String[]类型的 carrers 和 houseRegisters 绑定到 select 上，实现通过 option 标签对 select 添加选项，同时表单的 method 方法需指定为 post 来避免中文乱码问题。

在 userList.jsp 页面中使用 JSTL 标签遍历集合中的用户信息，JSTL 相关知识参见本书的相关内容。

userAdd.jsp 的代码如下:

```jsp
<%@ page language="java" contentType="text/html; charset=UTF-8" pageEncoding="UTF-8"%>
<%@ taglib prefix="form" uri="http://www.springframework.org/tags/form" %>
<!DOCTYPE html>
<html>
<head>
<meta charset="UTF-8">
<title>Insert title here</title>
</head>
<body>
<form:form modelAttribute="user"  method="post" action=" ${pageContext.request.contextPath }/user/save">
    <fieldset>
        <legend>添加一个用户</legend>
        <p>
            <label>用户名:</label>
            <form:input path="userName"/>
        </p>
        <p>
            <label>爱好:</label>
            <form:checkboxes items="${hobbys}"  path="hobby" />
        </p>
        <p>
            <label>朋友:</label>
            <form:checkbox path="friends" value="张三"/>张三
            <form:checkbox path="friends" value="李四"/>李四
            <form:checkbox path="friends" value="王五"/>王五
            <form:checkbox path="friends" value="赵六"/>赵六
        </p>
        <p>
            <label>职业:</label>
            <form:select path="carrer">
                <option/>请选择职业
                <form:options items="${carrers }"/>
            </form:select>
        </p>
        <p>
            <label>户籍:</label>
            <form:select path="houseRegister">
                <option/>请选择户籍
                <form:options items="${houseRegisters }"/>
            </form:select>
        </p>
```

```
        <p>
            <label>个人描述:</label>
            <form:textarea path="remark" rows="5"/>
        </p>
        <p id="buttons">
            <input id="reset" type="reset">
            <input id="submit" type="submit" value="添加">
        </p>
    </fieldset>
</form:form>
</body>
</html>
```

userList.jsp 的代码如下：

```
<%@ page language="java" contentType="text/html; charset=UTF-8" pageEncoding="UTF-8"%>
<%@ taglib uri="http://java.sun.com/jsp/jstl/core" prefix="c" %>
<!DOCTYPE html>
<html>
<head>
<meta charset="UTF-8">
<title>Insert title here</title>
</head>
<body>
    <h1>用户列表</h1>
    <a href="<c:url value="${pageContext.request.contextPath }/user/input"/>">继续添加</a>
    <table>
        <tr>
            <th>用户名</th>
            <th>兴趣爱好</th>
            <th>朋友</th>
            <th>职业</th>
            <th>户籍</th>
            <th>个人描述</th>
        </tr>
        <!-- JSTL 标签，请参考本书的相关内容 -->
        <c:forEach items="${users}" var="user">
            <tr>
                <td>${user.userName }</td>
                <td>
                  <c:forEach items="${user.hobby }" var="hobby">
                     ${hobby } 
                  </c:forEach>
                </td>
```

```
            <td>
                <c:forEach items="${user.friends }" var="friend">
                    ${friend } 
                </c:forEach>
            </td>
            <td>${user.carrer }</td>
            <td>${user.houseRegister }</td>
            <td>${user.remark }</td>
        </tr>
    </c:forEach>
    </table>
</body>
</html>
```

❸ 领域模型

应用中实现了 User 类属性和 JSP 页面中表单参数的绑定，User 类包含了和表单参数名对应的属性，以及属性的 set 和 get 方法。在 ch15SS 应用的 src 目录下创建包 pojo，并在该包中创建 User 类。

User 类的代码如下：

```
package pojo;
public class User {
    private String userName;
    private String[] hobby;//兴趣爱好
    private String[] friends;//朋友
    private String carrer;
    private String houseRegister;
    private String remark;
    //省略 setter 和 getter 方法
}
```

❹ Service 层

应用中使用了 Service 层，在 Service 层使用静态集合变量 users 模拟数据库存储用户信息，包括添加用户和查询用户两个功能方法。在 ch15SS 应用的 src 目录下创建包 service，并在该包中创建 UserService 接口和 UserServiceImpl 实现类。

UserService 接口的代码如下：

```
package service;
import java.util.ArrayList;
import pojo.User;
public interface UserService {
    boolean addUser(User u);
    ArrayList<User> getUsers();
}
```

UserServiceImpl 实现类的代码如下：

```java
package service;
import java.util.ArrayList;
import org.springframework.stereotype.Service;
import pojo.User;
@Service
public class UserServiceImpl implements UserService{
    //使用静态集合变量users模拟数据库
    private static ArrayList<User> users = new ArrayList<User>();
    @Override
    public boolean addUser(User u) {
        if(!"IT民工".equals(u.getCarrer())){//不允许添加IT民工
            users.add(u);
            return true;
        }
        return false;
    }
    @Override
    public ArrayList<User> getUsers() {
        return users;
    }
}
```

❺ Controller 层

在 Controller 类 UserController 中定义了请求处理方法，包括处理 user/input 请求的 inputUser 方法，以及处理 user/save 请求的 addUser 方法，其中在 addUser 方法中用到了重定向。在 UserController 类中，通过@Autowired 注解在 UserController 对象中主动注入 UserService 对象，实现对 user 对象的添加和查询等操作；通过 model 的 addAttribute 方法将 User 类对象、HashMap 类型的 hobbys 对象、String[]类型的 carrers 对象以及 String[]类型的 houseRegisters 对象传递给 View（userAdd.jsp）。在 ch15SS 应用的 src 目录下创建包 controller，并在该包中创建 UserController 控制器类。

UserController 类的代码如下：

```java
package controller;
import java.util.HashMap;
import java.util.List;
import org.apache.commons.logging.Log;
import org.apache.commons.logging.LogFactory;
import org.springframework.beans.factory.annotation.Autowired;
import org.springframework.stereotype.Controller;
import org.springframework.ui.Model;
import org.springframework.web.bind.annotation.ModelAttribute;
import org.springframework.web.bind.annotation.RequestMapping;
import pojo.User;
import service.UserService;
@Controller
```

```java
@RequestMapping("/user")
public class UserController {
    //得到一个用来记录日志的对象，这样打印信息的时候能够标记打印的是哪个类的信息
    private static final Log logger = LogFactory.getLog(UserController.class);
    @Autowired
    private UserService userService;
    @RequestMapping(value = "/input")
    public String inputUser(Model model) {
        HashMap<String, String> hobbys = new HashMap<String, String>();
        hobbys.put("篮球", "篮球");
        hobbys.put("乒乓球", "乒乓球");
        hobbys.put("电玩", "电玩");
        hobbys.put("游泳", "游泳");
        //如果model中没有user属性，userAdd.jsp会抛出异常，因为表单标签无法找到
        //modelAttribute属性指定的form backing object
        model.addAttribute("user", new User());
        model.addAttribute("hobbys", hobbys);
    model.addAttribute("carrers", new String[] { "教师", "学生", "coding 搬运工", "IT 民工", "其他" });
        model.addAttribute("houseRegisters", new String[] { "北京", "上海", "广州", "深圳", "其他" });
        return "userAdd";
    }
    @RequestMapping(value = "/save")
    public String addUser(@ModelAttribute User user, Model model) {
        if (userService.addUser(user)) {
            logger.info("成功");
            return "redirect:/user/list";
        } else {
            logger.info("失败");
            HashMap<String, String> hobbys = new HashMap<String, String>();
            hobbys.put("篮球", "篮球");
            hobbys.put("乒乓球", "乒乓球");
            hobbys.put("电玩", "电玩");
            hobbys.put("游泳", "游泳");
            //这里不需要model.addAttribute("user", new User()),
            //因为@ModelAttribute指定form backing object
            model.addAttribute("hobbys", hobbys);
    model.addAttribute("carrers", new String[] { "教师", "学生", "coding 搬运工", "IT 民工", "其他" });
            model.addAttribute("houseRegisters", new String[] { "北京", "上海", "广州", "深圳", "其他" });
            return "userAdd";
        }
    }
}
```

```
@RequestMapping(value = "/list")
public String listUsers(Model model) {
    List<User> users = userService.getUsers();
    model.addAttribute("users", users);
    return "userList";
}
}
```

❻ 测试应用

通过地址 http://localhost:8080/ch15SS/user/input 测试应用,添加用户信息页面效果如图 15.6 所示。

如果在图 15.6 中,"职业"选择"IT 民工",添加失败。失败后还回到添加页面,输入过的信息不再输入,自动回填(必须结合 form 标签)。自动回填是数据绑定的一个优点。失败页面如图 15.7 所示。

图 15.6　添加用户信息页面　　　　图 15.7　添加用户信息失败页面

在图 15.6 中输入正确信息,添加成功后,重定向到信息显示页面,效果如图 15.8 所示。

用户列表

继续添加

用户名	兴趣爱好	朋友	职业	户籍	个人描述
陈恒	游泳 乒乓球	张三 赵六	coding搬运工	其他	一个好人

图 15.8　信息显示页面

15.5 Spring 的数据库编程

视频讲解

数据库编程是互联网编程的基础,Spring 框架为开发者提供了 JDBC 模板模式,即 jdbcTemplate,它可以简化许多代码,但在实际应用中 jdbcTemplate 并不常用。更多应用中用的是 Hibernate 框架和 MyBatis 框架进行数据库编程。

本节仅简要介绍 Spring jdbcTemplate 的使用方法,而 MyBatis 框架的相关内容,读者可学习作者的另一本教程《Java EE 框架整合开发入门到实战——Spring+Spring MVC+MyBatis》。

15.5.1 Spring JDBC 的配置

本节主要使用 Spring JDBC 模块的 core 和 dataSource 包讲解 Spring 数据库编程。core 包是 JDBC 的核心功能包,包括常用的 JdbcTemplate 类;dataSource 包是访问数据源的工具类包。使用 Spring JDBC 操作数据库,需要对其进行配置。配置文件示例代码如下:

```xml
<!-- 配置数据源 -->
<bean id="dataSource" class="org.springframework.jdbc.datasource.DriverManagerDataSource">
    <!-- MySQL 数据库驱动 -->
    <property name="driverClassName" value="com.mysql.jdbc.Driver"/>
    <!-- 连接数据库的 URL -->
    <property name="url" value="jdbc:mysql://localhost:3306/test?characterEncoding=utf8"/>
    <!-- 连接数据库的用户名 -->
    <property name="username" value="root"/>
    <!-- 连接数据库的密码 -->
    <property name="password" value="root"/>
</bean>
<!-- 配置 JDBC 模板 -->
<bean id="jdbcTemplate" class="org.springframework.jdbc.core.JdbcTemplate">
    <property name="dataSource" ref="dataSource"/>
</bean>
```

上述示例代码中,配置 JDBC 模板时,需要将 dataSource 注入 jdbcTemplate;而在数据访问层(Dao 类)需要使用 jdbcTemplate 时,也需要将 jdbcTemplate 注入对应的 Bean 中。示例代码如下:

```
...
@Repository("testDao")
public class TestDaoImpl implements TestDao{
    @Autowired
    //使用配置文件中的 JDBC 模板
```

```
    private JdbcTemplate jdbcTemplate;
    …
}
```

15.5.2　Spring JdbcTemplate 的常用方法

在 15.5.1 节中获取 JDBC 模板后，如何使用它是本节将要讲述的内容。首先，需要了解 JdbcTemplate 类的常用方法：update()方法和 query()方法。

- **public int update(String sql,Object args[])**：该方法可以对数据表进行增加、修改、删除等操作。使用 args[]设置 SQL 语句中的参数，并返回更新的行数。示例代码如下：

```
String insertSql = "insert into user values(null,?,?)";
Object param1[] = {"chenheng1", "男"};
jdbcTemplate.update(sql, param1);
```

- **public List<T> query (String sql, RowMapper<T> rowMapper, Object args[])**：该方法可以对数据表进行查询操作。rowMapper 将结果集映射到用户自定义的类中（前提是自定义类中的属性要与数据表的字段对应）。示例代码如下：

```
String selectSql ="select * from user";
RowMapper<MyUser> rowMapper = new BeanPropertyRowMapper<MyUser>(MyUser.class);
List<MyUser> list = jdbcTemplate.query(sql, rowMapper, null);
```

下面通过一个实例演示 Spring JDBC 的使用过程，具体如下。

❶ **导入 JAR 包**

将 MySQL 数据库的驱动 JAR 包 mysql-connector-java-5.1.45-bin.jar、Spring JDBC 的 JAR 包 spring-jdbc-5.0.2.RELEASE.jar 以及 Spring 事务处理的 JAR 包 spring-tx-5.0.2.RELEASE.jar 复制到 ch15 应用的/ WEB-INF/lib 目录。ch15 应用所添加的 JAR 包如图 15.9 所示。

图 15.9　ch15 应用所添加的 JAR 包

第 15 章　Spring MVC 框架基础

❷ **完善配置文件**

在 ch15 的配置文件中，添加如下配置代码：

```xml
<!-- 配置数据源 -->
<bean id="dataSource" class="org.springframework.jdbc.datasource.DriverManagerDataSource">
    <!-- MySQL 数据库驱动 -->
    <property name="driverClassName" value="com.mysql.jdbc.Driver"/>
    <!-- 连接数据库的 URL -->
    <property name="url" value="jdbc:mysql://localhost:3306/springtest?characterEncoding=utf8"/>
    <!-- 连接数据库的用户名 -->
    <property name="username" value="root"/>
    <!-- 连接数据库的密码 -->
    <property name="password" value="root"/>
</bean>
<!-- 配置 JDBC 模板 -->
<bean id="jdbcTemplate" class="org.springframework.jdbc.core.JdbcTemplate">
    <property name="dataSource" ref="dataSource"/>
</bean>
```

❸ **创建实体类**

在 ch15 包 pojo 中创建实体类 MyUser。该类的属性与数据表 usertable 的字段一致。数据表 usertable 的结构如图 15.10 所示。

名	类型	长度	小数点	允许空值(	
uname	varchar	100	0	☐	🔑1
upwd	varchar	100	0	☑	

图 15.10　usertable 表的结构

实体类 MyUser 的代码如下：

```java
package pojo;
public class MyUser {
    private String uname;
    private String upwd;
    //此处省略 setter 和 getter 方法
    public String toString() {
        return "myUser [uname=" + uname +", upwd=" + upwd + "]";
    }
}
```

❹ **创建数据访问层 Dao**

在包 dao 中，创建 TestDao 接口和 TestDaoImpl 实现类。在实现类 TestDaoImpl 中使用

JDBC 模块 JdbcTemplate 访问数据库，并将该类注解为@Repository("testDao")。在配置文件中添加<context:component-scan base-package="dao"/>语句使注解生效。

TestDao 接口的代码如下：

```java
package dao;
import pojo.MyUser;
import java.util.List;
public interface TestDao {
    public int update(String sql, Object[] param);
    public List<MyUser> query(String sql, Object[] param);
}
```

TestDaoImpl 实现类的代码如下：

```java
package dao;
import java.util.List;
import org.springframework.beans.factory.annotation.Autowired;
import org.springframework.jdbc.core.BeanPropertyRowMapper;
import org.springframework.jdbc.core.JdbcTemplate;
import org.springframework.jdbc.core.RowMapper;
import org.springframework.stereotype.Repository;
import pojo.MyUser;
@Repository("testDao")
public class TestDaoImpl implements TestDao{
    @Autowired
    //使用配置文件中的 JDBC 模板
    private JdbcTemplate jdbcTemplate;
    /**
     * 更新方法，包括添加、修改、删除
     * param 为 sql 中的参数，如通配符?
     */
    @Override
    public int update(String sql, Object[] param) {
        return jdbcTemplate.update(sql, param);
    }
    /**
     * 查询方法
     * param 为 sql 中的参数，如通配符?
     */
    @Override
    public List<MyUser> query(String sql, Object[] param) {
        RowMapper<MyUser> rowMapper = new BeanPropertyRowMapper<MyUser>
            (MyUser.class);
```

```
            return jdbcTemplate.query(sql, rowMapper, param);
    }
}
```

❺ 创建测试类

在包 controller 中,创建控制器类 TestSpringJDBC。具体代码如下:

```
package controller;
import java.util.List;
import org.springframework.beans.factory.annotation.Autowired;
import org.springframework.stereotype.Controller;
import org.springframework.web.bind.annotation.RequestMapping;
import dao.TestDao;
import pojo.MyUser;
@Controller
public class TestSpringJDBC {
    @Autowired
    private TestDao testDao;
    @RequestMapping("/jdbcTest")
    public String test() {
        String insertSql = "insert into usertable values(?,?)";
        //数组 param 的值与 insertSql 语句中的"?"一一对应
        Object param1[] = {"chenheng1", "111"};
        Object param2[] = {"chenheng2", "222"};
        Object param3[] = {"chenheng3", "333"};
        Object param4[] = {"chenheng4", "444"};
        //添加用户
        testDao.update(insertSql, param1);
        testDao.update(insertSql, param2);
        testDao.update(insertSql, param3);
        testDao.update(insertSql, param4);
        //查询用户
        String selectSql ="select * from usertable";
        List<MyUser> list = testDao.query(selectSql, null);
        for(MyUser mu : list) {
            System.out.println(mu);
        }
        return "jdbcTest";
    }
}
```

通过地址 http://localhost:8080/ch15/jdbcTest 运行上述控制器,运行结果如图 15.11 所示。

```
    http://localhost:8080/ch15/jdbcTest
JDBC测试

  Markers  Properties  Servers  Data Source Explorer  Snipp
Tomcat v9.0 Server at localhost [Apache Tomcat] C:\Program Files\Java\jre1.8
myUser [uname=chenheng1, upwd=111]
myUser [uname=chenheng2, upwd=222]
myUser [uname=chenheng3, upwd=333]
myUser [uname=chenheng4, upwd=444]
```

图 15.11 Spring 数据库编程的运行结果

15.5.3 基于@Transactional 注解的声明式事务管理

@Transactional 注解可以作用于接口、接口方法、类以及类方法上。当作用于类上时，该类的所有 public 方法都将具有该类型的事务属性，同时，也可以在方法级别使用该注解来覆盖类级别的定义。虽然@Transactional 注解可以作用于接口、接口方法、类以及类方法上，但是 Spring 小组建议不要在接口或者接口方法上使用该注解，因为只有在使用基于接口的代理时它才会生效。@Transactional 注解使用步骤如下。

❶ 为数据源添加事务管理器

```
<!-- 为数据源添加事务管理器 -->
    <bean id="txManager"
        class="org.springframework.jdbc.datasource.DataSourceTransactionManager">
        <property name="dataSource" ref="dataSource" />
    </bean>
    <!-- 为事务管理器注册注解驱动 -->
<tx:annotation-driven transaction-manager="txManager" />
```

❷ 为 Service 层添加@Transactional 注解

在 Spring MVC 中，通常通过 Service 层进行事务管理，因此需要为 Service 层添加@Transactional 注解。添加@Transactional 注解的示例代码如下：

```
...
@Service("testService")
@Transactional
//加上注解@Transactional,就可以指定这个类需要受 Spring 的事务管理
//注意@Transactional 只能针对 public 属性范围内的方法添加
public class TestServiceImpl implements TestService{
}
...
```

15.6 基于 Spring MVC 框架的注册与登录

视频讲解

15.6.1 JSON 数据转换

为实现浏览器与控制器类之间的 JSON 数据交互，Spring MVC 提供了 MappingJackson2HttpMessageConverter 实现类默认处理 JSON 格式请求响应。该实现类利用 Jackson 开源包读写 JSON 数据，将 Java 对象转换为 JSON 对象和 XML 文档，同时也可以将 JSON 对象和 XML 文档转换为 Java 对象。

Jackson 开源包及其描述如下所示。

- jackson-annotations-2.9.4.jar：JSON 转换注解包。
- jackson-core-2.9.4.jar：JSON 转换核心包。
- jackson-databind-2.9.4.jar：JSON 转换的数据绑定包。

以上 3 个 Jackson 的开源包在编写本书时，最新版本是 2.9.4，读者可通过地址 http://mvnrepository.com/artifact/com.fasterxml.jackson.core 下载得到。

在使用注解开发时，需要用到两个重要的 JSON 格式转换注解，分别是@RequestBody 和@ResponseBody。

- @RequestBody：用于将请求体中的数据绑定到方法的形参中，该注解应用在方法的形参上。
- @ResponseBody：用于直接返回 return 对象，该注解应用在方法上。

15.6.2 注册与登录

创建 Spring MVC 应用 ch15SSS，实现以下功能。

- 未注册用户请运行注册页面进行注册，注册时检查用户名是否已注册（要求使用 Ajax 异步提交检查）。
- 已注册的用户运行登录页面进行登录，登录成功的用户可以查看所有用户姓名。

实现应用 ch15SSS 的步骤如下。

❶ 创建应用 ch15SSS，并导入相关 JAR 文件

除了导入 ch15 应用所需要的 JAR 文件，还需要导入 JSTL 和 JSON 需要的 JAR 文件。具体如图 15.12 所示。

❷ 创建相关 JSP 文件

应用中需要的 js 和图片，请参见提供的源代码，这里只说明 JSP 相关代码。

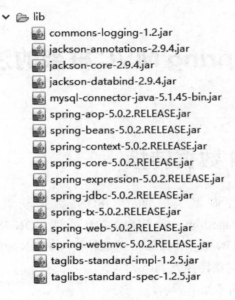

图 15.12 相关 JAR 文件

1）注册页面 register.jsp 的代码

```
<%@ page language="java" contentType="text/html; charset=UTF-8" pageEncoding=
"UTF-8"%>
<%@ taglib prefix="c" uri="http://java.sun.com/jsp/jstl/core"%>
<%
String path = request.getContextPath();
String basePath = request.getScheme()+"://"+request.getServerName()+":"
+request.getServerPort()+path+"/";
%>
<!DOCTYPE html>
<html>
<base href="<%=basePath%>">
<head>
<meta charset="UTF-8">
<style type="text/css">
    .textSize{
        width: 100pt;
        height: 15pt;
    }
</style>
<title>注册页面</title>
<script type="text/javascript" src="js/jquery-3.2.1.min.js"></script>
<script type="text/javascript">
    //输入姓名后，调用该方法，判断用户名是否可用
    function nameIsNull(){
        $.ajax({
```

```
                //请求路径
                url : "${pageContext.request.contextPath }/user/isUse",
                //请求类型
                type : "post",
                //data 表示发送的数据
                data : JSON.stringify({uname:$("#uname").val()}),
                //定义发送请求的数据格式为 JSON 字符串
                contentType : "application/json;charset=utf-8",
                //定义回调响应的数据格式为 JSON 字符串，该属性可以省略
                //dataType : "json",
                //成功响应的结果
                success : function(data){
                    if(data == "no"){
                        $("#isExit").html("<font color=red size=5>×</font>");
                        alert("用户已存在，请修改！");
                    }else{
                        $("#isExit").html("<font color=green size=5>√</font>");
                        alert("用户可用");
                    }
                },
                error : function() {
                    alert("处理异常！");
                }
            });
        }
    </script>
</head>
<body style="background-image: url('images/bb.jpg');">
    <form action="user/regist" method="post" name="registForm">
        <table>
            <tr>
                <td>姓名：</td>
                <td>
        <input class="textSize" type="text" id="uname" name="uname" onblur=
        "nameIsNull()" />
                    <span id="isExit"></span>
                </td>
            </tr>
            <tr>
                <td>密码：</td>
                <td><input class="textSize" type="password" name="upass"/></td>
            </tr>
            <tr>
                <td>确认密码：</td>
                <td><input class="textSize" type="password" name="reupass"/></td>
```

```
                </tr>
                <tr>
    <td colspan="2" align="center"><input type="button" value="注册" onclick=
    "allIsNull()"/></td>
                </tr>
        </table>
    </form>
</body>
</html>
```

2）登录页面 login.jsp 的代码

```
<%@ page language="java" contentType="text/html; charset=UTF-8" pageEncoding=
"UTF-8"%>
<%
String path = request.getContextPath();
String basePath = request.getScheme()+"://"+request.getServerName()+
":"+request.getServerPort()+path+"/";
%>
<!DOCTYPE html>
<html>
<head>
<base href="<%=basePath%>">
<meta charset="UTF-8">
<title>后台登录</title>
<script type="text/javascript">
    //确定按钮
    function gogo(){
        document.forms[0].submit();
    }
    //取消按钮
    function cancel(){
        document.forms[0].action = "";
    }
    </script>
</head>
<body>
    <form action="user/login" method="post">
    <table>
        <tr>
            <td colspan="2"><img src="images/login.gif"></td>
        </tr>
        <tr>
            <td>姓名：</td>
            <td><input type="text" name="uname" class="textSize"/></td>
        </tr>
```

```html
            <tr>
                <td>密码: </td>
                <td><input type="password" name="upass" class="textSize"/></td>
            </tr>
            <tr>
                <td colspan="2">
                    <input type="image" src="images/ok.gif" onclick="gogo()" >
                    <input type="image" src="images/cancel.gif" onclick="cancel()">
                </td>
            </tr>
        </table>
    </form>
  </body>
</html>
```

3) 主页面 main.jsp 的代码

```jsp
<%@ page language="java" contentType="text/html; charset=UTF-8" pageEncoding="UTF-8"%>
<%@ taglib prefix="c" uri="http://java.sun.com/jsp/jstl/core"%>
<%
String path = request.getContextPath();
String basePath = request.getScheme()+"://"+request.getServerName()+":"+request.getServerPort()+path+"/";
%>
<!DOCTYPE html>
<html>
<head>
<base href="<%=basePath%>">
<meta charset="UTF-8">
<title>后台主页面</title>
</head>
<body>
    查询出所有用户:
    <br>
    <c:forEach items="${allUsers}" var="map">
        ${map.uname}<br>
    </c:forEach>
</body>
</html>
```

❸ **创建相关配置文件**

这里只提供 springmvc-servlet.xml 的代码，web.xml 的内容请读者参考 ch15 应用。
springmvc-servlet.xml 的核心代码如下：

```xml
<!-- 附加 props 配置文件 -->
    <context:property-placeholder location="classpath:db.properties" />
```

```xml
<!-- scan the package and the sub package -->
<context:component-scan base-package="com.controller"/>
<context:component-scan base-package="com.service"/>
<context:component-scan base-package="com.dao"/>
<mvc:annotation-driven />
<!-- 静态资源需要单独处理,不需要dispatcher servlet -->
<mvc:resources location="/images/" mapping="/images/**"></mvc:resources>
<mvc:resources location="/js/" mapping="/js/**"></mvc:resources>
<!-- 配置视图解析器 -->
<bean class="org.springframework.web.servlet.view.InternalResourceViewResolver"
      id="internalResourceViewResolver">
    <!-- 前缀 -->
    <property name="prefix" value="/WEB-INF/jsp/" />
    <!-- 后缀 -->
    <property name="suffix" value=".jsp" />
</bean>
<bean id="dataSource" class="org.apache.tomcat.dbcp.dbcp2.BasicDataSource">
    <property name="driverClassName" value="${db.driverClassName}" />
    <property name="url" value="${db.url}" />
    <property name="username" value="${db.username}" />
    <property name="password" value="${db.password}" />
    <!-- 读取classpath:db.properties属性文件 -->
</bean>
<bean id="jdbcTemplate" class="org.springframework.jdbc.core.JdbcTemplate">
    <property name="dataSource" ref="dataSource"/>
</bean>
<!-- 添加事务支持 -->
<bean id="txManager"
      class="org.springframework.jdbc.datasource.DataSourceTransactionManager">
    <property name="dataSource" ref="dataSource" />
</bean>
<!-- 注册事务注解处理器-->
<tx:annotation-driven transaction-manager="txManager" />
```

❹ 创建控制器类

UserController.java 的代码如下:

```java
package com.controller;
import org.springframework.beans.factory.annotation.Autowired;
import org.springframework.stereotype.Controller;
import org.springframework.ui.Model;
import org.springframework.web.bind.annotation.RequestBody;
import org.springframework.web.bind.annotation.RequestMapping;
import org.springframework.web.bind.annotation.ResponseBody;
import com.service.UserService;
```

```java
@Controller
@RequestMapping("/user")
public class UserController {
    @Autowired
    private UserService userService;
    /**
     * 验证用户名是否可以
     */
    @RequestMapping("/isUse")
    @ResponseBody
    public String isUse(@RequestBody String uname) {
        //返回 JSON 字符串
        return userService.isUse(uname);
    }
    /**
     * 实现注册功能
     */
    @RequestMapping("/regist")
    public String regist(String uname, String upass) {
        return userService.regist(uname, upass);
    }
    /**
     * 实现登录功能
     */
    @RequestMapping("/login")
    public String login(Model model, String uname, String upass) {
        return userService.login(model, uname, upass);
    }
}
```

❺ 创建 Service 层

Service 层有接口和接口的实现类。

UserService.java 的代码如下:

```java
package com.service;
import org.springframework.ui.Model;
public interface UserService {
    public String isUse(String uname);
    public String regist(String uname, String upass);
    public String login(Model model, String uname, String upass);
}
```

UserServiceImpl.java 的代码如下:

```java
package com.service;
import java.util.List;
import java.util.Map;
```

```java
import org.springframework.beans.factory.annotation.Autowired;
import org.springframework.stereotype.Service;
import org.springframework.transaction.annotation.Transactional;
import org.springframework.ui.Model;
import com.dao.UserDao;
@Service
@Transactional
public class UserServiceImpl implements UserService{
    @Autowired
    private UserDao userDao;
    @Override
    public String isUse(String uname) {
        List<Map<String, Object>> allU = userDao.isUse(uname);
        if(allU.size() > 0) {
            return "no";
        }
        return "ok";
    }
    @Override
    public String regist(String uname, String upass) {
        if(userDao.regist(uname, upass)) {
            return "login";
        }
        return "register";
    }
    @Override
    public String login(Model model, String uname, String upass) {
        if(userDao.login(uname, upass).size() > 0) {
            model.addAttribute("allUsers", userDao.selectAll());
            return "main";
        }
        model.addAttribute("errorMsg", "用户名密码错误！");
        return "login";
    }
}
```

❻ 创建数据访问层

该层有两个 dao 类，一个是 BaseDao，一个是 UserDao。在 src 目录下还有一个数据库信息文件 jdbc.properties。

jdbc.properties 的代码如下：

```
db.driverClassName=com.mysql.jdbc.Driver
db.url=jdbc:mysql://localhost:3306/test?characterEncoding=utf8
db.username=root
db.password=root
```

BaseDao.java 的代码如下：

```java
package com.dao;
import java.sql.ResultSet;
import java.sql.ResultSetMetaData;
import java.sql.SQLException;
import java.util.HashMap;
import java.util.List;
import java.util.Map;
import org.springframework.beans.factory.annotation.Autowired;
import org.springframework.jdbc.core.JdbcTemplate;
import org.springframework.jdbc.core.RowMapper;
public class BaseDao {
    @Autowired
    private JdbcTemplate jdbcTemplate;
    /**
     * @description 添加、修改、删除操作
     * @param String sql 代表SQL文件
     * @param 根据参数进行增删改，注意参数Object[] arg0元素的个数和顺序一定与
     * String sql 中的通配符？一一对应，如果SQL语句中没有通配符，请传递null
     * @return boolean 更新成功返回true，否则false
     */
    public boolean updateByParam(String sql, Object[] arg0) {
        //SQL语句中没有通配符
        if(arg0 == null || arg0.length == 0){
            if(jdbcTemplate.update(sql)>0)
                return true;
            else
                return false;
        }
        if(jdbcTemplate.update(sql,arg0)>0)
            return true;
        else
            return false;
    }
    /**
     * @description support select SQL Statement by param
     * @param String sql 代表SQL文件
     * @param 根据参数进行查询，注意参数Object[] arg0元素的个数和顺序一定与String sql
     * 中的通配符"？"——对应，如果SQL语句中没有通配符，请传递null
     * @return List<Map<String, Object>>：查询结果
     */
    @SuppressWarnings("unchecked")
    public List<Map<String, Object>> findByParam(String sql,Object arg0[]){
        if(arg0 == null || arg0.length == 0){
            return jdbcTemplate.query(sql, new IRowMapper());
        }
```

```java
        return jdbcTemplate.query(sql, new IRowMapper(),arg0);
    }
    /**
     *
     * 数据库表的列名与map的key相同
     *
     */
    @SuppressWarnings("rawtypes")
    public class IRowMapper implements RowMapper{
        @Override
        public Map<String, Object> mapRow(ResultSet rs, int rowNum) throws SQLException {
            Map<String, Object> row = new HashMap<String, Object>();
            ResultSetMetaData rsMetaData = rs.getMetaData();
            for(int i=1,size=rsMetaData.getColumnCount(); i<=size; i++){
                //查询中，数据表的字段名和页面中的要一模一样
                row.put(rsMetaData.getColumnLabel(i).toLowerCase(), rs.getObject(i));
            }
            return row;
        }
    }
}
```

UserDao.java 的代码如下：

```java
package com.dao;
import java.util.List;
import java.util.Map;
import org.springframework.stereotype.Repository;
@Repository("userDao")
public class UserDao extends BaseDao{
    public List<Map<String, Object>> isUse(String uname){
        String sql = "select * from usertable where uname = ?";
        String param[] = {uname};
        return findByParam(sql, param);
    }
    public boolean regist(String uname, String upass) {
        String sql = "insert into usertable values(?,?)";
        String param[] = {uname, upass};
        return updateByParam(sql, param);
    }
    public List<Map<String, Object>> login(String uname, String upass) {
        String sql = "select * from usertable where uname = ? and upwd = ?";
        String param[] = {uname, upass};
        return findByParam(sql, param);
```

```
    }
    public List<Map<String, Object>> selectAll() {
        String sql = "select * from usertable";
        return findByParam(sql, null);
    }
}
```

15.7 本章小结

本章简单介绍了 Spring MVC 框架基础，包括 Spring MVC 的工作流程、控制器、表单标签与数据绑定、数据库编程等内容。

习题 15

1．在开发 Spring MVC 应用时，如何部署 DispatcherServlet？又如何创建 Spring MVC 的配置文件？
2．简述 Spring MVC 的工作流程。
3．举例说明数据绑定的优点。
4．Spring MVC 有哪些表单标签？其中，可以绑定集合数据的标签有哪些？
5．@ModelAttribute 可实现哪些功能？

附录 A 案例 1：基于 Struts 2 框架的名片管理系统的设计与实现

本案例系统使用 Struts 2 框架实现各个模块，JSP 引擎为 Tomcat 9.0，数据库采用的是 MySQL 5.5，集成开发环境为 Eclipse IDE for Java EE Developers。

❶ 系统功能需求

名片管理系统是针对注册用户使用的系统，提供如下功能：

① 非注册用户可以注册为注册用户（使用 Ajax 技术验证注册名称的唯一性）。
② 成功注册的用户可以登录系统。
③ 成功登录的用户可以添加、修改、删除以及浏览自己客户的名片信息。
④ 成功登录的用户可以在个人中心查看自己的基本信息和修改密码。

❷ 系统模块划分

注册用户登录成功后，进入管理主页面（main.jsp），可以对自己的客户名片进行管理。系统模块划分如图 A.1 所示。

图 A.1　名片管理系统模块划分图

读者扫描右侧二维码，可阅读本系统详细的设计文档、观看视频讲解以及下载源代码。

源码+视频

案例 2：基于 Spring MVC 框架的名片管理系统的设计与实现

本案例系统使用 Spring MVC 框架实现各个模块，JSP 引擎为 Tomcat 9.0，数据库采用的是 MySQL 5.5，集成开发环境为 Eclipse IDE for Java EE Developers。

❶ 系统功能需求

名片管理系统是针对注册用户使用的系统，提供如下功能：
① 非注册用户可以注册为注册用户（使用 Ajax 技术验证注册名称的唯一性）。
② 成功注册的用户可以登录系统。
③ 成功登录的用户可以添加、修改、删除以及浏览自己客户的名片信息。
④ 成功登录的用户可以在个人中心查看自己的基本信息和修改密码。

❷ 系统模块划分

注册用户登录成功后，进入管理主页面（main.jsp），可以对自己的客户名片进行管理。系统模块划分如图 B.1 所示。

图 B.1 名片管理系统模块划分图

读者扫描右侧二维码，可阅读本系统详细的设计文档、观看视频讲解以及下载源代码。

源码+视频

案例3：Spring MVC 整合 MyBatis

本案例首先讲解 Spring MVC 如何整合 MyBatis，然后详细介绍 MyBatis 的映射文件及动态 SQL 的编写，最后使用 Spring MVC + MyBatis 实现名片管理系统。

MyBatis 是主流的 Java 持久层框架之一，它与 Hibernate 一样，也是一种 ORM（Object/Relational Mapping，即对象关系映射）框架。MyBatis 本是 apache 的一个开源项目 iBatis，2010 年这个项目由 apache software foundation 迁移到了 google code，并且改名为 MyBatis。MyBatis 提供的持久层框架包括 SQL Maps 和 Data Access Objects（DAO），它消除了几乎所有的 JDBC 代码和参数的手工设置以及结果集的检索。MyBatis 使用 XML 或注解进行配置和原始映射，将接口和 Java 的 POJOs（Plain Old Java Objects，普通的 Java 对象）映射成数据库中的记录。因其性能优异，且具有高度的灵活性、可优化性、易于维护以及简单易学等特点，受到了广大互联网企业和编程爱好者的青睐。

读者扫描下方二维码，可观看视频学习 Spring MVC 整合 MyBatis 的流程、MyBatis 的映射文件及动态 SQL 的编写。同时，还可下载基于 Spring MVC + MyBatis 实现名片管理系统的源代码。

源码+视频

案例 4：在 Eclipse 中使用 Maven 整合 Spring MVC+MyBatis 框架

Apache Maven 是一个软件项目管理工具。基于项目对象模型（Project Object Model，POM）的理念，通过一段核心描述信息来管理项目构建、报告和文档信息。在 Java 项目中，Maven 主要完成两项工作：① 统一开发规范与工具；② 统一管理 jar 包。

Maven 统一管理项目开发所需要的 jar 包，但这些 jar 包将不再包含在项目内（即不在 lib 目录下），而是存放于仓库中。仓库主要包括以下内容。

（1）中央仓库：存放开发过程中的所有 jar 包，例如 JUnit，这些包都可以通过互联网从中央仓库中下载，仓库地址：http://mvnrepository.com。

（2）本地仓库：本地计算机中的仓库。

下载 Maven 的本地仓库，配置在"%MAVEN_HOME%\conf\settings.xml"文件中，找到 localRepository 即可；Eclipse 中自带 Maven 的默认本地仓库地址在"{user.home}/.m2/repository/settings.xml"文件中，同样找到 localRepository 即可。

Maven 项目首先会从本地仓库中获取所需要的 jar 包，当无法获取指定 jar 包时，本地仓库会从远程仓库（中央仓库）中下载 jar 包，并放入本地仓库以备将来使用。

如何在 Eclipse 中使用 Maven 整合 Spring MVC + MyBatis 框架，请扫描下方二维码，查看配置说明文档。

源码+视频

参 考 文 献

[1] QST 青软实训. Java Web 技术及应用[M]. 北京：清华大学出版社，2015.
[2] Budi Kurniawan，Paul Deck. Servlet JSP 和 Spring MVC 初学指南[M]. 北京：人民邮电出版社，2016.
[3] 杨开振，等. Java EE 互联网轻量级框架整合开发——SSM 框架（Spring MVC + Spring + MyBatis）和 Redis 实现[M]. 北京：电子工业出版社，2017.
[4] 黑马程序员. Java EE 企业级应用开发教程（Spring+Spring MVC+MyBatis）[M]. 北京：人民邮电出版社，2017.

图书资源支持

感谢您一直以来对清华版图书的支持和爱护。为了配合本书的使用,本书提供配套的资源,有需求的读者请扫描下方的"书圈"微信公众号二维码,在图书专区下载,也可以拨打电话或发送电子邮件咨询。

如果您在使用本书的过程中遇到了什么问题,或者有相关图书出版计划,也请您发邮件告诉我们,以便我们更好地为您服务。

我们的联系方式:

地　　址: 北京市海淀区双清路学研大厦 A 座 701

邮　　编: 100084

电　　话: 010-62770175-4608

资源下载: http://www.tup.com.cn

客服邮箱: tupjsj@vip.163.com

QQ: 2301891038(请写明您的单位和姓名)

用微信扫一扫右边的二维码,即可关注清华大学出版社公众号"书圈"。

资源下载、样书申请

书圈

扫一扫,获取最新目录